Lecture Notes in Morphogenesis

Lecture Notes in Morphogenesis is an interdisciplinary book series which aims to face the questions of emergence, individuation and becoming of forms from several different points of view: those of pure and applied mathematics, of computational algorithms, of biology, of neurophysiology, of cognitive and social structures. The set of questions above concerns all the manifestations of Being, all the manifestations of Life. At the heart of contemporary embryogenesis lies an essential question: How can form emerge from the constant, chaotic flow? How can a sequence of purely informational elements—an a-signifying combination of chemical substances organized in the DNA molecule—evolve into the highly complex and structured forms of the living organism? A similar question can be asked when we deal with the morphogenesis of vision in neural systems and with the creation of evolving synthetic images, since digital technology makes possible the simulation of emergent processes both of living bodies and of visual forms. Finally the very idea that abstract structures of meaning could be captured in terms of morphodynamic evolution opens the door to new models of semiolinguistics, semiotic morphodynamics, and cognitive grammars. An entire heritage of ideas and concepts has to be reconsidered in order to face new and challenging problems: the theoretical framework opened by Goethe with the introduction of the word "Morphogenesis" is developed by D'Arcy Thompson in "On Growth and Form", it is reorganized with new theoretical insights by the classical structuralism of Levi-Strauss and formalized by the dynamical structuralism of René Thom. The introduction of the post-structuralists ideas of individuation (in Gilbert Simondon and Gilles Deleuze) and plasticity of structures builds a bridge to contemporary problems of morphogenesis at a physical, biological, social and transindividual level. The objective of this book series is to provide suitable theoretical and practical tools for describing evolutionary phenomena at the level of Free boundary problems in Mathematics, Embryogenesis, Image Evolution in Visual Perception, Visual Models of Morphogenesis, Neuromathematics, Autonomy and Self-Organization, Morphogenetic Emergence and Individuation, Theoretical Biology, Cognitive Morphodynamics, Cities Evolution, Semiotics, Subjectivation processes, Social movements as well as new frontiers of Aesthetics. To submit a proposal or request further information, please use the PDF Proposal Form or contact directly: *Dr. Thomas Ditzinger (thomas.ditzinger@springer.com)*

More information about this series at http://www.springer.com/series/11247

Denise Pumain
Editor

Theories and Models of Urbanization

Geography, Economics and Computing Sciences

Springer

Editor
Denise Pumain
Department of Geography
University Paris I Pantheon Sorbonne
Paris, France

ISSN 2195-1934 ISSN 2195-1942 (electronic)
Lecture Notes in Morphogenesis
ISBN 978-3-030-36655-1 ISBN 978-3-030-36656-8 (eBook)
https://doi.org/10.1007/978-3-030-36656-8

This Springer imprint is published by the registered company Springer Nature Switzerland AG
The registered company address is: Gewerbestrasse 11, 6330 Cham, Switzerland

Contents

Contributors

Elsa Arcaute Centre for Advanced Spatial Analysis (CASA), University College London, London, UK

Natacha Aveline-Dubach CNRS UMR Géographie-cités, Aubervilliers, France

Solène Baffi Department of Geography and Environmental Studies, Stellenbosch University, Stellenbosch, South Africa;
Université Paris 1 Panthéon-Sorbonne, UMR 8504 Géographie-Cités, Paris, France

Marc Barthelemy Institut de Physique Théorique (IPhT/CEA-CNRS), Gif-sur-Yvette, France;
Centre d'Analyse et de Mathématique Sociales (CAMS/EHESS), Paris, France

Michael Batty University College London, Centre for Advanced Spatial Analysis (CASA), London, UK

Mehdi Bida University of Lausanne, Lausanne, Switzerland

Olivier Bouba-Olga Université de Poitiers, Poitiers, France

Clémentine Cottineau Université Paris 1 Panthéon-Sorbonne, UMR 8504 Géographie-Cités, Paris, France;
CNRS, UMR 8097 Centre Maurice Halbwachs, Paris, France;
Centre for Advanced Spatial Analysis, University College London, London, UK

Eric Denis CNRS, UMR Géographie-Cités, Paris, France

Olivier Finance UMR 7362 Laboratoire Image Ville Environnement, Université de Strasbourg, Strasbourg, France;
UMR 8504 Géographie-cités, Paris, France

Mauricio Franco-Cisterna Laboratorio de Ecoinformática, Instituto de Conservación, Biodiversidad y Territorio, Universidad Austral de Chile, Valdivia, Los Ríos, Chile

Michel Grossetti CNRS, EHESS, Université de Toulouse, Toulouse, France

Erez Hatna College of Global Public Health, New York University, New York, USA

Denise Pumain Université Paris I, UMR Géographie-Cités, Paris, France;
Université Paris I Panthéon Sorbonne, UMR Géographie-Cités, Paris, France

Juste Raimbault Center for Advanced Spatial Analysis (University College London), London, UK;
Institut des Systèmes Complexes Paris-Ile-de-France (CNRS 3611), Paris, France;
UMR Géographie-Cités (CNRS 8504), Paris, France

Céline Rozenblat University of Lausanne, Lausanne, Switzerland

Horacio Samaniego Laboratorio de Ecoinformática, Instituto de Conservación, Biodiversidad y Territorio, Universidad Austral de Chile, Valdivia, Los Ríos, Chile;
Instituto de Ecología y Biodiversidad (IEB), Santiago, Chile;
Instituto de Sistemas Complejos de Valparaíso (ISCV), Valparaíso, Chile

Lena Sanders CNRS, UMR Géographie-cités, Paris, France

Boris Sotomayor-Gómez Laboratorio de Ecoinformática, Instituto de Conservación, Biodiversidad y Territorio, Universidad Austral de Chile, Valdivia, Los Ríos, Chile

Elfie Swerts FORS—Swiss Centre of Expertise in the Social Sciences, Université de Lausanne, Lausanne, Switzerland;
UMR 8504 Géographie-cités, Paris, France

Isabelle Thomas Université Catholique de Louvain, Louvain-la-Neuve, Belgium

Céline Vacchiani-Marcuzzo CNRS, UMR Géographie-cités, Paris, France;
Université de Reims Champagne-Ardenne, UMR Géographie-cités, Reims, France

Fulong Wu Bartlett School of Planning, University College, London, UK

Chapter 1
Introduction

Denise Pumain

> *I, too, have thought of a model of cities from which I deduce all the others. It is a city which is made of exceptions, impossibilities, contradictions, incongruities, and misinterpretations. If a city thus made is all that is most improbable, by lowering the number of abnormal elements the probability grows that the city really exists. Therefore, it suffices that I take exceptions from my model, and in whatever way I proceed I will arrive at one of the cities which, though yet by exception, exists. But I cannot push my operation further than a certain limit: I would get cities too plausible to be true.*
> Italo Calvino, *Le città invisibili*, Giulio Einaudi Editore, 1972 (my translation).

This book aims to offer a broad vision of the current state of knowledge on theories and explanatory models of the urbanization process. In a century when human populations are becoming predominantly urban, it is crucial that the science of cities can understand the meaning of this evolution and help manage this tremendous transition in our way of inhabiting the Earth. The authors in this book bring different points of view, reflecting the necessary plurality of evolving science. But all of them aim to clearly articulate their understanding of urban dynamics and discuss their perspectives on the future of cities. The debates between theories from geography, archeology, sociology and history and those from economics have been re-energized recently by the intervention of physicists and mathematicians in the quantitative description of urban properties. While showing the interest of rethinking the dynamics of cities in relation to the theories and models of complex systems, the book provides useful nuances to the possibility of transferring certain aspects of these theories and the models associated with them to urban systems. This reflection is an essential prerequisite in the current race to develop cities that are more "intelligent", but still struggle to offer a better respect for their environment and the rights of each person to a decent life in the city.

D. Pumain (✉)
Université Paris I Panthéon Sorbonne, UMR Géographie-Cités, Paris, France
e-mail: pumain@parisgeo.cnrs.fr

D. Pumain (ed.), *Theories and Models of Urbanization*, Lecture Notes in Morphogenesis, https://doi.org/10.1007/978-3-030-36656-8_1

1.1 Why an Open Discussion of Major Urban Issues Is an Urgent Matter

If, like Italo Calvino, you like paradoxes, if you are passionate about great challenges, consider for a moment the case of cities: everyone knows for him or herself what a city is, what it was and can recognize a city when they see one, but among the multiple definitions that are proposed, none can be agreed on by everyone, and this despite the more or less sophisticated delineations provided by statistical offices and the availability of powerful earth observation instruments to identify built-up surfaces and to capture digital traces given actively or passively by people and objects now connected to the Internet; the concern for cities is one of the great businesses of the 21st century, because before it ends two-thirds or three-quarters of humanity will live in cities, and yet, apart from wanting them even more "intelligent", no consensus is emerging about whether to make them larger or smaller, denser or more dispersed, high raised or more spread out, filled with more technologies or filled with more natural elements…; cities are places where many new sources of wealth are often created or adopted, but in parallel they seem to exacerbate situations of increasing poverty, social segregation and between group tensions…; cities have existed as physical objects and as a concept for some ten thousand years and among them a large number have existed at the same location and under the same name for several centuries or even millennia. Only a very small number of cities have disappeared completely since the 15th century, yet each day new questions arise about how making them sustainable and resilient…; the cities have communicated with each other since the dawn of time, their destinies are linked and intertwined within networks and systems of cities, yet each continues to think of itself as a unique, autonomous center, which can choose its future more or less independently from other cities…

Fortunately many scholars are tackling these conundrums. Whatever their discipline, many work under the umbrella of complex systems, of which cities are a most interesting case. But here the Babel tower is a major risk as well, since most of the required trans-disciplinary tools for safely building a cumulative knowledge have not yet been clearly recognized or invented, starting with a basic common vocabulary, let alone a common ontology. Many premature generalizations emerge, ignoring whole swathes of concepts or observations that were made earlier or in neighboring research fields.

Thus it has become urgent to provide clues about real scientific advances that may have emerged from such new perspectives. As often in social sciences, a plurality of theories is useful and exists about cities, and the authors in this book especially will discuss which regularities and processes still justify the existence of different theories. They are also keen to try finding the best way for combining the innovations brought from different sciences. In particular, they analyze which features urbanization theories share with complexity theories and which do not. Part of the discussion is about how an evolutionary (geo-historical) perspective that characterises social sciences can be composed with supposedly "universal" principles from physical sciences or economics and how mathematic and computational methods can be integrated within urban models. Another important contribution of this pool of authors is about checking the meaningfulness of theories in different parts of the world and assessing where different interpretative frameworks are necessary—without emphasizing excessively regional idiosyncrasies. Especially, do we have to develop a specific theory for urban systems in regions that are sometimes lazily referred as "the global South" or can we modulate the explanations by introducing elements of their geo-historical context within general theories of urbanization?

The most obvious reason for discussing urban theories and models is because any particular conception of what a city is or should be, as well as what can be expected as future urban dynamics, potentially has immense and durable consequences on urban practical problems. The matter is particularly urgent in the context of the contemporary context of energetic and broader ecological transition urged by the climatic change. That will require a better adjustment between urban development and the quantity and quality of available resources. Demography also is a problem linked with the future of cities because the gap in urban concerns seem to widen between wealthier cities whose population tend to stabilize or even start shrinking and poorer cities where demographic growth is still booming and urban quality of life and environment are much more threatened. Within such a challenging context, decisions are made each day by investors and mayors for engaging the future of transportation systems, changing the degree of convenience of urban layouts for urban citizens, adapting to new conditions under climate change. Such decisions contribute to improve and sometimes deteriorate the income level, employment rate and quality of life in the city, its attractiveness and social cohesion. A variety of solutions is always possible, that can be explored with the help of urban theories and models.

This book is organized into five major parts, each of which addresses one of the main questions currently posed to explanatory theories of cities and their modeling. In each part, from two to four chapters thoroughly present and discuss points of view, sometimes complementary and sometimes contradictory, in order to clearly explain the arguments and the stakes of the ongoing debates.

1.2 Urban Systems as Complex Systems

Complexity theories have percolated into the field of science of cities many times in recent decades. The *cybernetic inspiration* of Forrester's (1969) models led to the search for the explanation of the autonomous development of a city in its environment, thought as a system articulating three major functionalities (habitat, production, social life) interacting with each other. The model was intended to better understand the sometimes counterintuitive consequences of urban planning policies. On this basis, the geographer Berry (1964) proposed to conceive a two-scale model of 'cities as systems within systems of cities' according to their functional complementarity on a regional or national territory. The *self-organization* theories of the 1980s, derived from chemistry and physics, proposed to explain the regularities observed at the macro-geographical level by interactions at the micro-geographical level between often uncoordinated and seemingly random actions of many actors. These theories proposed an interpretation of the "bifurcations" observed during the history of these systems, these unpredictable structural changes being associated with small modifications of certain parameters of the models during phases of instability (Prigogine and Stengers 1979; Allen 2012; Haken 1983; Weidlich and Haag 1988). The advanced *theories of complex systems* more recently attempt to explain the processes of emergence, that is, the appearance of novelty in urban systems, in terms of innovation and creativity, as well as in their socio-spatial and political organization.

The first two chapters of this book written by Michael Batty and Juste Raimbault are devoted to assessing the relevance of these explanatory contributions. Both are concerned with explaining how the apparent chaos of individual decisions, actions and representations may be composed for explaining the "order" i.e. regularities that are detected when observing cities at various spatial and temporal scales. Which definition of complexity is more relevant for framing urban theories? Are cities and systems of cities, which are objects of the social sciences, amenable to the same basic explanatory principles as those constructed for the natural sciences? If one distinguishes quite clearly between *dynamics*, sometimes similar to those of certain physical processes, and *evolutions* that proceed more from history, how far can one identify the similarities between biological evolution and urban history? On what basis can we build a specific vision of the complexity of urban systems? After enlightening the challenging question of urban system boundary Michael Batty in Chap. 2 demonstrates how to connect major properties of urban systems with concepts of scale, fractality, allometry, entropy and information theory. He provides analytical representations of these important notions adapted to urban objects. Juste Raimbault in Chap. 3 introduces from a broad review of literature three types of complexities: emergence, computational complexity, and informational complexity. He also examines which linkages can be established between them and suggests an enlarged view for better integrating various disciplinary approaches and developing new modelling methods. He emphasizes the usefulness of considering reflexivity in the process of tackling with complexity in social sciences.

1.3 Geographical Scales and Scaling Laws

Following Michael Batty's encouragement to measure the importance of spatiality and scales in the interpretation of urban morphogenesis, a special place is reserved in the next three chapters to the question of transposing the invariant scaling laws established in physics to understand the morphogenesis of the physical and social structures of cities. Reliability and significance of the measurements, quality of the transposition of statistical models to geographical problems are at the heart of the critical analysis of this important contribution to the science of the cities envisaged as complex systems. The "scaling laws", whether formulated in terms of power function like Zipf's law, or as relations of allometry or in the form of fractal models, are indeed often considered as "signatures" of the complexity of the systems. Chapter 4 written by Elsa Arcaute and Erez Hatna discusses the interest and limits of this approach, from a statistical perspective that must take into account the geographical dimension of objects. The authors place in the forefront of the discussion the question of the definition and delineation of "cities", too often neglected in the literature by authors who are not very concerned with the social and material realities involved in these denominations. They provide a sensitivity analysis of diverse statistical methods that are used for measuring the exponents summarizing quantitatively the degree of inequalities between cities for some attribute within a whole system of cities. Questioning the solidity of statistical procedures that are used when transferring the scaling concepts from physics and biology toward urban science lead them to build new hypothesis including regional processes. Which uncertainties are arising from the measurement of hierarchical and spatial distribution for a variety of urban activities? To what extent can we confirm the often supposed universality of such scaling laws? In Chap. 5 Olivier Finance and Elfie Swerts continue this reviewing assessment of scaling laws from the point of view of urban geography. They discuss alternative interpretations that were proposed in the light of the spatial functioning of urban systems and their historical evolution. How are scaling laws linked to spatial concentration processes, to socio-economic transformation of urban settlements and to the theory of the hierarchical diffusion of innovation? Relying on their own experience with original data, investment flows in France and employment in China, they are able to provide guidelines for a safe way of handling scaling laws as a new method applicable to the making of urban science. On the case of brand new mobile phone data among Chilian cities, Horacio Samaniego and his colleagues Mauricio Franco-Cisterna and Boris Sotomayor-Gómez in Chap. 6 show how the networks that structure the interactions of interpersonal communication within cities can consolidate the explanatory framework provided by scaling laws. They bring forward useful connections between the new science of complex networks and urban science, recalling in passing the precise definitions of all the elements of a formal description of the networks.

1.4 Urban Development: Economical Versus Geographical Theories

Economic activities in their growth and the diversity of their expressions in the urban realm are often considered as the main engine of urban development. The correlation between economic growth and urbanization is undeniable in the long term, despite the significant fluctuations and reversals of the dynamics that were generated locally, either through over exploitation of the resources or the obsolescence of certain specializations, or at broader scales when countries of very contrasted income levels and types of productive organizations became connected, for instance by colonization. But can economic science suffice to theorize the expansion of cities in their extent and diversity? The economic science considers cities first and foremost as a source of 'agglomeration economies', and this supplementary return to investment is narrowly linked to the concentration and diversity of activities that are gathered in densely populated areas. Undoubtedly agglomeration economies forms the heart of economic theory of cities. That position reviewed in the light of the complexity sciences stipulates that it is the interactions between people that determine the functioning and evolution of cities, as systems of articulation and concentration of productive activities of all kinds. Chapter 7 written by Marc Barthélémy proposes a precise investigation of these local individual interactions, first empirical according to mobile phone data, then formalized in a predictive model. How are the polycentric hierarchized patterns of urban activities, which lead since long to develop a geographical concept of centrality, to be derived from mapping the hotspots of mobility patterns in a city? More generally, can stochastic models from statistical physics improve the current state of predictive power of the classical economic models?

However, the dominant mainstream principles of market economy that are sometimes represented as "natural" and "universal" may hide the associated social organizational processes which maintain inter-urban competition and reinforce urban concentrations, sometimes probably beyond their actual collective benefit. Urban theories may well be linked with theories of political economy, as argued since long by clever observers of the practical dynamics of urban systems (to my knowledge, Botero (1588) would be among the first urban theoreticians in this perspective) and nowadays often denunciated as a side effect of neo-liberalism (Harvey 2010; Davis 2006). Indeed, some economists question the support given by public policies to large urban concentrations, which are supposed to generate more growth through agglomeration economies, and which therefore push for metropolization (World Bank 2009). The forms that urban development has taken in the world of advanced capitalism, finance and digital technologies are bringing new problems in terms of territorial inequalities. Olivier Bouba Olga and Michel Grossetti in Chap. 8 criticize a set of principles that are summarized in the "CAGE" acronym (Competitiveness, Attractiveness, Globalization, Excellence) and widely enter in the justifications of a diversity of public policies in France. Considering them as part of a collective "myth" for the French case they challenge these principles first according to their empirical validity and second to their consequences that mostly tend to increase the territorial inequalities. In Chap. 9 written by Lena Sanders, Isabelle Thomas and Céline Vacchiani-Marcuzzo another concern emerges from the epistemological posture defended by economists, compared to geographers. Interestingly the authors replace the epistemological debate into the sociological context of each discipline to build a better understanding of the divergences and sometimes lack of communication between these two domains of research. The question of methodological individualism that dominates in economics and is classically opposed to multi-scalar approaches in geography is not the only source of dispute. Will this still durably hamper a fruitful dialogue for building sounder theories and models?

1.5 Universality or Local and Temporal Specificity of Urban Theories

The authors of Chap. 9 show that interdisciplinary misunderstandings do not come most often from philosophically opposed beliefs, but especially from the choice of objects and questions considered as the most important by each discipline. Economists and physicists share some interest in identifying and modeling universal processes. Geographers wish to analyze as well the more or less decisive interferences occurring between urbanization processes and the cultural and political organization of the territories where they take place. Are processes and forms of urbanization really comparable and similar in time and space? Should we formalize different theories and models to account for developments in various parts of the world? These important questions are discussed in the third part of the book. China is an interesting case because of its very rapid urban growth of the recent decades that is narrowly coupled with the extraordinary take-off of the economic development. Elevation of the income level is materialized and somehow visually summarized as a metaphor in the surge of series of high rise buildings with very high population densities, while concerns for migrant's status, air pollution and financial risks linked to this development are arising. About this case of China's urbanization, Fulong Wu in Chap. 10 recalls the debates that have recently agitated mainly Anglo-Saxon geography under the cover of contesting urban theories that would be too adjusted to a North Western perspective, sometimes considered as colonial, and which would have forgotten the diversity of the world. These debates have not taken the same scale in Europe, and the fact that Fulong Wu reports on these debates on the case of China makes it possible to check the degree of relevance of these arguments. His careful analysis of similarities and differences is together illuminating the limits to comparison and providing new insights in measuring the powerfulness of cultural and political features that are inherent to each different society. Natacha Aveline in Chap. 11 also explores the Chinese model of urbanization according to the financing of housing. She does not provide another illustration of an economic bubble linked with the housing market, but "intends to highlight the challenges posed by China's residential market dynamics to the bubble theory". She excels at dismantling the misconceptions and revising the debates related to the financialization of the construction of Chinese cities, by analyzing very precisely the role of banking systems, local governments and households in this process. Is an alternative model to urban construction cycles possible to avoid further large urban crisis? In Chap. 12 Solène Baffi and Clémentine Cottineau challenge another series of models of urban development that are supposed to be universal, at two geographical scales. Discussing the meaning of "emergent" as applied to countries or to complex system's dynamics, they try to identify a series of urban models that would be relevant and useful for poor countries entering the urban development. At local scale, the BRICS states provide really original models of policies in favor of the "compact city" through "transit-oriented development" by introducing entirely new solutions for managing the urban mobility in difficult socio-economic contexts. At the scale of systems of cities, they develop forms of spatial organization that are different from those experimented long ago in Europe and that were formalized under central places theory. These national cases are meaningful examples of how the urban complexity is linked with the territorial diversity, of which Eric Denis provides in Chap. 13 a fascinating global map. Relying on the best harmonized data bases recently made available on the surfaces covered by urban constructions and the populations they host, Eric Denis analyses for the first time at world level the diversity of the relationship between urban physical and demographic expansion between 1960 and 2015. He explores the diversity in trends of the spatial expansion of cities around the world, as well as the variety of the level of correlation with indicators of economic development. Is urban sprawl always correlated with urban population growth? How do really evolve the economic indicators that are usually considered as the best translations, both as causes and consequences, of the urban spatial and demographic expansion?

1.6 Computational Turn in Science for Urban Models

Among the great diversity of existing urban models, this book focuses particularly on those related to urban theories because they encapsulate in mathematical or computer forms processes that have been the subject of a very large number of observations in different urban contexts. Most are dynamic models, designed sometimes to explain and sometimes to predict (or at least frame) likely changes. As Michael Batty briefly points out in Chap. 2 , urban theories use the concepts of complex systems because cities and systems of cities are never in equilibrium but instead are always in open evolution. Explanation and understanding of forms and processes therefore always involve observations and conceptualizations that are both spatial and temporal (Pumain and Sanders 2013). The scales of time and space are variable depending on the problem, as can be seen for example in the examples presented in Chaps. 7 or 9. Urban models may be designed for many purposes, but two kinds of use are dominant: sometimes they help to embody theoretical hypothesis under manageable form in order to test them, and most of times they are conceived for preparing decisions at different levels of certainty according to their predictive powerfulness and quantified precision of their outputs. Chapter 14 by Juste Raimbault introduces a complete change in that modeling landscape through promoting major advances in multi-agent modelling thanks to evolutionary algorithms and intensive computing. Entirely new modelling methods are applied to co-evolution processes in urban systems. The classical approaches in urban planning for simulating interactions between cities and transport networks (coined as LUTI models for instance) are totally renewed from those concepts and the method may be used at any geographical scale. In Chap. 15, another advance in modelling is proposed by Mehdi Bida and Céline Rozenblat that may help solving some of the difficulties mentioned in Chap. 9. They suggest that it is now possible to model urban development from the bottom up, following the ways economic firms develop their network within and between cities. Thus they provide a micro-founded economic model that is able to reproduce the two major emerging geographical properties at the higher level of the systems of cities, i.e. their hierarchical organization of city sizes and the diversity of functional economic specialization of cities.

As a whole, the chapters of this book have been conceived by their authors and previously discussed in the framework of a colloquium.[1] Their approaches to theories and models are diverse but agree on two very important points: the theoretical construction is based on empirical observations taken in a variety of geographical contexts, and taking into account this diversity is of great importance, even in operations of abstract construction or experimentation. This makes their comparison possible and their cross-talk fascinating.

References

Allen, P.M.: Cities and Regions as Self-Organizing Systems: Models of Complexity. Routledge (2012)

Berry, B.J.L.: Cities as systems within systems of cities. Pap. Reg. Sci. **13**(1), 147–163 (1964)

Botero, G.: Della cause della grandezza e magnificenza delle città. Tre libri. (Re-edited in 1598 with the book "La Ragion di Stato" (7ª ed.) in Venise, and a cura di Luigi Firpo by Unione Tipografico Editrice Torinese in 1941) (1588)

Davis, M.: Planet of Slums. Verso, London and New York (2006)

Forrester, J.: Urban Dynamics. MIT Press, Cambridge (1969)

Haken, H.: Synergetics, an Introduction: Nonequilibrium Phase Transitions and Self-Organization in Physics, Chemistry, and Biology, 3rd rev. enl. edn. Springer, New York (1983)

Harvey, D.: Social Justice and The City. University of Georgia Press (2010)

Prigogine, I., Stengers, I.: La nouvelle alliance: métamorphose de la science. Gallimard (1979)

Pumain, D., Sanders, L.: Theoretical principles in inter-urban simulation models: a comparison. Environ. Plann. A **45**, 2243–2260 (2013)

Weidlich, W., Haag, G. (eds.): Interregional Migrations. Dynamic Theory and Comparative Analysis. Springer, Berlin, (1988)

World Bank.: Annual Report (2009)

[1]Closing conference of the ERC GeoDiverCity program organized at the Institute of Complex Systems of Paris-Ile-de-France on October 12th and 13th 2017. http://geodivercity.parisgeo.cnrs.fr/blog/. We acknowledge ERC for partly financing the colloquium.

Part I
Theories of Complexity and Urban Systems

Chapter 2
Defining Complexity in Cities

Michael Batty

Abstract This chapter defines a series of measures of complexity that pertain to the spatial structure of cities. First the development of complexity theory is sketched with specific reference to the ways in which it has and is being exploited in applied areas such as urban science and city planning. The key signatures of a complex system are outlined, with a focus on the various subsystems of the city manifesting self-similarity over a range of scales, thus invoking ideas about fractal geometry. A series of scaling relations are then defined that are adopted as these signatures. These scaling relations are power laws and they emerge in many different contexts with respect to cities. Here our focus is on spatial scale which serves to define these relationships. As well as basic fractal scaling, these include allometry, spatial interaction, mass, distance and area relations, and rank-size rules that pertain to city size and related distributions. Once these functions have been introduced, we examine ways in which the classic entropy or information formula first stated by Shannon (1948) can be used to measure the degree of complexity in a city. These measures focus on the shape of these scaling distributions such as population that define a city and the number of objects or components that count the size of such distributions. We conclude with some challenges for defining complexity further.

2.1 Articulating Complexity

At the turn of the millennium, Hawking (2000), the astrophysicist known for his portrayal and popularization of the phenomenon of the black hole, told the Newspaper *The San Jose Mercury*: "… that the 21st century would be the century of complexity". From the mid-20th century onwards, the idea that any system embedded in the real world manifested a degree of complexity that was beyond our scientific understanding, grew in importance. This did not mean that we could no longer abstract systems to the point where we might consider them closed and tractable but it did mean that as soon as we considered any system in its human context, it was complex beyond our

M. Batty (✉)
University College London, Centre for Advanced Spatial Analysis (CASA), London, UK
e-mail: m.batty@ucl.ac.uk

D. Pumain (ed.), *Theories and Models of Urbanization*, Lecture Notes in Morphogenesis, https://doi.org/10.1007/978-3-030-36656-8_2

wider comprehension. By the end of the 19th century, two main approaches to science had developed, the first based on the laws of mechanics which sought to understand how systems were organised in the simplest way possible using well-defined physical regularities, and the second, theories developed more than two centuries after the first, based on the laws of 'organics'—of evolution—which sought to understand how systems changed according to how their elements adapted to one another and to their wider context. At the risk of oversimplification, the first approach sought to see systems as *machines* where physics represented the dominant paradigm whereas the second sought to see them as *organisms* where biology was to the fore. A related difference is reflected in the fact the science of machines is the science of negative feedback whereas the science of organisms is the science of positive feedback.

Complexity theory grew slowly but surely during the first half of the last century. A somewhat prescient statement by Weaver (1948) in an address to the Rockefeller Foundation divided science into three domains: simple systems with few variables such as those in classical mechanics which were largely deterministic in their predictions, systems that contained extensive elements that appeared disorganised but were predictable only in a statistical sense; and systems that were largely unpredictable in which their elements adapted to one another and in which order could emerge from what appeared to be random behaviour at the most individual level, systems of organized complexity. In this sense, these latter systems were more like organic systems rather than mechanical ones, unpredictable even in the statistical sense but systems that enabled innovation, novelty and surprise in that their unpredictability provided scope for future inventions. Weaver's paper remained buried for some years but it was picked up by Jacobs (1961) who argued, in her seminal book *The Death and Life of Great American Cities*, that cities were not problems of simplicity, or of disorganised complexity like statistical systems but problems of organized complexity in which diversity and choice were key forces in their definition and structure.

Besides Weaver's paper and Jacobs's book, there were other attempts at thinking about systems that, implicitly at least, questioned predictability. Simon (1961) in his paper "The Architecture of Complexity" argued that systems needed to be thought of as nested, as hierarchical with the components of their hierarchy reflecting the way the parts could be added up to form the whole. In fact, the mantra of complexity theory which is originally accredited to Aristotle and which was widely acknowledged in gestalt psychology, is that "the whole is greater than the sum of the parts". This embraces the notion that it is not possible to predict the structure of the system from its individual elements and it is reflected in other statements about complexity ranging from the architect Mies van der Rohe's cliché that "less is more" with respect to contemporary buildings of the modern movement to Anderson's (1971) statement about complexity as "more is different" that marked the beginning of the field in the 1970s.

In some respects, both models of science—classical mechanics and evolutionary theory—were used to provide analogies to human systems almost from the time they first emerged. It was mechanics however that provided the first model and by the middle of the 20th century the so-called systems approach which articulated systems as being centrally organised, often in an equilibrium which was maintained through negative feedbacks, held sway in a variety of disciplines that required some overarching theoretical structure. Many of the social sciences fell into this camp while biology and some applied engineering were also articulated in similar terms. In fact, physics and economics remained aloof but in the event, it was a critique of these disciplines that led to the systems approach falling into disrepute. The idea that the systems to which the approach was applied functioned centrally was clearly problematic and the idea that such systems operated from the bottom up rather than the top down began to take hold.

Thus the characteristic feature of complex systems is that they function from the bottom up. This does not mean that there is a complete absence of processes that operate at different scales above the bottom for most real systems are a mixture of top down and bottom up. But as human systems operate at the level of the individual, then it is this base on which the evolution of the system rests. Individuals acting as groups and collectivities and governments and many other agencies work at different levels but at the end of the day, everything happens from the bottom up. The notion that human systems function as being predominantly top down is a notion that has become passé, largely because human systems start small and become big; this means that individuals drive the system at every level. Such systems can decline from large to small but their genesis is always small, evolving from the bottom up. When systems grow in this fashion, the big question is how their parts are coordinated because what appears are patterns which are ordered at higher levels.

In other words, order seems to emerge from chaos although on further reflection the action of the individual is never chaotic. It might appear random in one sense but that is only because we cannot reconstruct the overall decision-making process of each individual in question. In effect, physical constraints on the process of decision-making as well as long-term collective goals tend to drive the system towards certain structures that appear ordered and explicable. Moreover, the precise form of the system that emerges also depends on historical accidents. Random growth reflects history but, in this sense, random does not mean chaotic: it only means that in explaining a system, we are unable to reconstruct the parts that lead to the whole. In fact, the great challenge of complexity theory is to describe systems from the bottom up and to articulate the processes that enable this order to emerge.

We could continue introducing properties of complex systems almost indefinitely for there is no exhaustive definition, nor will there ever be. But to progress, we need to examine different features that pertain to cities and to begin, we will examine how the definition of the city reflects its complexity in terms of spatial-geographical, and then non-spatial boundaries which determine the system of interest. As we will see, the boundary issue is one of the most difficult problems of analysis for how we interpret the city depends on how we define where cities begin and end. Once we have noted these ideas, we will introduce the idea of self-similar objects that define the spatial organisation of cities—fractals—showing how various scaling relations emerge quite naturally from ideas about how cities fill space. We will then introduce an almost nihilistic model of urban evolution that enables us to generate elements of cities with respect to their shape and size as well as boundaries. The competitive equilibrium that we describe serves to generate signatures of complexity that recur again and again in the various contributions to this book. These are essentially generic distributions of the elements that make up a city which are fashioned as power laws, and this lets us see a series of scaling relations that characterise the properties of the city with respect to its complexity. This leads to measures of complexity based on information that our definitions imply. One of the features that is still lacking when it comes to cities and complexity are insightful statements of the processes that enable cities to evolve. Although for a long time, we have all agreed that cities can never be in equilibrium, the processes that drive their evolution are still hard to pin down and the great challenge of complexity theory is to define these processes so that we have some way of thinking about how cities might evolve in the future.

2.2 Boundaries to City Systems

The systems approach takes as its central construct the division between the system itself and its wider context which is usually called the environment. It is assumed that there is a clear distinction between the two and that our understanding of the system does not depend on us having to explain what happens in the environment. In short, it is assumed that the density of interactions between component parts of the system—through its subsystems and their elements for example—is much greater than any interactions that take place between the system and its environment. In fact, a common assumption is that the line between the system and its environment is drawn in such a way as to minimise these interactions, both physically and functionality. The most obvious spatial example in terms of a city system and its wider environment (or hinterland) occurs where the flows between the environment and system are as small as possible and the flows within the system as large as possible, or in some sense maximal. The classic exemplar of such a division is between the solar system and the rest of the universe. In general, Newton's laws apply to a very fine approximation to this system but break down when considered for the whole universe. Thus, it is possible to produce very good predictions of what happens if we simply study the solar system regardless of anything else. To an extent, it is scale that makes this possible in that the solar system is a cluster of related masses that is many light years away from anything else that might interfere with its local gravitation.

In many systems, particularly those which are dominated by human actions, the biggest problem with setting boundaries in this way is that it is often impossible to minimise the flows between and maximise those within. In a world where cities now have a global role, key economic interactions are no longer restricted by any physical boundaries that are essentially artificial. Local flows of course remain local but even at the finest spatial scale there are interactions that are generated by global forces. If we cannot divide our world of cities into discrete entities, then it is increasingly difficult to isolate properties that can be studied as if they are invariant in some way. This is also part and parcel of the problem that cities are getting more complex as they evolve. There is little doubt that since the industrial revolution began that cities have become ever more complex as more and more opportunities for interaction have been invented. Now that our society and cities are almost entirely digital at one level, this complexity is increasing at a faster rate than we are able keep up with, largely because the ingestion time for new digital technologies is getting shorter and shorter.

Physical boundaries need to be well-defined but non-physical can be conceptual. Most definitions with respect to cities are physical in some sense, but there can be city systems separated from their environment that appear entirely artificial. The model of urban development for a typical US city developed by Forrester (1969) divided a hypothetical inner city (moulded on his impression of Boston from its wider environment) from the rest of the city so that suburbs of the city were not defined and simply absorbed into the wider environment. In such a model, the division was quite arbitrary and meant that the inner city could not interact with its environment in a way that generated realistic predictions. The model was highly critiqued because of this artificial separation although there was a logic to divide the system from its environment in this way so that the plight of the inner city could be highlighted, thus demonstrating how difficult it was to regenerate a more economically viable form. In fact, what Forrester showed was that as the boundary changed, the results from the model changed: the fact that the predictions of the model changed dramatically as the boundary between inner and outer city changed, starkly pointed to the dilemma facing policy makers in US cities in the late 1960s. In terms of complexity, this is a clear example of the intrinsic difficulty of capturing the right balance between the system and its environment.

The usual problem with defining boundaries between the city and its wider environment relate to defining the points or areas where the influence of the city is weakest. The problem is that this is a forever shifting boundary for as the city grows or declines, waxes or wanes, the degree of influence also varies. For example, in a project on defining London's periphery, it appears that a good division from work by Arcaute et al. (2014) is based on including all those areas that are at least 14 persons per hectare in population density and which contribute more than 30% of their flows to areas around the area in question. The problem however is that as London is growing by additional population around its edge, then this boundary is continually changing and if we were to add other key features that define the city, for example with respect to the myriad of other flows of energy, people and money, and much else pertaining to more qualitative features, it becomes almost impossible to find any 'best' definition. Definitions of the city thus depend on what the purpose is for which the definition is required. Thus it is easy to sink into a black hole of relativism unless one is very careful about what is included by any definition for whatever purpose.

There is another intriguing property of geographical boundaries that serves to introduce the notion of infinitely complex systems. In 1967, Benoit Mandelbrot wrote a famous paper in the journal *Science* entitled "How Long is the Coast of Britain?" to which he gave the answer 'infinity'. Imagine you are measuring the length of a coastline. You first choose a scale S and then between two points that define the line, you count the number of segments $N(S)$ at that scale S that defines the line. You then choose another scale—let us say a finer scale $S' < S$ and count the number of segments $N\left(S'\right)$ that define this line. Now we need to find the lengths of the line $L(S)$ and $L\left(S'\right)$ for each of these scales where we define length as the number of segments multiplied by the scale. For the first scale, $L(S) = N(S)S$ and for the next scale $L(S') = N(S')S'$. Now if these lengths are the same, then it is obvious that the line is straight, that is $L(S) = L(S') = N(S)S = N(S')S'$, where the number of segments must go up linearly as the scale gets finer. This means that no more detail is being picked up as the scale changes, and that $N(S') = N(S)S/S'$. Now imagine that more detail is actually picked up as the scale changes, then the line gets longer and we can write the equation for the number of segments as a non-linear function of the scale, that is $N(S) \sim 1/S^\alpha$ where if the number of segments gets bigger, then $\alpha > 1$ and if smaller, $\alpha < 1$. When $\alpha = 1$, no more detail is being picked up and the line is linear.

The question of course is whether or not a scale change changes the nature of the phenomenon, in this case, the coastline. Mandelbrot (1967) following earlier observations by Richardson (1961) demonstrated that this indeed was the case. In short, as the scale gets finer, the length of the coastline gets proportionately greater in length and in the limit, the length of the coastline becomes infinite. Only when you begin to scrutinise the nature of the measurement does this makes sense because one continually downsizes to below the scales we can experience and ultimately one ends up measuring around every atom and molecule that makes up the material that is the coastline. We need to say a little more about the power law that emerges with respect to the number of segments and length. We can now write the equation for length as $L(S) = N(S)S \sim 1/S^{1-\alpha}$ and from this, it is easy to see that when $\alpha = 1$, the line is straight but when the power differs from one, the line increases more or less than proportionately in length. To relate this to our treatment of scaling below, let us note that we could consider the number of segments $N(S)$ as the frequency $f(S)$ and the scale as the size; thus the number of segments is the frequency at a given size, that is $f(S) \sim 1/S^\alpha$ which can easily be converted to the counter-cumulative which is a kind of rank-size rule. The key issue of course is that this is a power law, the signature of a fractal, and that the power is related to the fractal dimension. From the pure relationship $N(S) \sim 1/S^\alpha$, we can derive the parameter as $\alpha = -\log(N(S))/\log(S)$ which is in fact the fractional or fractal dimension. In general, it differs from 1 which is the Euclidean dimension of the straight line. More on this later as it is key to the morphology of cities and is a measure in and of itself of complexity.

We can also extend our idea of the boundary to the area of the space in an obvious but non-rigorous way by taking the square of the length $L(S)^2$. This might be fashioned as $A(S) = L(S)^2 \sim S^{2(1-\alpha)} = S^\beta$ which is an equivalent power law with a different fractal dimension. Just as we pick up more detail from the line as we make the scale finer, we can do the same with area, on the assumption of course that measuring area in this way is meaningful. In cities, boundaries may reflect increasing detail but only over very limited orders of magnitude. When it comes to development that fills areas, then again, the notion of more detail being picked up is relevant, as for example in ways space is filled at the city, district and neighbourhood levels using self-similar modules such as principles of vehicular-pedestrian segregation in housing layouts and other modular constructions. But in terms of literal area, then one has to be careful. In biological systems where more and more dendritic detail is observed as one looks at more microscopic levels, then area does increase in the same way as that when one observes a coastline. But for built environments, the comparison is rather blunt and it is the analogy with ideas about self-similarity that is important. There are, in fact, other ways of measuring space-filling than scaling the boundary equations which we have explored elsewhere (Batty and Longley 1994).

In this book, there are many examples dealing with cities as fractal structures, where urban form consists of clusters that have a degree of self-similarity often measured using statistics associated with their fractal dimension. Fixing the orders of magnitude, that is the scale variations over which we measure fractality, is difficult in that the nature of the phenomena—total development, network structure, different land uses, different demographic areas—all of these elements of urban structure generate different fractal dimensions and there is even the possibility that the entire variation of phenomena in the city is multi-fractal in some sense (Salat, Murcio and Arcaute 2017). But in general, most cities appear to manifest a fractal dimension towards the Euclidean dimension of the plane—likely to be 1.71 or greater—although much depends on where we draw the boundary (Batty and Longley 1994). Moreover, as the city grows and spreads (or even declines), the boundary changes with respect to where we consider the periphery to be and there is no theory that tells us where the best boundary might be and what we might expect the fractal dimension to be or any other properties that pertain to structure such as density. Thus, we are still bogged down in this sea of relativism—where is the boundary? And how do the various properties of the city such as density vary with respect to city size? To explore these issues further, we will now introduce a series of scaling relations that illustrate how cities are structured with respect to their scale, shape, size and the interactions between the elements that define their morphology.

2.3 Scaling and Complexity

There are many scaling relationships that define various properties of cities but there is no integrated theory that ties all these together. However, we can classify them into at least four types. First, we will present ways in which various attributes of different cities change with their size under the heading of *allometric relations*. Second, we will characterise *spatial interactions* between cities in terms of their size and the spacing between them and illustrate how these define the forces that bind entire cities to one another. Third we will define the relationship between city size and the spatial extent of the city which relate to how the size of the city grows with respect to its area which we call the *mass-area* or *mass-radius* relations. And last but not least, we will define relationships between cities of different sizes in terms of their frequency distributions. We will call these *rank-size distributions* although as we will see, we can relate these properties of scaling to each other in diverse ways. This might imply that they can all be derived from each other but this is not the case in strict formal terms, although there are many tantalising links that need further work in generating a more integrated theory of the complex city in terms of scaling relations. Arcaute (2018, this volume) extends our treatment of scaling to networks and other features of cities in a later chapter.

Before we introduce these relations, we need to define some standard variables. We will begin with population P_i in city i which we will also normalise as a probability p_i which makes its use more generic. Wherever we define P_i, there is a simple transformation to probability form as $p_i = P_i / \sum_k P_k$ and from probability form back to the population as $p_i \sum_k P_k = P_i$. Using these definitions, first *allometry* is in some respects the most obvious of all scaling relationships. Usually an allometric relation is defined between one attribute of the city say, Y_i and some measure of size as $Y_i \sim P_i^{\theta}$ where θ is the allometric coefficient. If $\theta > 1$, then this implies that the variable Y_i increases more than proportionately and one of the most important relationships which has been tested in recent years is where these variables are income and population (Bettencourt 2013; West 2017). If there is positive allometry where $\theta > 1$, then this implies economies of scale where income increases more than proportionately with city size which is consistent with ideas in macroeconomics. The opposite where $\theta < 1$ implies the existence of diseconomies.

Diseconomies are also functions of the increasing separation or distance between places. We can develop a simple spatial interaction that encompasses both allometry and distance deterrence where we measure the attraction between two cities i and j in terms of their allometry, that is P_i^{θ} and P_j^{γ}, which combines to produce $P_i^{\theta} P_j^{\gamma}$. We can then simply factor the deterrent effect of distance into the relationship as $T_{ij} \sim P_i^{\theta} P_j^{\gamma} / d_{ij}^{\mu}$ where θ, γ, and μ are the allometric parameters controlling the economies or diseconomies of scale. Note that the signs of these relationships can be positive or negative and the values greater than 1 or less than 1. There is thus complete flexibility in using relationships such as these to reflect the properties of city systems. There is a third relation that in essence comes from allometry and this is the mass-radius or mass-area relationship which we define as relating the area of the city to its mass as measured by its population. This is for an individual city so we can drop the locational referent although different cities are likely to have different mass-area relations. The relations can be stated as $P \sim A^{\theta}$ where θ is the coefficient that converts the area into the mass and it is related to the fractal dimension as we defined it in the previous section. Then if we define the area as $A = \pi d^2$, and use this in the mass-area equation, we see that $P \sim A^{\theta} = (\pi d^2)^{\theta}$. With a little bit of judicious manipulation, it is easy to demonstrate that $P \sim d^D$ where D is the fractal dimension.

So far, we have mixed these scaling relations which pertain to both the intra-urban and inter-urban domains. Allometry can relate to both but most of the recent work has been upon examining systems of cities, the distribution of city sizes rather than components within the city itself. There is still allometry to be worked out here but the predominant focus has been on cities of different sizes. Spatial interaction is relevant to forces within and between cities while the mass-area relations pertain to individual cities but can be applied to many cities so that comparisons can be made with respect to how different cities fill space (Batty and Longley 1994). Our last relation is perhaps the most ubiquitous and to an extent nihilistic because all it does is to portray size distributions themselves. This is the rank-size relation in its various forms that was first demonstrated over a century or more ago (Auerbach 1913). This can pertain to populations and other attributes of locations within cities as well as between cities and in this sense, it is generic because all the components that we have hinted at so far, populations and their various types, distances, areas, densities and so on can be examined in terms of the shape of their distributions and how these are formed.

The rank-size relation is easy to state and we begin by simply noting that the relationship between the frequency f with which cities of different sizes P exist which we call $f(P)$ is a simple power law of the form $f(P) \sim 1/P^{\alpha} = P^{-\alpha}$. We will justify this in a moment but this essentially is from observations that were first popularized for cities and word frequency distributions by Zipf (1949). If we first produce the cumulative frequency distribution as $F(P) \sim P^{-\alpha+1}$ and then define the rank as $r(P) = 1 - F(P)$ which is the counter-cumulative, Zipf's Law is the inverse of this which is $P \sim r^{1/(1-\alpha)}$. Now the pure Zipf law which was fitted by Zipf to US city populations using 1930 Census data and then again by Krugman (1996) to 1990 data gives a Zipf parameter $\alpha = 2$. This means that the size of any city $P(r)$ at rank r is $1/r$ of the population of the first ranked city $P(1)$. The second city's population is thus half the first city's population and so on down the hierarchy. In fact, this appears to be an extreme case of the rank size relation although under fairly minimal conditions, Gabaix (1999) shows that a stochastic process can generate this pure rank size rule.

The key link between all these scaling relations is that fact that they are all power laws, and thus scale independent. In short, if we scale the relationship by multiplying the relevant size variable by a scalar Φ, we can easily demonstrate that the relevant equations predict proportionate change and in this sense is self-similar. First we look at the fractal relations pertaining to boundaries and areas that we introduced above. We can write these as

$$\text{For fractal boundaries}: L_i^{'} \sim (\Phi S_i)^{1-\alpha} = \Phi^{1-\alpha} S_i^{1-\alpha} \sim L_i$$

$$\text{For fractal areas}: A_i^{'} \sim (\Phi S_i)^{\beta} = \Phi^{\beta} P_i^{\beta} \sim A_i$$

Then for each of the four scaling relations, we scale and simplify as follows:

$$\text{For allometry}: T_i' \sim (\Phi P_i)^{\theta} = \Phi^{\theta} P_i^{\theta} \sim Y_i$$

$$\text{For spatial interactions}: T_{ij}' \sim (\Phi P_i)^{\theta} P_j^{\gamma} / d_{ij}^{\mu} = \Phi^{\theta} P_i^{\theta} P_j^{\gamma} / d_{ij}^{\mu} \sim T_{ij}$$

$$\text{For mass} - \text{area relations}: P' \sim (\Phi d)^{D} = \Phi^{D} d^{D} \sim P$$

$$\text{For the rank size rule}: P \sim r^{1/(1-\alpha)} \text{in the form of } r = P^{1-\alpha} \text{became}$$

$$r' \sim (\Phi P)^{1-\alpha} = \Phi^{1-\alpha} P^{1-\alpha} \sim r$$

Note that we can scale population size for allometry, and any or all the independent variables in the spatial interaction model, that is populations and/or distances, distance as a proxy for area in the mass-area relations, and population in the rank size rule. In short, what all this implies is that in terms of these properties or signatures, the city or system of cities is fractal in that as we scale, the underlying relations remain invariant. There is a good deal more we can say about these ideas but we prefer to leave the reader to explore their many implications which are sketched in various chapters in the rest of this book.

There is a particularly simple explanation for the emergence of city size distributions that follow power laws, where there are few very big cities and very many small ones. The essence of this is a competitive process, articulated at the level of whole cities which is limited from below with respect to the size cities can drop to and operates using random growth rates for all cities. If for example there are n cities, then the probability of one of these becoming very large is much less than the probability of it staying small. If cities grow or decline with random growth rates, then the probability of a string of positive growth rates for one city decreases through time as the system of cities evolves. Cities will grow and decline randomly but if there are enough cities, then some will certainly get large but most will remain small. To make sure there are no negative cities, then cities cannot fall below a lower bound which may actually be zero and if a city does so, it disappears from the map. Usually this process replaces any cities which fall to the minimum threshold on the assumption that the city disappears and a new one is introduced. In some variants of the model, the number of cities expands with new ones continually being introduced, implying positive growth rates overall.

It is easy to demonstrate that a power law of city size distributions results from this process (Batty 2013), and it is possible to introduce various constraints that enable the ultimate distribution to mirror distributions such as the lognormal, the stretched exponential, the double Pareto and so on. In fact, in many versions of the model, the power law only exists for the heavy tail which we define as the upper tail of the rank size distribution where the largest cities exist. The reason why we consider this model to be nihilistic is that other than competition which we enable through the choice of successive randomly chosen growth rates, there is nothing in the model that ensures that the system of cities produces anything other than a power law. There is no locational specificity in the model and although this can be introduced to provide a spacing for cities in their wider system, this does not make any difference to the ultimate size distribution. If the system of cities is constrained in some way as real systems tend to be, then the parameter defining the Zipf distribution is likely to vary from unity (see Cristelli et al. 2012). At this point, we have introduced a range of scaling laws that define the way elements of cities at different scales nest within one another and we are now in a position to develop measures of complexity that enable us to demonstrate how complex systems betray a degree of complexity almost beyond our comprehension that will continue to challenge our interpretations.

2.4 Measuring Complexity Using Information Theory

An obvious measure of complexity is the number of distinct elements that is contained within the system. This of course in complex systems can be many-fold but in terms of our discussion of fractal structures such as the coastline boundary, the number of elements is infinite. So one clear measure is simply this number which we will call n. How these n elements relate to one another is another aspect of complexity and a good measure of this is their distribution—how each object relates to all the others. This covers their interactions as we noted above as well as their relative position in terms of the shape of their distribution. We can measure the distribution of objects by their size or probability which as before for each object we define as p_i where $p_i = P_i / \sum_k P_k$ and $\sum_i p_i = 1$. Now the classic measure of complexity which is called information is the entropy $H = -\sum_i p_i \log p_i$ which varies from a minimum value of 0 to a maximum value of $\log n$ where n is the number of objects or elements comprising the system (Shannon 1948). The two features—the number of objects and their distribution—have very simple properties from this measure. Ignoring the distribution, it is clear that the entropy measure which we define as one measure of complexity increases without bound as the system gets bigger. This is what we might expect as cities get bigger and as the population of the planet increases. In terms of the distribution, the greatest complexity occurs when the distribution of probabilities (or population) is uniform, that is when $p_i = 1/n$ and the entropy in this case is clearly $H = \log n$. If the distribution is at the other extreme which we might define as $p_k = 1,\ p_i = 0\ for\ all\ i \neq k$, then $H = 0$. The interpretation of the shapes of these distributions is important. When the entropy is zero, all the action is concentrated on one object or element and arguably this is a very simple system. When all the probabilities are the same, then the energy required to hold such a system together is much greater than for the case where only one object is significant and this also seems to accord with our notion of what complexity is.

All our scaling relations that we argue are the signatures of any complex system are functions based on power laws. Rather than examine all six of the relations noted above, we will take a generic power law and examine its properties with respect to the entropy formula. Assume the power law is the allometric relationship $Y_i \sim P_i^\theta$ which is generic where the probability is defined as $p_i = P_i^\theta / \sum_k P_k^\theta = P_i^\theta / Z$. If we use this in the entropy equation, the complexity is defined as $H = -\sum_i p_i \log p_i = -\sum_i p_i(\theta \log P_i - \log Z)$. Now if $\theta > 0$, then as the allometric coefficient increases, then the system of cities becomes ever more unequal and at the limit, the entropy or complexity is zero, like the extreme distribution noted above. If the coefficient falls to zero, the distribution becomes more even as it approaches the uniform distribution. If the coefficient is negative, then inverse scaling is invoked but the implications for the shape of the distribution are the same.

In fact when we assume a distribution and use this to define its entropy, we could also derive the distribution by maximising the entropy subject to the constraints that we know the distribution must meet. In the case of the allometric relation, we can maximise H subject to the constraint on the increase in population which we can write as $\sum_i p_i \log P_i = Y$ which is a constraint on total income expressed logarithmically and a normalisation constraint on the total probability, that is $\sum_i p_i = 1$. Now we can write the entropy-maximising equation as a Lagrangean as $\max L = -\sum_i p_i \log p_i - \theta(\sum_i p_i \log P_i - Y) - \lambda(\sum_i p_i - 1)$ and from this, we can derive the distribution as $p_i = P_i^\theta / Z = P_i^\theta / \sum_k P_k^\theta$. Although we have assumed a distribution so that we can measure its complexity using the entropy equation, the entropy-maximising method has been widely used to generate models in the spatial domain, so-called spatial interaction models, analogous to gravity models in social physics and discrete choice models in economics which are used to predict flows and activity at different locations (Wilson 1970).

There is one last twist in the tail that we will make with respect to defining and measuring complexity. Our focus here is largely on spatial complexity in that scale and size are key to our definitions. So far, we have defined the distribution of elements in a city system as being discrete probabilities but we could rework our analysis using probability densities. All this means is that our probabilities are defined as $p_i = \rho_i \Delta x_i$ where ρ_i is the density and Δx_i is the interval over which the density is measured. Substitution of one for the other makes no difference to the measurements apart from the degree to which density is approximated and if the distribution is on a regular lattice with identical areas for each zone, then distribution is essentially the same as density. There are however some simplifications that can be made to the entropy and by way of conclusion it is worth noting these. First we substitute p_i as $\rho_i \Delta x_i$ in the entropy equation which gives $H = -\sum_i \rho_i \Delta x_i \log \rho_i \Delta x_i$ and this simplifies to $H = -\sum_i p_i \log p_i / \Delta x_i + \sum_i p_i \log \Delta x_i$. The two components reflect the distributional and the number of elements properties of our measure of complexity. The first component $-\sum_i p_i \log p_i / \Delta x_i$ approximates the continuous entropy which measures the shape of the density and which we have called the spatial entropy (Batty 1974). It converges to the continuous entropy measure as the scale or size of the unit over which the object is measured converges to zero and in this sense, it is independent of the number of objects. The second component $\sum_i p_i \log \Delta x_i$ is the scale itself that measures the complexity of the number of objects and as this converges to zero, the complexity increases without bound. In fact, we argue that this is a much more satisfactory definition of complexity because the two components are isolated and it is thus possible to explore the trade-off between density of the objects or elements and their number (Batty et al. 2014).

2.5 Widening Complexity

In the next chapter, Raimbault (2020, this volume) takes a very different view of complexity from that presented here. His view is wider, more expansive and less technically orientated than ours and he broadens the discussion to questions of emergence and computation as well as information which this chapter has been mainly focused upon. Nevertheless, throughout this book wherever complexity is touched upon, the general view is that there is no single definition; consistent with the notion that systems that are complex are infinitely so, there is no definition or even philosophy that is more appropriate than any other. As Raimbault suggests, a pluralistic view which he calls 'applied perspectivism' is relevant. Our view here is that complexity is as much a point of view as an approach although as we have seen there are many specific features of the approach in an applied context.

There are many types of emergence, information and computation which characterize cities. This is because there are many characterizations of city systems and thus there is no one generic template for measuring any city's complexity. For example, modular physical principles that layout urban development in a certain patterned manner at the most local scale, often reflect themselves at coarser scales simply as a consequence of applying these modules at the most local scale. Similar patterns emerge at the macro-scale and this is the essence of fractal structure. Patterns can also emerge at a single scale as simple principles relating to whether or not objects favour co-location with other objects. Segregated populations might emerge in this way and highly clustered patterns can result from a sequence of seemingly random decisions that ultimately add up to what is a highly ordered spatial system. Complexity theory however does differ with respect to the system it is applied to. For physical systems without any human operations, much simpler patterns might emerge whereas in human systems, human actions tend to be more unpredictable and the methods whereby patterns scale from the micro to macro are more convoluted.

Much of this pertains to the degree to which a system is predictable. Complex systems are in essence unpredictable in the large although they may be predictable to an extent in the small. It is impossible in advance to say what degree of predictability might be expected for one of the features of complex systems is that they exhibit novelty and surprise. Complex human systems operating at the level of the individual are subject to all the whims and nuances of human decision-making and what is urgently required is a much more detailed catalogue of different urban forms that are associated with different varieties of complexity. In this chapter, we have defined a series of signatures that characterize spatial systems whose activities are composed of numbers of objects or elements arranged across some spatial distribution. What we need is much more integrated thinking so that we can embrace many different types of complexity in the quest to produce a much more rounded view. We will end where we came in with Hawking's (2000) response. His argument that this century will be the century of complexity was in response to the question: "Some say that while the twentieth century was the century of physics, we are now entering the century of biology. What do you think of this?" And he responded, and I paraphrase, we should go beyond disciplinary boundaries and define new approaches, new sciences if you like, where the great challenge is to cope with the kind of complexity that we see everywhere around us, particularly in cities.

References

Anderson, P.W.: More is different. Science **177**(4047), 393–396 (1971)
Arcaute, E.: Scaling Laws: Insights and Limitations (2018)
Arcaute, E., Hatna, E., Ferguson, P., Youn, H., Johansson, A., Batty, M.: Constructing cities, deconstructing scaling laws. J. Royal Soc. Interface **12**, 20140745. http://dx.doi.org/10.1098/rsif.2014.0745 (2014)
Auerbach, F.: Das Gesetz der Bevölkerungskonzentration, Petermann's Geographische Mitteilungen, **59**, 74–77 (1913)

Batty, M.: Spatial Entropy. Geogr. Anal. **6**, 1–31 (1974)
Batty, M.: The New Science of Cities. The MIT Press, Cambridge, MA (2013)
Batty, M., Morphet, R., Masucci, P., Stanilov, K.: Entropy, complexity, and spatial information. J. Geogr. Syst. **16**, 363–385. https://doi.org/10.1007/s10109-014-0202-2 (2014)
Batty, M., Longley, P.A.: Fractal Cities: A Geometry of Form and Function. Academic Press, London and San Diego, CA. Available at www.fractalcities.org (1994)
Bettencourt, L.M.A.: The origins of scaling in cities. Science **340**, 1438 (2013). https://doi.org/10.1126/science.1235823
Cristelli, M., Batty, M., Pietronero, L.: There is more than a power law in Zipf. Sci. Rep. **2**, 812. https://doi.org/10.1038/srep00812 (2012)
Forrester, J.W.: Urban Dynamics. The MIT Press, Cambridge, MA (1969)
Gabaix, X.: Zipf's law for cities: an explanation. Quart. J. Econ. **114**(3), 739–767 (1999)
Hawking, S.: Unified theory. In: Is Getting Closer, Hawking Predicts. Interview in San Jose Mercury News, 29A (2000)
Jacobs, J.: The Death and Life of Great American Cities. Random House, New York (1961)
Krugman, P.A.: Confronting the mystery of urban hierarchy. J. Jpn. Int. Econ. **10**, 399–418 (1996)
Mandelbrot, B.: How long is the coast of Britain? Science **156**(3775), 636–638 (1967)
Raimbault, J.: Relating complexities for the reflexive study of complex systems (2020)
Richardson, L.F.: The problem of contiguity: an appendix to statistics of deadly quarrels. Gen. Syst. Yearb. **6**, 139–187 (1961)
Salat, H., Murcio, R., Arcaute, E.: Multifractal methodology. Physica A Stat. Mech. Its Appl. **473**, 467–487 (2017)
Shannon, C.E.: A Mathematical Theory of Communication. Bell Syst. Tech. J. **27**(3), 379–423 (1948)
Simon, H.A.: The architecture of complexity. Proc. Am. Philos. Soc. **106**(6), 467–482 (1961)
Weaver, W.: Science and complexity. Am. Sci. **36**(4), 536–544 (1948)
West, G.: Scale: The Universal Laws of Growth, Innovation, Sustainability, and the Pace of Life in Organisms, Cities, Economies, and Companies. Penguin Press, New York (2017)
Wilson, A.G.: Entropy in Urban and Regional Modelling. Pion Press, London (1970)
Zipf, G.K.: Human Behavior and the Principle of Least Effort. Addison Wesley, Reading, MA (1949)

Chapter 3
Relating Complexities for the Reflexive Study of Complex Systems

Juste Raimbault

Abstract Several approaches and corresponding definitions of complexity have been developed in different fields. Urban systems are the archetype of complex socio-technical systems concerned with these different viewpoints. We suggest in this chapter some links between three types of complexity, namely emergence, computational complexity and informational complexity. We discuss the implication of these links on the necessity of reflexivity to produce a knowledge of the complex, and how this connects to the interdisciplinarity of approaches in particular for socio-technical systems. We finally synthesize this positioning as a proposal of an epistemological framework called applied perspectivism, and discuss the implications for the study of urban systems.

3.1 Introduction

The various disciplines or approaches concerned with the study of urban systems have the common point of having postulated their *complexity*: Batty's "new science of cities" is rooted within complexity science paradigms (Batty 2007, 2013); Pumain's evolutionary urban theory makes the fundamental assumption of urban systems as co-evolving complex systems (Pumain 1997, 2017); Haken's synergetics have application in urban space cognition for example (Haken and Portugali 2016); recent works by physicists have applied tools from statistical physics to urban systems (West 2017); architecture has a long tradition in integrating complexity in the urban fabric (Alexander 1977); the new economic geography is aware of the complexity of its objects of study despite its reductionism in methods used (Krugman 1994) and has developed an evolutionary branch (Cooke 2018); artificial life has introduced

J. Raimbault (✉)
Institut des Systèmes Complexes Paris-Ile-de-France (CNRS 3611), Paris, France
e-mail: juste.raimbault@polytechnique.edu

UMR Géographie-Cités (CNRS 8504), Paris, France

Center for Advanced Spatial Analysis (University College London), London, UK

D. Pumain (ed.), *Theories and Models of Urbanization*, Lecture Notes in Morphogenesis, https://doi.org/10.1007/978-3-030-36656-8_3

complex simulation models with application to urban problems (Raimbault et al. 2014); to give a few examples.

What is meant by complexity remains however not well defined, and as (Chu 2008) recalls, several definitions and characterizations exist and there is a priori no reason to think that these could converge. Interestingly, a reductionist view on this matter would aim at some unified definition, whereas most complexity approaches will take this diversity as an asset to be developed (take for example the complex thinking advocated by Morin (1991) which relies on the progressive integration of multiple disciplines and thus viewpoints on the world). Manson (2001) develops different notions of complexity and their potential in the study of geographical systems. In the case of urban form and design, Boeing (2018) gives an overview of existing measures of urban form and how they relate to various complexity approaches. The previous chapter (Batty 2020) has investigated this issue by providing a broad historical overview of complexity approaches to urban systems, and details several measures of complexity linked for example to fractals and Shannon information. Contrary to (Ladyman et al. 2013) which aims at a unified approach of complexity, we will embrace here a diversity of views and highlight their complementarity.

We propose here a theoretical entry to an effective integration of some approaches to complexity, inducing a framework to study complex socio-technical systems. More precisely, we illustrate the links between three types of complexities (emergence, computational complexity, and informational complexity, which choice will be discussed below), and develop the implication of these links on the production of a knowledge of the complex. This work also has more general epistemological implications, since we introduce a development modestly contributing (i.e. in our context of the study of urban systems) to the *knowledge of knowledge* (Morin 1986). An implicit aim is thus to interrogate the links between complexity and processes of knowledge production.

Our argument will therefore (i) establish some links between different types of complexities; (ii) explore the necessity of reflexivity possibly as a consequence of these links; (iii) suggest a practical framework to apply these principles to the construction of integrative approaches to complex systems.

The rest of this chapter is organized as follows: we first introduce more precisely the types of complexities we consider; we then detail some links which are currently not obvious in the literature (i.e. links of computational complexity or informational complexity with emergence). After recalling reasons suggesting a necessity of reflexivity, we sketch how the production of knowledge on complex systems falls at the intersection of different complexities and how this implies reflexivity. We finally develop implications for interdisciplinarity, introduce the corresponding applied perspectivism framework and discuss the implications for the study of urban systems.

3.2 Complexity and Complexities

What is meant by complexity of a system often leads to misunderstandings since it can be qualified according to different dimensions and visions. We distinguish first the complexity in the sense of weak emergence and autonomy between the different levels of a system, and on which different positions can be developed as in (Deffuant et al. 2015). We will not enter a finer granularity, the vision of social complexity giving even more nightmares to the Laplace daemon, and since it can be understood as a stronger emergence (in the sense of weak and strong emergence as viewed by Bedau (2002)).

We thus simplify and assume that the nature of systems plays a secondary role in our reflexion, and therefore consider complexity in the sense of an emergence. This choice answers the rationale that a certain level of complexity can already be present in the *structure* of a system (in the sense of the interaction between its elements) whatever the complexity of its elements themselves. This idea is in line with the "sociophysics" approach to socio-technical and human issues as claimed by Caldarelli et al. (2018), in the sense of extracting patterns and modeling stylized processes[1] We do not tackle here the question of relating complexities between levels of a system, which remains open (a system composed of complex agents can sometimes be simple and reciprocally) and is out of the reach of this work. This issue is related to finding the appropriate design of levels in modeling and the appropriate ontology to answer a given question (Roth 2009). Wolf et al. (2018) suggest for example that biological complexity has intrinsic physical roots, implying a transfer of complexity between two distinct levels.

Beyond the view of complexity as emergence, we distinguish two other "types" of complexity, namely computational complexity and informational complexity, that can be seen as measures of complexity, but that are not directly equivalent to emergence. We can for example consider the use of a simulation model, for which interactions between elementary agents translate as a coded message at the upper level: it is then possible by exploiting the degrees of freedom to minimize the quantity of information contained in the message. The different languages require different cognitive efforts and compress the information in a different way, having different levels of measurable complexity (Febres et al., 2013). In a similar way, architectural artefacts are the result of a process of natural and cultural evolution, and witness more or less this trajectory.

Numerous other conceptual or operational characterizations of complexity exist, and it is clear that the scientific community has not converged on a unique definition. Indeed, (Chu 2008) proposes to continue exploring the different existing approaches, as proxies of complexity in the case of an essentialism, or as concepts in themselves

[1] We however do not endorse the idea that importing methods and tools from physics to study social problems would consist in a "physics of human" since the objects studied are not physical objects anymore—this relates to the debate wether disciplines can be defined by methods or by objects of study; we believe that the emergence of disciplines is far more complex and also involves social phenomena, as illustrated by the imperialist positioning of (Caldarelli et al. 2018).

otherwise. This approach is in a certain way reflexive, since the complexity should emerge naturally from the interaction between these different approaches studying complexity, hence the reflexivity. An example of approach not taken into account here is chaos in dynamical systems, although numerous links for example with computational complexity (Prokopenko et al. 2019) naturally exist. We propose to focus only on the three concepts described above in particular, for which the relations are already not obvious.

Indeed, links between these three types of complexity are not systematic, and depend on the system considered. Epistemological links can however be introduced. We will develop the links between emergence and the two other complexities, since the link between computational complexity and informational complexity is relatively well explored, and corresponds to issues in the compression of information and signal processing, or moreover in cryptography.

We can furthermore note that complexity is not the only concept for which different approaches diverge for complex enough systems: (Thurner et al. 2017) detail three approaches to entropy, linked to information theory, thermodynamics and statistical inference, and show that the corresponding measures do not coincide for some examples of complex systems.

3.3 Computational Complexity and Emergence

Different clues suggest a certain necessity of computational complexity to have emergence in complex systems, whereas reciprocally a certain number of adaptive complex systems have high computational capabilities.

A first link where computational complexity implies emergence is suggested by an algorithmic study of fundamental problems in quantum physics. Indeed, Bolotin (2014) shows that the resolution of the Schrödinger equation with any Hamiltonian is a NP-hard and NP-complete problem, and thus that the acceptation of $\boldsymbol{P} \neq \boldsymbol{NP}$ implies a qualitative separation between the microscopic quantum level and the macroscopic level of the observation. Therefore, it is indeed the complexity (here in the sense of their computation) of interactions in a system and its environment that implies the apparent collapse of the wave function, what rejoins the approach of (Gell-Mann and Hartle 1996) by quantum decoherence, which explains that probabilities can only be associated to decoherent histories (in which correlations have led the system to follow a trajectory at the macroscopic scale). The *Quantum Measurement Problem* arises when considering a microscopic wave function giving the state of a system that can be the superposition of several states, and consists in a theoretical paradox, on the one hand the measures being always deterministic whereas the system has probabilities for states, and on the other hand the issue of the non-existence of superposed macroscopic states (collapse of the wave function). As reviewed by Schlosshauer (2005), different epistemological interpretations of quantum physics are linked to different explanations of this paradox, including the "classical" Copenhagen one which attributes to the act of observation the role of collapsing the wave function. Gell-Mann and Hartle (1996) recalls that this interpretation is not absurd since it is indeed the correlations between the quantum object and the world that product the decoherent history, but that it is far too specific to consider the observer only, and that the collapse happens in the emergence itself: the cat is either dead or living, but not both, before we open the box. The paradox of the Schrödinger cat appears then as a fundamentally reductionist perspective, since it assumes that the superposition of states can propagate through the successive levels and that there would be no emergence, in the sense of the constitution of an autonomous upper level. To summarize, the work of Bolotin (2014) suggests that computational complexity is sufficient for the presence of emergence. From a contrapositive point of view, (Elliott and Gu 2017) show that quantum computation reduces drastically the memory needed to simulate stochastic processes, highlighting also the role of the memory space dimension in computational complexity. Backing up these ideas, (Davies 2004) suggests the existence of a minimal computational complexity for a system to exhibit emergence, based on a theoretical information capacity of the universe.

Another approach proposed by Frauchiger and Renner (2018) demonstrates that quantum theory is inconsistent with its use to describe itself. More precisely, it is shown through a thought experiment that complex macroscopic systems that would answer to quantum rules and would themselves use quantum mechanics to describe the world lead to an intrinsic contradiction. This suggests another entry to emergence and the decoherence problem, and gives empirical support to this link between emergence and computational complexity.

This effective separation of scales does not a priori imply that the lower level does not play a crucial role, since Vattay et al. (2015) prove that the properties of quantum criticality are typical of molecules of the living, without a priori any specificity for life in this complex determination by lower scales: (Verlinde 2017) has recently introduced a new approach linking quantum theories and general relativity in which it is shown that gravity could be an emergent phenomenon and that path-dependency in the deformation of the original space introduces a supplementary term at the macroscopic level, which allows explaining the deviations in observational data attributed up to now to *dark matter*.

Reciprocally, the link between computational complexity and emergence is revealed by questions linked to the nature of computation (Moore and Mertens 2011). Cellular automatons, that are moreover crucial for the understanding of several complex systems, have been shown as Turing-complete, i.e. are able to compute the same functions than a Turing machine, commonly accepted as all what is "computable" (CHURCH's thesis). The Game of Life is such an example (Beer 2004). There even exists a programming language allowing to code in the *Game of Life*, available at https://github.com/QuestForTetris. Its genesis finds its origin in a challenge posted on *code golf* aiming at the conception of a Tetris game simulated by the game of life, and ended in an extremely advanced collaborative project. This property of the game of life to be used as a computing device can be used to simulate "meta-pixels", i.e. a cellular automaton at an upper level (Todesco 2013), which behavior weakly emerges from the lower cell but for which rules can be autonomously stated, illustrating a system that could in theory exhibit weak emergence at an arbitrary number of levels. This also suggests an importance of reflexivity, on which we will come back below.

Some organisms without a central nervous system are capable of solving difficult decisional problems (Reid et al. 2016). An ant-based algorithm is shown by Pintea et al. (2017) as solving a Generalized Travelling Salesman Problem (GTSP), problem which is NP-difficult. This fundamental link had already been conceived by TURING, since beyond his fundamental contributions to contemporary computer science, he studied morphogenesis and tried to produce chemical models to explain it (Turing 1952) (that were far from actually explaining it but which conceptual contributions were fundamental, in particular for the concept of reaction-diffusion). We moreover know that a minimum of complexity in terms of constituting interactions in a particular case of agent-based system (models of Boolean networks), and thus in terms of possible emergences, implies a lower bound on computational complexity, which becomes significant as soon as interactions with the environment are added (Tošić and Ordonez 2017).

3.4 Informational Complexity and Emergence

Informational complexity (see DeDeo (2016) for a smooth introduction to information theory), or the quantity of information contained in a system and the way it is stored, also bears some fundamental links with emergence. Information is equivalent to the entropy of a system and thus to its degree of organisation—this what allows to solve the apparent paradox of the Maxwell Daemon that would be able to diminish the entropy of an isolated system and thus contradict the second law of thermodynamics: it indeed uses the information on positions and velocities of molecules of the system, and its action balances to loss of entropy through its captation of information (the Maxwell Daemon is more than an intellectual construction: (Cottet et al. 2017) implements experimentally a daemon at the quantum level).

This notion of local increase in entropy has been largely studied by CHUA under the form of the *Local Activity Principle*, which is introduced as a third principle of thermodynamics, allowing to explain with mathematical arguments the self-organization for a certain class of complex systems that typically involve reaction-diffusion equations (Mainzer and Chua 2013).

The way information is stored and compressed is essential for life, since the ADN is indeed an information storage system, which role at different levels is far from being fully understood. Cultural complexity also witnesses of an information storage at different levels, for example within individuals but also within artefacts and institutions, and information flows that necessarily deal with the two other types of complexities. Information flows are essential for self-organization in a multi-agent system. Collective behaviors of fishes or birds are typical examples used to illustrate emergence and belong to the canonic examples of complex systems. We only begin to understand how these flows structure the system and what are the spatial patterns of information transfer within a *flock* for example: Crosato et al. (2017) introduce first empirical results with transfer entropy for fishes and lay the methodological basis of this kind of studies. Similarly, (Lecheval et al. 2018) show that nonlinear interaction between fishes are essential for the propagation of information in the school during a collective U-turn. Lizier et al (2008) introduce a local measure of transfer entropy to characterize information diffusion in cellular automatons. In the field of artificial life, open-ended evolution is an open research area closely link to emergence, and (Corominas-Murtra et al. 2018) introduce a definition of open-ended evolution based on algorithmic information theory.

Furthermore, different theoretical approaches of complex systems suggest a strong link between informational complexity and self-organization. For example, (Haken and Portugali 2016) develop an overview from the point of view of synergetics on the role of different types of information in self-organizing systems ranging from neuroscience and cognition to urban dynamics. The theoretical approach of multi-scale information to complex systems proposed by Allen et al. (2017), defining an information profile across scales which shape will be linked to the complexity of the system, is another crucial entry into complexity from the viewpoint of information theory. Gershenson (2012) proposes to interpret complex systems as evolving information, and introduces rules for the behavior of information which are particularly suited to the understanding of cognition and life. Finally, (Hoel 2017) shows that the causal structure of some systems in terms of information theory can only be captured by considering emerging levels.

3.5 Reflexivity in the Study of Complex Systems

Furthermore, one aspect of knowledge production on complex systems which seems to be recurrent and even inevitable, is a certain level of reflexivity (and that would be inherent to complex system in comparison to simple systems, as we will develop further). We mean by this term both a practical reflexivity, i.e. a necessity to increase the level of abstraction, such as the need to reconstruct in an endogenous way the disciplines in which a reflection is positioned as proposed by Raimbault (2019), or to reflect on the epistemological nature of modeling when constructing a model, but also a theoretical reflexivity in the sense that theoretical apparels or produced concepts can recursively apply to themselves.

The practical inevitability of reflexivity is well-known in social sciences and humanities, but (Bourdieu 2004) postulated this would be more generally linked to the nature of scientific knowledge which is inherently social (Maton 2003). In the case of a strong constructivist approach to science, knowledge is by nature reflexive but highly contingent to the social structure producing it. The question remains open for disciplines in which the object of study is not social or anthropologic, but as soon as producers of knowledge are potential subjects to be studied, the discipline becomes reflexive. There are some clues that reflexivity may be more generally needed in the study of complex systems and not only social systems: in the context of simulating open-ended evolution, what is a grand challenge in the field of artificial life, (Banzhaf et al. 2016) suggest that a reflexive programming language, in the sense that it can write or embed parts of its own code recursively, would be a crucial feature to simulate open-ended evolution.

This practical observation can be related to old epistemological debates questioning the possibility of an objective knowledge of the universe that would be independent of our cognitive structure, somehow opposed to the necessity of an "evolutive rationality" implying that our cognitive system, product of the evolution, mirrors the complex processes that led to its emergence, and that any knowledge structure will be consequently reflexive. We naturally do not pretend here to bring a response to such a broad and vague question as such, but we propose a potential link between this reflexivity and the nature of complexity.

3.6 Production of Knowledge

3.6.1 From Complexity to Reflexivity

We now have enough material to come to the importance of reflexivity in studying complexity. It is possible to position knowledge production at the intersection of interactions between types of complexity developed above. First of all, knowledge as we consider it cannot be dissociated from a collective construction, and implies thus an encoding and a transmission of information: this relates at another level to all issues linked to scientific communication. The production of knowledge thus necessitates this first interaction between computational complexity and informational complexity. The link between informational complexity and emergence is introduced if we consider the establishment of knowledge as a morphogenetic process. It is shown by Antelope et al. (2016) that the link between form and function is fundamental in psychology: we can interpret it as a link between information and meaning, since semantics of a cognitive object cannot be considered without a function. Hofstadter (1980) recalls the importance of symbols at different levels for the emergence of a thought that consists in signals at an intermediate level. Finally, the last relation between computational complexity and emergence is the one allowing us a positioning in particular on knowledge production on complex systems, the previous links being applicable to any type of knowledge.

Therefore, any *knowledge of the complex* embraces not only all complexities and their relations in its content, but also in its nature as we just showed. The structure of knowledge in terms of complexity is analog to the structure of systems it studies. We postulate that this structural correspondence implies a certain recursivity, and thus a certain level of *reflexivity* (in the sense of knowledge of itself and its own conditions). In other words and as we will detail more below, understanding the complex requires complexity, what is intrinsically reflexive.

3.6.2 *The Complexity of Interdisciplinarity*

We can try to extend to reflexivity in terms of a reflection on the disciplinary positioning: following (Pumain 2005), the complexity of an approach is also linked to the diversity of viewpoints that are necessary to construct it. Some links with the previous types of complexities naturally appear: for example, (Gell-Mann 1995) considers the effective complexity as an *Algorithmic Information Content* (close to Kolmogorov complexity) of a Complex Adaptive System *which is observing another* Complex Adaptive System, what gives their importance to informational and computational complexities and suggests the importance of the observational viewpoint, and by extension of their combination. This furthermore paves the way for the perspectivist approach of complex sciences we will introduce below.

This "interdisciplinary complexity" would be a supplementary dimension linked to the knowledge of complex systems. To reach this new type of complexity, reflexivity must be at the core of the approach. Read et al. (2009) recall that innovation has been made possible when societies reached the ability to produce and diffuse innovation on their own structure, i.e. when they were able to reach a certain level of reflexivity. The *knowledge of the complex* would thus be the product and the support of its own evolution thanks to reflexivity which played a fundamental role in the evolution of the cognitive system: we could thus suggest to gather these considerations, as proposed by PUMAIN, as a new epistemological notion of *evolutive rationality*.

It is highly likely that these approaches could not be tackled in a simple way. Indeed, we can remark that given the law of *requisite complexity*, proposed by Gershenson (2015) as an extension of *requisite variety* (Ashby 1991). One of the crucial principles of cybernetics, the *requisite variety*, postulates that to control a system having a certain number of states, the controller must have at least as much states. Gershenson (2015) proposes a conceptual extension to complexity, which can be theoretically justified for example by Allen et al. (2017) which introduce the multiscale *requisite variety*, showing its compatibility with a theory of complexity based on information theory. Therefore the *knowledge of the complex* will necessarily have to be a *complex knowledge*. This other point of view reinforces the necessity of reflexivity, since following MORIN (see for example Morin (1991) on the production of knowledge), the *knowledge of knowledge* is central in the construction of a complex thinking.

3.7 Discussion

3.7.1 *Towards an Epistemological Framework of Applied Perspectivism*

The view of complexity we just gave can be formulated as a research program, for the development of an *applied perspectivism*, which has already been sketched for example by Banos et al. (2018). This epistemological positioning is based on Giere's perspectivism (Giere 2010b) and aims at sketching practical research guidelines from its viewpoint. We recall that perspectivism has been commented as a "third way" beyond the constructivism-realism debate, which stipulates that any scientific knowledge construction process is a perspective by an agent to answer a purpose with a media, which is called a model. This cognitive approach to science (Giere 2010a) is compatible with the view of cognitive systems by Gershenson (2012) as agents processing information and with the view of (Gell-Mann 1995) of effective complexity as already mentioned.

This approach is particularly relevant for the study of complex multidimensional systems: (Muelder and Filatova 2018) show for example that even different formalizations of the same theory can lead to highly different outcomes. This does not imply that one is more relevant than others (this point being validated by the internal consistence and external validations), but on the contrary that they inform on different dimensions of the system.

Raimbault (2017a) has introduced a knowledge framework to study complex systems, based on complementary *knowledge domains*, which are the theoretical, empirical, modeling, data, methods and tools domains. Any scientific perspective on a complex system would then be a co-evolutive system of cognitive agents and the knowledge domains. Applied perspectivism would then consist in the reflexive quantification of these dynamics during the production of knowledge itself, and a proactive engagement of agents in the explicit positioning of their perspectives and the construction of integrative coupled perspectives.

The main principles of applied perspectivism would in particular include, echoing several of the golden rules by Banos (2013) for simulation models in social sciences:

- Actively foster the consistence of several perspectives when studying a common object, and foster their communication and integration. This point is exactly the virtuous spiral between disciplinarity and interdisciplinarity introduced by Banos (2013).
- To ease the coupling of perspectives, each view to be included must be well self-aware of its own positioning in the scientific landscapes, of its own strengths and weaknesses, and what it can bring to an integrated perspective. To achieve this, an increased reflexivity of each discipline, including the ones which are traditionally not reflexive by nature, is crucial.
- The new simulation model exploration methods, which have recently provided new ways of producing knowledge (Pumain and Reuillon 2017), help for a higher integration of knowledge domains, and thus of the components of perspectives. This is in line of considering computer simulation as genuine experiments themselves (Boge 2019).
- Assuming some kind of "transfer hypothesis" between the models and the perspectives of which they are the media, an approach to couple perspectives (and for example theories) is achieved through the coupling of corresponding models. How to deal with possible ontological incompatibilities and how to couple models

in general are open research questions that still need to be investigated for this point to be relevant.

At this stage, this framework remains at a proposal stage, and several directions must be developed to make it more robust: (i) as the framework acts in a way as a "model of scientific activities", we believe that a formalization into a modeling or logic framework would ensure a greater consistence (although several approaches would be possible, recalling the reflexivity through the recursion in the framework itself that would need several perspectives to be described); (ii) in that spirit, specific simulation models can be developed to answer specific questions such as finding the appropriate compromise in the disciplinary-interdisciplinary interplay, or including more social aspects in the framework (incentives, relations between individuals); (iii) following these preliminary experiments, different practical implementations could be proposed and tested on real cases, such that these experimentations would inform back the framework itself and the different models.

In the concluding chapter of this book (Pumain and Raimbault 2020), a sketch of reflexive analysis of urban theories is proposed, through a citation network analysis of the scientific neighborhood of cited references of all chapters in the book. The resulting knowledge maps provide an overview of involved disciplines and their relative positioning. Each approach could then use this knowledge on itself to refine research directions or interactions with other approaches.

3.7.2 *Implications for Urban Theories*

Our tour on complexities and reflexivity has direct implications for urban theories, since as we already developed, this kind of socio-technical system is the archetype of a complex system on which numerous complementary views can be formulated. We suggest that these aspects may transfer to the models themselves if they become relatively independent: (White et al. 2017) speculates that future urban modeling approaches will have to be themselves self-adaptive and be based on artificial intelligence to dynamically develop their responses to changing contexts of application. The development of multi-scale territorial models for sustainable policies, recalled as a crucial issue by Rozenblat and Pumain (2018), will indeed necessarily require first a practical epistemological reflexivity to integrate the several disciplines concerned (in particular geography and economics which have still much to do to communicate, and secondly a reflection on the emergence of intelligence in territorial systems themselves.

Our work suggests the crucial role of reflexivity to study urban systems, and some links remain also to be established with other types of reflexivity and methods to reach it, such as the one proposed by Anzoise (2017) to enhance reflexivity of all actors in anthropological research, including not only researchers but also others stakeholders and the people studied. Also, our view of integrated knowledge domains, goes as already suggested by Raimbault (2017c) in a direction beyond the arbitrary "qualitative-quantitative" opposition which is a recurrent issue in social sciences, and echoes previous proposals in the literature such as (Shah and Corley 2006) suggesting that more robust theories shall get rid of this opposition.

Finally, we can try to summarize from these epistemological developments some "practical" implications that could be used as useful guides for complex urban research:

1. As suggested by the applied perspectivist positioning, the coupling of models should play a crucial role in the capture of complexity.
2. Similarly, the inspiration for complex urban approaches shall essentially be interdisciplinary and aim at combining different points of view.
3. Different knowledge domains cannot be dissociated for any complex approach, and should use them in a strongly dependent way.
4. These approaches should imply a certain level of reflexivity.
5. The construction of a complex knowledge (Morin 1991) is neither inductive nor deductive, but constructive in the idea of a morphogenesis of knowledge: it can be for example difficult to clearly identify precise "scientific deadlocks" since this metaphor assumes that an already constructed problem has to be unlocked, and even to constrain notions, concepts, objects or models in strict analytical frameworks, by categorizing them following a fixed classification, whereas the issue is to understand if the construction of categories is relevant. Doing it a posteriori is similar to a negation of the circularity and recursion of knowledge production. The elaboration of ways to report knowledge that would translate the diachronic character and its evolutive properties remains an open problem.

3.8 Conclusion

We started a journey into concepts of complexity by illustrating numerous approaches from diverse fields considering urban systems as complex and introducing corresponding concepts. This suggested to investigate the links between different types of complexities. We then suggested that the production of knowledge on the complex is at the intersection of these links between complexities, and hence that reflexivity of this knowledge could be necessary. This naturally brought us to complexity as a diversity of viewpoints on a system. These considerations are summarized into a proposal of an epistemological framework of *applied perspectivism* that we believe to be a useful tool for future complex urban theories and models. A broader integration with other views of complexity and existing approaches in social sciences remains to be done, such as evidence-based foundations of the approach, achieving its own reflexivity.

Acknowledgements This work is part of Dynamicity, a FUI project funded by BPI France, Auvergne-Rhône-Alpes region, Ile-de France region and Lyon metropolis. This work was also funded by the Urban Dynamics Lab grant EPSRC EP/M023583/1.

References

Alexander, C.: A Pattern Language: Towns, Buildings, Construction. Oxford University Press (1977)
Allen, B., Stacey, B.C., Bar-Yam, Y.: Multiscale information theory and the marginal utility of information. Entropy **19**(6), 273 (2017)
Antelope, C., Hubatsch, L., Raimbault, J., Serna, M.J.: An interdisciplinary approach to morphogenesis. In: SFICSSS 2016 Proceedings (2016)
Anzoise, V.: Perception and (re) framing of urban environments: a methodological reflection toward sentient research. V. Anthropol. **30**(3), 191–205 (2017)
Ashby, W.R.: Requisite variety and its implications for the control of complex systems. In: Facets of Systems Science, pp. 405–417. Springer (1991)
Banos, A.: Pour des pratiques de modélisation et de simulation libérées en géographies et shs, p. 1. Université Paris, HDR (2013)
Banos, A., Chasset, P.-O., Commenges, H., Cottineau, C., Pumain, D., and Raimbault, J.: A spatialised bibliometrics approach of a scientific journal production. ArXiv E-Prints (2018)
Banzhaf, W., Baumgaertner, B., Beslon, G., Doursat, R., Foster, J.A., McMullin, B., et al.: Defining and simulating open-ended novelty: requirements, guidelines, and challenges. Theory Biosci. **135**(3), 131–161 (2016)
Batty, M.: Cities and Complexity: Understanding Cities with Cellular Automata, Agent-Based Models, and Fractals. MIT press (2007)
Batty, M.: The New Science of Cities. MIT Press (2013)
Batty, M.: Which complexity for urban systems? In Theories and Models of Urbanization. Springer lecture notes in morphogenesis (2020)
Bedau, M.: Downward causation and the autonomy of weak emergence. Principia: Int. J. Epistem. **6**(1), 5–50 (2002)
Beer, R.D.: Autopoiesis and cognition in the game of life. Artif. Life **10**(3), 309–326 (2004)
Boeing, G.: Measuring the complexity of urban form and design. Urban Des. Int. **23**(4), 281–292 (2018)
Boge, F.J.: Why computer simulations are not inferences, and in what sense they are experiments. Euro. J. Philos. Sci. **9**(1), 13 (2019)
Bolotin, A.: Computational solution to quantum foundational problems. arXiv Preprint arXiv:1403.7686 (2014)
Bourdieu, P.: Science of Science and Reflexivity. Polity (2004)
Caldarelli, G., Wolf, S., Moreno, Y.: Physics of humans, physics for society. Nat. Phys. **14**(9), 870 (2018)
Chu, D.: Criteria for conceptual and operational notions of complexity. Artif. Life **14**(3), 313–323 (2008)
Cooke, P.: Evolutionary complexity geography and the future of regional innovation and growth policies. In: Resilience and Regional Dynamics, pp. 11–30. Springer (2018)
Corominas-Murtra, B., Seoane, L F., Solé, R.: Zipf's law, unbounded complexity and open-ended evolution. J. R. Soc. Interface **15**(149), 20180395 (2018)
Cottet, N., Jezouin, S., Bretheau, L., Campagne-Ibarcq, P., Ficheux, Q., Anders, J., Huard, B.: Observing a quantum maxwell demon at work. Proc. Natl. Acad. Sci. **114**(29), 7561–7564 (2017)
Crosato, E., Jiang, L., Lecheval, V., Lizier, J.T., Wang, X.R., Tichit, P., Prokopenko, M.: Informative and misinformative interactions in a school of fish. arXiv Preprint arXiv:1705.01213 (2017)
Davies, P.: Emergent biological principles and the computational properties of the universe: explaining it or explaining it away: essays & commentaries. Complexity **10**(2), 11–15 (2004)
DeDeo, S.: Information theory for intelligent people (2016)
Deffuant, G., Banos, A., Chavalarias, D., Bertelle, C., Brodu, N., Jensen, P., Varenne, F.: Visions de la complexité. Le démon de laplace dans tous ses états. Nat. Sci. Sociétés **23**(1), 42–53 (2015)
Elliott, T.J., Gu, M.: OCCAM'S vorpal quantum razor: memory reduction when simulating continuous-time stochastic processes with quantum devices. ArXiv E-Prints (2017)

Febres, G., Jaffé, K., Gershenson, C.: Complexity measurement of natural and artificial languages. arXiv Preprint arXiv:1311.5427 (2013)
Frauchiger, D., Renner, R.: Quantum theory cannot consistently describe the use of itself. Nat. Commun. **9**(1), 3711 (2018)
Gell-Mann, M.: The Quark and the Jaguar: Adventures in the Simple and the Complex. Macmillan (1995)
Gell-Mann, M., Hartle, J.B.: Quantum mechanics in the light of quantum cosmology. In Foundations of Quantum Mechanics in the Light of New Technology: Selected Papers from the Proceedings of the First Through Fourth International Symposia on Foundations of Quantum Mechanics, pp 347–369. World Scientific (1996)
Gershenson, C.: The world as evolving information. In: Unifying Themes in Complex Systems vii, pp. 100–115. Springer (2012)
Gershenson, C.: Requisite variety, autopoiesis, and self-organization. Kybernetes **44**(6/7), 866–873 (2015)
Giere, R.N.: Explaining Science: a Cognitive Approach. University of Chicago Press (2010a)
Giere, R.N.: Scientific Perspectivism. University of Chicago Press (2010b)
Haken, H., Portugali, J.: Information and self-organization: a unifying approach and applications. Entropy **18**(6), 197 (2016)
Hoel, E.: When the map is better than the territory. Entropy **19**(5), 188 (2017)
Hofstadter, D.R.: Gödel, Escher, Bach. Vintage. New York (1980)
Krugman, P.: Complex landscapes in economic geography. Am. Econ. Rev. **84**(2), 412–416 (1994)
Ladyman, J., Lambert, J., Wiesner, K.: What is a complex system? Euro. J. Philos. Sci. **3**(1), 33–67 (2013)
Lecheval, V., Jiang, L., Tichit, P., Sire, C., Hemelrijk, C.K., Theraulaz, G.: Social conformity and propagation of information in collective u-turns of fish schools. Proc. R. Soc. B **285**(1877), 20180251 (2018)
Lizier, J.T., Prokopenko, M., Zomaya, A.Y.: Local information transfer as a spatiotemporal filter for complex systems. Phys. Rev. E **77**(2), 026110 (2008)
Mainzer, K., Chua, L.O.: Local Activity Principle. World Scientific (2013)
Manson, S.M.: Simplifying complexity: a review of complexity theory. Geoforum **32**(3), 405–414 (2001)
Maton, K.: Reflexivity, relationism, & research: Pierre Bourdieu and the epistemic conditions of social scientific knowledge. Space Cult. **6**(1), 52–65 (2003)
Moore, C., Mertens, S.: The Nature of Computation. Oxford University Press (2011)
Morin, E.: La méthode 3. La connaissance de la connaissance, Essais, Seuil (1986)
Morin, E.: La méthode, tome 4. Les idées (1991)
Muelder, H., Filatova, T.: One theory—many formalizations: testing different code implementations of the theory of planned behaviour in energy agent-based models. J. Artif. Soc. Soc. Simul. **21**(4), 5 (2018)
Pintea, C.-M., Pop, P.C., Chira, C.: The generalized traveling salesman problem solved with ant algorithms. Complex Adapt. Syst. Model. **5**(1), 8 (2017)
Prokopenko, M., Harré, M., Lizier, J., Boschetti, F., Peppas, P., Kauffman, S.: Self-referential basis of undecidable dynamics: from the liar paradox and the halting problem to the edge of chaos. Phys. Life Rev. (2019)
Pumain, D.: Pour une théorie évolutive des villes. Espace Géographique **26**(2), 119–134 (1997)
Pumain, D.: Cumulativité des connaissances. Revue Européenne Des Sciences Sociales. Euro. J. Soc. Sci. **(XLIII-131)**, 5–12 (2005)
Pumain, D.: Geography far from equilibrium. In Non-Equilibrium Social Science and Policy, pp. 71–80. Springer (2017)
Pumain, D., Raimbault, J.: Perspectives on urban theories. In Theories and Models of Urbanization. Springer Lecture Notes in Morphogenesis (2020)
Pumain, D., Reuillon, R.: Urban Dynamics and Simulation Models. Springer International Publishing (2017)

Raimbault, J.: An applied knowledge framework to study complex systems. In Complex systems Design & Management, pp. 31–45 (2017a)
Raimbault, J.: Invisible bridges? Scientific landscapes around similar objects studied from economics and geography perspectives. In: 20th European Colloquium in Theoretical and Quantitative Geography 2017 (2017b)
Raimbault, J.: Un cadre de connaissances pour une géographie intégrée. In: Journée des jeunes chercheurs de l'Institut de géographie de Paris (2017c)
Raimbault, J.: Exploration of an interdisciplinary scientific landscape. Scientometrics **119**(2), 617–641 (2019)
Raimbault, J., Banos, A., Doursat, R.: A hybrid network/grid model of urban morphogenesis and optimization. In 4th International Conference on Complex Systems and Applications, pp. 51–60 (2014)
Read, D., Lane, D., Van der Leeuw, S.: The innovation innovation. In Complexity Perspectives in Innovation and Social Change, pp. 43–84. Springer (2009)
Reid, C.R., MacDonald, H., Mann, R.P., Marshall, J.A., Latty, T., Garnier, S.: Decision-making without a brain: how an amoeboid organism solves the two-armed bandit. J. R. Soc. Interface **13**(119), 20160030 (2016)
Roth, C.: Reconstruction failures: questioning level design. In: Epistemological Aspects of Computer Simulation in the Social Sciences, pp. 89–98. Springer (2009)
Rozenblat, C., Pumain, D.: Conclusion: toward a methodology for multi-scalar urban system policies. Int. Transnat. Perspect. Urban Syst. 385 (2018)
Schlosshauer, M.: Decoherence, the measurement problem, and interpretations of quantum mechanics. Rev. Mod. Phys. **76**(4), 1267 (2005)
Shah, S.K., Corley, K.G.: Building better theory by bridging the quantitative–qualitative divide. J. Manage. Stud. **43**(8), 1821–1835 (2006)
Thurner, S., Corominas-Murtra, B., Hanel, R.: Three faces of entropy for complex systems: Information, thermodynamics, and the maximum entropy principle. Phys. Rev. E **96**(3), 032124 (2017)
Todesco, G.M.: Cellular automata: the game of life. In: Imagine Math 2, pp. 231–243. Springer (2013)
Tošić, P.T., Ordonez, C.: Boolean network models of collective dynamics of open and closed large-scale multi-agent systems. In International Conference on Industrial Applications of Holonic and Multi-Agent Systems, pp. 95–110. Springer (2017)
Turing, A.M.: The chemical basis of morphogenesis. Philos. Trans. Royal Soc. London B: Biol. Sci. **237**(641), 37–72 (1952)
Vattay, G., Salahub, D., Csabai, I., Nassimi, A., Kaufmann, S.A.: Quantum criticality at the origin of life. In: Journal of Physics: Conference Series, vol. 626, p. 012023. IOP Publishing (2015)
Verlinde, E.: Emergent gravity and the dark universe. SciPost Phys. **2**(3), 016 (2017)
West, G.: Scale: The Universal Laws of Growth, Innovation, Sustainability, and the Pace of Life in Organisms, Cities, Economies, and Companies. Penguin (2017)
White, R., Recio, G., Banzhaf, W.: The necessity of disequilibrium. In: European Colloquium on Theoretical and Quantitative Geography 2017 (2017)
Wolf, Y.I., Katsnelson, M.I., Koonin, E.V.: Physical foundations of biological complexity. Proc. Natl. Acad. Sci. **115**(37), E8678–E8687 (2018)

Part II
Scaling Laws: Universal Solutions for Urban Planning or Contingent Products of Circumstances?

Chapter 4
Scaling Laws: Insights and Limitations

Elsa Arcaute and Erez Hatna

Abstract This chapter revisits rudimentary concepts pertaining to power laws and self-similarities, to illustrate the insights and limitations of scaling laws applied to urban systems. As a first step, we look at the role of city boundaries and the definition of the urban system. We portray different ways to delineate cities and metropolitan areas and discuss the effects of this choice on the system of cities. In particular, we look at the different results obtained for the Zipf's exponent, and we show that Zipf's law is a necessary but not a sufficient condition to define a system of cities. We sketch some of the statistical difficulties arising when measuring power laws, and we show how these can lead to contradictory results when looking at scaling laws if not addressed properly. We describe some of the more rigorous approaches aiming at tackling this, such as the methods developed by Clauset and colleagues at the SFI on the one hand, and Leitão and colleagues on the other. In addition, we show that the aggregated level at which scaling laws are analysed does not provide any information about the heterogeneities present in urban systems. We briefly touch on the role of hierarchies and unexplored self-similarities to better understand these.

4.1 Introduction

Urban systems are complex systems that present emergent properties that are the result of interactions taking place at the very local scales and feeding back in a multi-scalar way across different spatio-temporal scales. These so-called emergent patterns can be seen as regularities that are ubiquitous, that is, regularities that are present regardless of the geographical or historical context of each individual country. Some of these are: the distribution of city sizes, Zipf's law; the growth of cities,

E. Arcaute (✉)
Centre for Advanced Spatial Analysis (CASA), University College London, London, UK
e-mail: e.arcaute@ucl.ac.uk

E. Hatna
College of Global Public Health, New York University, New York, USA
e-mail: eh109@nyu.edu

D. Pumain (ed.), *Theories and Models of Urbanization*, Lecture Notes in Morphogenesis, https://doi.org/10.1007/978-3-030-36656-8_4

Gibrat's law; morphological properties of cities, such as fractal properties; a hierarchical organisation, such as different geographical levels of regional organisation; and economies of agglomeration, that can be seen in terms of the so-called urban scaling laws. Encompassing these regularities under the same framework can be seen as an effort towards formulating a science of cities (Batty 2013).

In this chapter, we will focus on urban scaling laws, which have had a lot of attention in the last decade. Although this surge was initiated by the influential PNAS paper (Bettencourt et al. 2007), economists have been looking at this problem from the perspective of productivity in cities from almost a century (Marshall 1890). They observed that clustering of economic activity produces a more than proportional outcome per capita. The tension between mechanisms of division of labour and competition giving rise to economies and diseconomies of scale, has long been advanced by Adam Smith since the 18th century (Smith 1778). Outside of the economic realm, within the urban context, the earliest attempts can be traced back to the 70s (Nordbeck 1971; Woldenberg 1973). From the perspective of the morphological structure of cities, the ideas of Mandelbrot (1983) inspired two groups to model growth and self-similarity in cities using fractals: in the UK (Batty et al. 1989; Batty and Longley 1994), and in France (Frankhauser 1998). On the other hand, in other disciplines, such as physics and biology, scaling and self-similarities in critical phenomena have been widely studied (Bak 1996). Among these, Geoffrey West pursued the task to derive the mechanisms that gave rise to the sub-linear scaling of the metabolic rate with respect to its mass, i.e. Kleiber's law of efficiency of the metabolic rate (Kleiber 1947). In a paper in *Science* (West et al. 1997), the authors explained how a branching mechanism was responsible for the observed phenomenology, and derived the expected exponent from such a mechanism, obtaining ¾. Within the context of urban systems, Denise Pumain developed a working paper in 2004 during a visit to the Santa Fe Institute (Pumain 2004), which was the initiating seed for this *renaissance* of urban scaling laws. In that working paper, she discussed many important intervening aspects that were later on overlooked, such as the role of the speed of transportation for the definition of cities and functional areas, the sensitivity of the measures to the different boundary delimitations, in addition to the growth of cities and their fractal properties. It is outside the scope of this small chapter to give a full historical review of scaling laws, for an overall picture the book by Geoffrey West is recommended (West 2017), and for a technical perspective a forthcoming book on urban scaling (Arcaute and Hatna 2020).

4.2 Which System of Cities?

Within the spirit of the quest in complexity science of finding generic mechanisms leading to the observed patterns in urban systems, the first emerging challenge is the definition of the system itself as pointed out in Pumain (2004). To which extent are the observed patterns constrained to urban cores or metropolitan areas? Does the system need to have a critical mass to present specific behaviours? That is, does a city need to have a minimum number of individuals to be considered a city, or to present specific characteristics? The definition of cities has changed throughout the years, from ad hoc and size-independent assumptions, such as the presence of a cathedral, to assigning a random minimum population size, say of 100,000 individuals, to the more complicated aspect of considering economic functional areas, where urban cores are extended according to commuting zones (see definitions given by the Eurostat bureau, such as Larger Urban Zones, and others derived by the OECD).

In this section, we will explore different ways to construct coherent definitions across systems, from urban cores to metropolitan areas.

4.2.1 Defining Urban Cores

Different approaches have been explored in order to construct definitions of cities which are free from administrative constraints, and which can be developed in a systematic way, so that a standard framework for different countries can be formulated. These are based on street networks (Jiang and Jia 2011; Strano et al. 2012; Masucci et al. 2015; Arcaute et al. 2016; Molinero et al. 2017), satellite images (Fluschnik et al. 2016), percolation (Rybski et al. 2013), or aggregations of geographical units (Arcaute et al. 2015), to name a few. The finer granular methods, such as the former ones, can be more accurate with respect to defining the boundaries of cities, nevertheless, data might have to be imputed from the administrative geographical units to the newly defined boundaries. This issue can be avoided if the units containing the data are considered as the most granular level.

When working with systems of cities, we would like to capture as many cities as possible, instead of selectively retaining the largest ones only. This matter is of extreme importance if we wish to investigate the role of city size in manifesting certain patterns. In this sense, we would like to make use of a method to define cities that can be automatized to the extend in which a coherent system of cities is derived, and not only the most representative ones, defined in an ad hoc way. Many of the above-mentioned methods do this, where a single parameter is used to define all the settlements in a country, instead of carefully finding the boundaries for each of them. For example, in Arcaute et al. (2015), we cluster census units according to population density, and define a global threshold for all settlements in the UK. Figure 4.1 shows the variation of population density in space, and how the threshold determines the transversal cut at which cities could be defined. For very high thresholds, only the urban cores, corresponding to the red colours, will be part of the system of cities, while for low thresholds, large extents, where settlements are merged into each other, will appear.

Looking carefully at the figure, the spatial distribution of population is such that two important cities, Liverpool and Manchester, are almost merging into one another. By selecting a threshold before the merging takes place, one ensures some degree of validation, see Fig. 4.2.

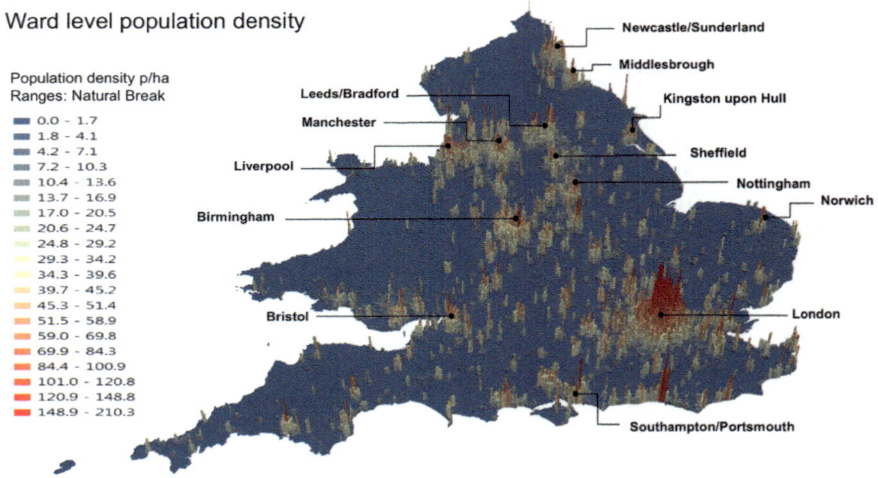

Fig. 4.1 3D plot of population density at the ward level for England and Wales

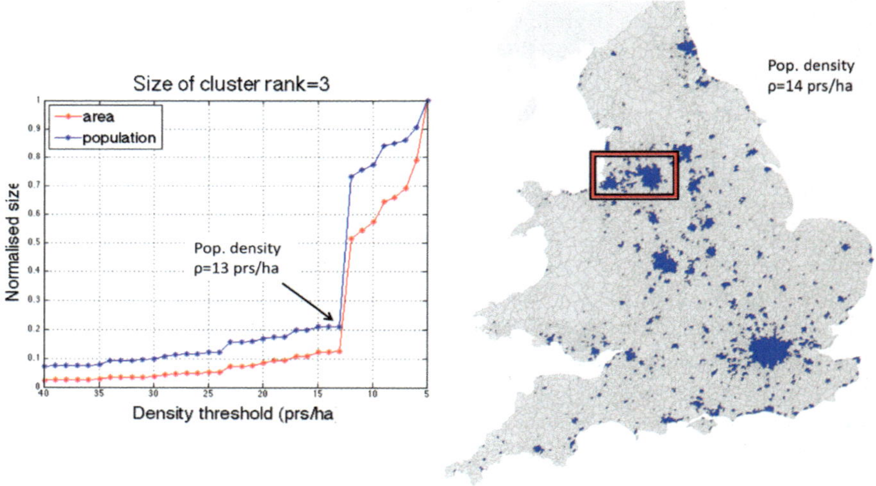

Fig. 4.2 At a density of 12 prs/ha, Manchester and Liverpool, pictured inside the red rectangle, are merged into one cluster. We choose a density before this happens: 14 prs/ha

In Arcaute et al. (2015), we use satellite maps to validate the choice for the population density threshold of 14 prs/ha, which immediately shows the limitation of such a methodology if one is to include economic functional areas, since these cannot be defined through satellite images.

4.2.2 Metropolitan Areas

Comparative analyses between countries and cities, are mainly focused on economic indicators. In order to draw meaningful conclusions, one needs to consider economic functional areas in a systematic and coherent way. Given the importance in understanding the drivers for economic growth, considerable effort has been done in this direction. In Europe, the Eurostat developed in 2004 a harmonised definition for metropolitan areas called Larger Urban Zones (LUZ). These were later on improved and adapted by the OECD. In France, in addition to these definitions, one finds Aire Urbaines, while in the US, economic functional areas are encompassed within the term Metropolitan Statistical Areas (MSAs). The lack of automatization of such methods, results in producing only a small sample of metropolitan areas for the whole of individual countries. This limitation is important, since when looking at generic patterns in urban systems, we expect these to hold for the whole urban system, and not only for the biggest cities, or metropolitan areas. Scaling laws in particular, relate urban indicators from one scale to another. If only a handful is selected, two different problems emerge. On the one hand, mathematically one does not have enough orders of magnitude to properly assess the validity of the scaling laws. On the other, the characteristics are only valid for cities/metropolitan areas that have a minimum population size. The extrapolation of the observed phenomenology for all cities cannot be made, and hence, policy arguments need to be put into the context of the limitations of the analysis with respect to city sizes.

There are many definitions of metropolitan areas, incorporating a set of different functionalities defined according to the needs and background of each country. In the context of the urban scaling laws, we focus on the functional areas given by commuters. This will limit the variables needed to generate the set of metropolitan areas across countries. One way to proceed, is to start with an urban core defined following the techniques described in 2.1, in general considered to be densely populated, and then to incorporate other areas from which individuals commute to work. The percentage needs to be consistent across the country, and in general it is considered to be 30%. Within the spirit of looking at the sensitivity of the models or measures to city definitions, different ranges should be considered so that deviations can be observed. Figure 4.3 shows some definitions for France (Cottineau et al. 2018) for different density thresholds and commuting flows.

Although we have simplified the definition of metropolitan areas for the purpose of creating a consistent laboratory of city definitions, this discourse cannot be properly addressed without taking into consideration the intervening economic opportunities related to transport and linkages of various forms (Bretagnolle et al. 2002). Extensive work has been pursued in this direction by Denise Pumain and her group (Pumain 1992, 1997, 2003). This is a crucial point in order to understand how city growth can be driven by the effects of *metropolisation* (Pumain and Rozenblat 2018), where cities are conceived as interacting entities, which cannot be reduced to independent entities characterised only by their size, and removed from a framework of a system of cities. Furthermore, urban scaling laws have city size as their pillar, and as we will

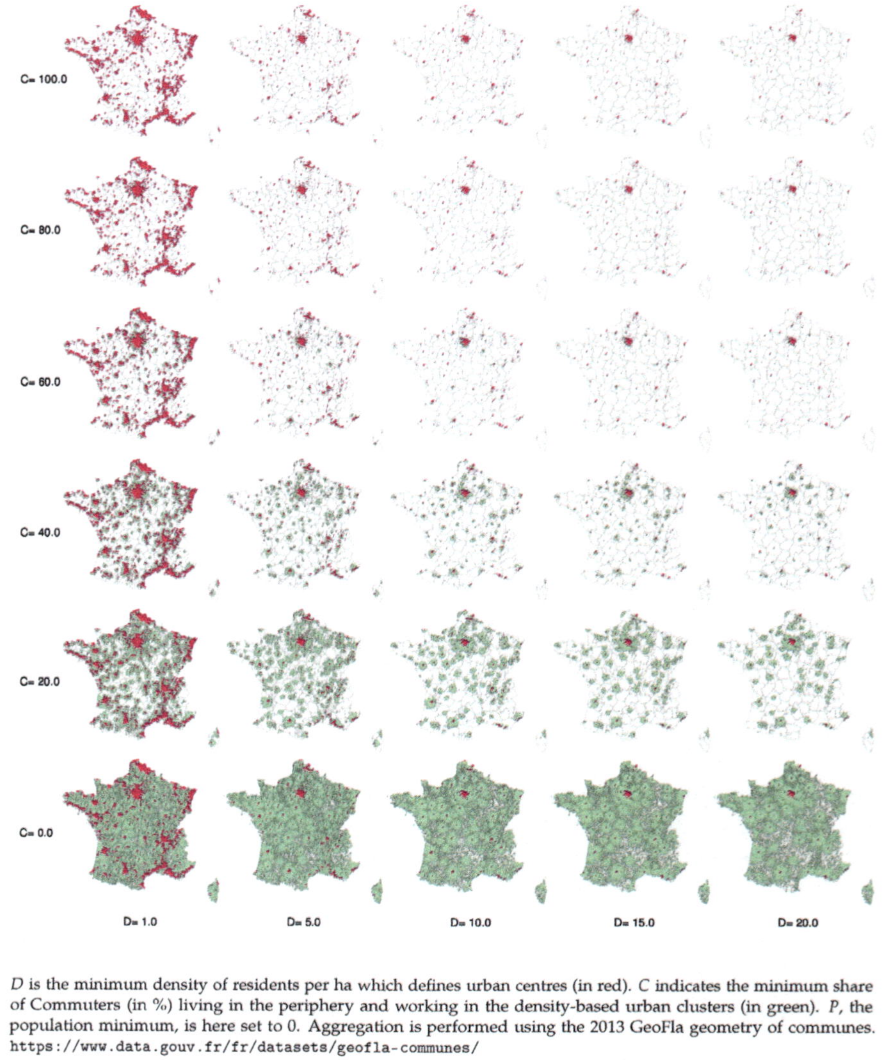

Fig. 4.3 Urban cores and functional areas, figure taken from Cottineau et al. (2018)

see in the next section, such a pillar is a fragile structure, given the large fluctuations encountered when measuring rank size. The clear delineation of the system under consideration cannot be overlooked when considering urban scaling laws.

There are other methods that propose different levels of urbanisation at the local level to integrate urban areas, hence providing a range of boundaries for the sensitivity analysis of the models. These are based on the fractal properties of the system at the local and global level (Sémécurbe et al. 2018) to name a few.

4.2.3 Zipf's Law

Countries are defined by different political and historical trajectories, which create specific conditions for certain functions to thrive and succeed among others. It would then be sensible to think that some cities would grow more rapidly than others, and that growth would be governed by those conditions, creating different distributions for city sizes. Nevertheless, it is observed that in spite of this diversity of trajectories and opportunities, there is a regularity in the way cities grow, creating a consistent distribution of city sizes across countries following Zipf's law (Zipf 1949). In such a distribution, it is observed that if we denote by N_k the size of the city of rank k, where N_1 corresponds to the largest one, N_2 to the second largest one and so on, then $N_1 = 2N_2 = 3N_3 = 4N_4 = ... = k\,N_k$. This regularity is valid for systems of cities in a country, and it has been observed to hold worldwide, withstanding time. This is an extraordinary empirical law first observed for cities by Auerbach in 1913 (Auerbach 1913).

There are however a few caveats to note. On the one hand, this regularity does not hold for the full set of cities within a country, but only *for the tail of the distribution,* as is the case of most power laws (Newman 2005). This means that there is a minimum population size for cities from which Zipf's law is verified. Nevertheless, it is observed that this population cut-off is not the same for all countries, indicating that generalisations for city sizes across countries should not be made. For example, when considering cities larger than a certain size, say typically 100,000 individuals, this has different implications for countries in Europe compared to countries in Asia or America.

This lack of universality prompts to a careful exploration of the following questions when describing generic patterns, or when trying to identify mechanisms for regularities worldwide: what is the system under consideration? Is it defined in a consistent way across countries such that comparisons can be drawn? Is the phenomenon valid for ALL cities within a country or only for a subset? If the latter, how is this integrated within policy making schemes? Do different regimes need to be identified for small and for large cities, such that different frameworks need to be developed?

The second caveat refers to disagreement with respect to the validity of Zipf's law. Given that not all cities are included in the power law distribution, some authors have proposed a lognormal distribution instead (Eeckhout 2004). Others have argued, that two different distributions intervene: a power law for the largest cities and a lognormal for the smaller ones (Levy 2009). Malevergne et al. (2011) provides a full statistical test to be able to distinguish between the two and concludes that there are two different distributions for small and large cities. In this sense, this caveat links back to the problem pointed out in the preceding paragraph. If within a system of cities there are different generational mechanisms giving rise to the observed population size for small and large cities, how can we encompass within a single law all cities? In particular, within the context of urban scaling laws, where city size is the main variable at play, do we need to exclude small cities?

The lack of rigour when measuring the exponent further obscures the question of the validity of Zipf's law as pointed out by Malevergne et al. (2011). In Cottineau (2017), the author shows how for the same country different sets of exponents get reported in the literature, due to the different methods applied on the one hand, and to the inconsistency in the way the system of cities is defined on the other. In Pumain (2004), the author illustrates this point through the variation in population size and ranking of cities according to the definition chosen: e.g. for central communes, Paris would have a population of around 2 million; while for urban systems, Paris would account for 11 million people.

In spite of the complications pointed out, it is typically expected that a system of cities obeys Zipf's law. Testing for the different boundary definitions, it is observed that the exponent fluctuates in a way that does not allow us to use it as a proxy to establish which definition is the best one. In Fig. 4.4, we used the method developed in Clauset et al. (2009) to compute the Zipf exponent for the definitions of cities at different population density thresholds as outlined in Sect. 2.1. Note that the exponent

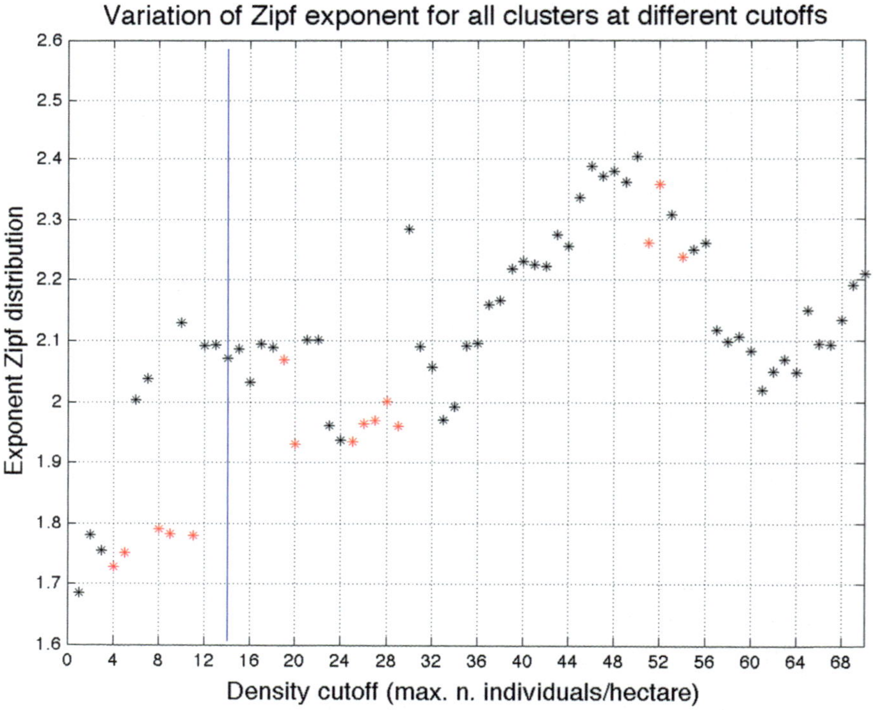

Fig. 4.4 Power law exponent for the definitions of cities given by different density thresholds. This corresponds to the exponent for the cumulative distribution: $\tau = \alpha + 1$, where α is the usual Zipf exponent. Red dots indicate that the power law is rejected, and hence that this specific configuration does not follow Zipf's law. The blue line indicates the population density threshold at which cities have been verified to be well-defined.

τ is computed for the cumulative distribution: $\tau = \alpha + 1$, where α is the Zipf exponent commonly found in the literature. The method provides a test indicating whether the power law hypothesis can be rejected or not, and hence allowing us to identify which configurations do not obey Zipf's law (marked in red in the figure).

Looking at the figure, and recalling the findings in Cottineau (2017), we can see that the value of the exponent does not necessarily give an indication of how good the definition of cities is. Zipf's law then becomes a necessary but not a sufficient condition for urban systems to properly represent cities.

4.3 Testing Universality: Scaling Laws in England and Wales

In order to understand what is meant by *universality*, let us depart from the results obtained in Bettencourt et al. (2007), where a taxonomy for urban indicators was derived according to the regime they were expected to belong to. This is dictated by the value of the exponent β in $A \sim N^{\beta}$, where A is the urban indicator, and N denotes the population size: (1) $\beta > 1$, superlinear regime, i.e. increasing returns are expected for outputs resulting from interactions between individuals such as wealth, crime, innovation, etc.; (2) $\beta \sim 1$, outputs are proportional to population, these correspond to basic needs such as electricity consumption (at the household level), number of households, etc.; (3) $\beta < 1$, sublinear regime, outputs resulting from services and infrastructure such as the length of roads, the number of gas stations, etc., are expected to grow slower than population size. A summary of the table in Bettencourt et al. (2007) can be seen in the following Fig. 4.5.

Note that the taxonomy is not constrained to the exact value of β, but to the interval in which it belongs to.

In Arcaute et al. (2015) we tested these empirical findings for the case of England and Wales using the 2001 census, and we found that key variables did not show the expected scaling behaviour according to the classification given above. Income in particular, although expected to belong to the superlinear regime, varied linearly with the population regardless of the choice of definition of cities or metropolitan areas. Furthermore, the lack of superlinearity is still present whether we consider income, net income, and control for housing, see Fig. 4.6.

Any individual living in London will intuitively tell you that a linear fit seems incorrect, since London is an expensive place where wealth tends to accumulate in comparison to other places. In order to better understand what is going on, one needs to look closely at the scatter plots, and realise that the distribution of residuals confirm such an intuition. Figure 4.7 shows the residuals for two different definitions of cities given by a density cut-off of 14 prs/ha on the left and of 30 prs/ha on the right. These maps show the shear difference of wealth accumulated on the South versus the North. Such a division has been known for centuries, and it is referred to as the North-South divide.

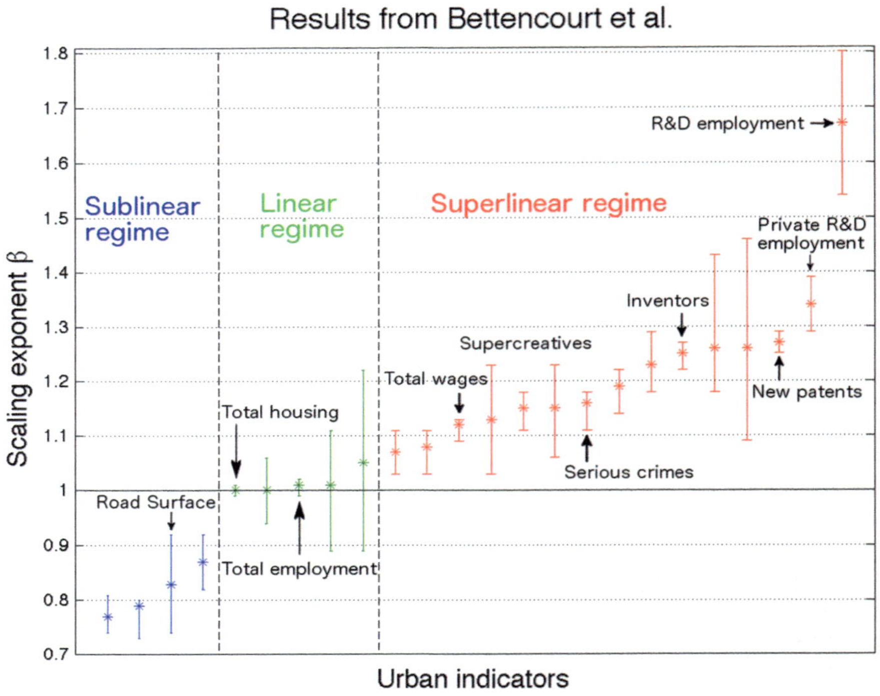

Fig. 4.5 Exponents with 95% CI for different urban indicators found for the USA, Germany and China in reference (Bettencourt et al., 2007). These are colour-coded according to their regime. Figure taken from Arcaute et al. (2015)

The residuals show that income in London is higher than in the post-industrial cities of the north, confirming the general perception. Nevertheless, it is important to note that the lack of superlinearity in the system is also generated through the high income concentrated in the small cities, in particular in those surrounding London. In this sense, the particular geography of the North-South divide, in which the largest city (London) and many small ones belong to the wealthy South, while the rest of the largest cities are part of the less wealthy North, is key to obtaining a linear exponent. If on the other hand, we were to consider a situation in which a large number of small cities would have a very low income, then the exponent would be superlinear. We will see in the next sections, that a more careful statistical approach is needed to properly assess the exponent, since a fit in the log-log scale is insufficient, regardless of the apparent large R^2. In any case, from this simple exercise, we observe that the exponent itself does not hold information with respect to the real distribution of wealth. For the same exponent, we could have a very different set of residuals. In Bettencourt et al. (2010), the authors use the residuals to measure the performance of cities, relating the observations at the local level to the behaviour at the system level.

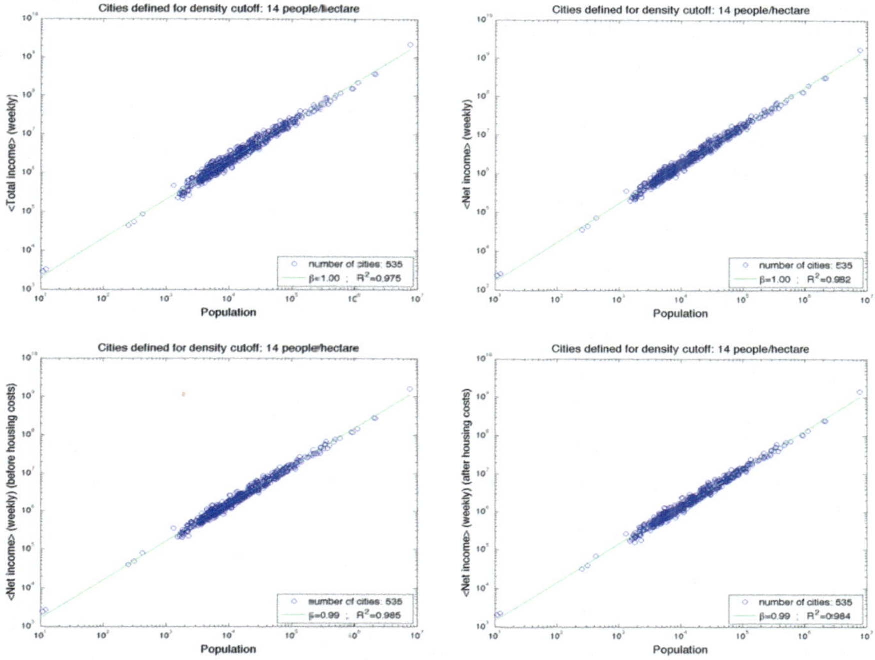

Fig. 4.6 Scatter plots for weekly mean income, considering before and after taxes and housing. The plots show a linear fit for all of them and a very high R^2

Nevertheless, given that the exponent is sensitive to the subset of cities under consideration (Cristelli et al. 2012; Cottineau et al. 2018), this could lead to misleading results, since a city could be over-performing for a specific population cutoff, and then under-performing for a different one. This is illustrated in Fig. 4.8, where for the same system of cities, different exponents are found according to the minimum population size under consideration. The variations can lead to a change in regime, e.g. from linear to sublinear (Fig. 4.8a), or even more dramatically from superlinear to sublinear (Fig. 4.8b), or from sublinear to superlinear (Fig. 4.8c). The change in regime is sometimes only observed for the highly dense areas (Fig. 4.8d).

In Fig. 6 of Arcaute et al. (2015), we showed that this is also the case when measuring patenting activity. If the system of cities is defined for cities larger than 10,000 individuals, then the system presents the expected superlinear behaviour, and the 2nd and 3rd largest cities in the UK systematically under-perform, while London presents the expected behaviour for its size. If on the other hand, the population cut-off is increased to 50,000 individuals, then the system no longer exhibits superlinearity, the 2nd and 3rd largest cities perform according to their size, and this time London over-performs.

At this point it is important to note that these inconsistencies arise from a lack of statistical rigour in measuring the exponent, in particular given that the fit changes

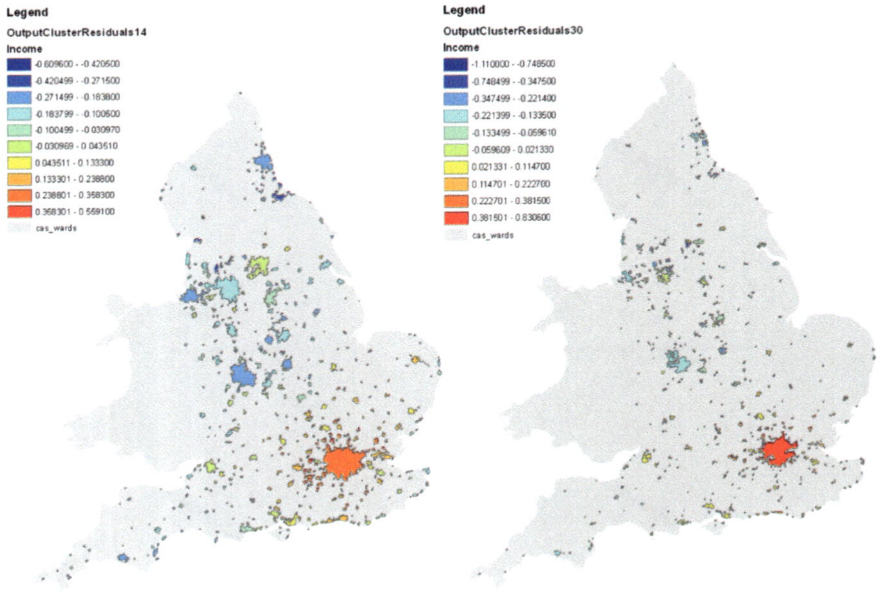

Fig. 4.7 Residuals from the scatter plot of income versus population in a log scale for two definitions of cities given at two different population density cutoffs: 14 prs/ha on the left and 30 prs/ha on the right

according to the number of cities under consideration (Cristelli et al. 2012). For example, there are 535 settlements within the definition of cities in the UK at 14 prs/ha if no population cut-off is applied, while after applying a population cut-off of 100,000, only 46 remain. On the other hand, Fig. 4.8c presents a change in regime, from sublinear to superlinear in some cases, when a minimum population size of 10,000 individuals is imposed. With such a constraint, there are still 300 cities in the system, which in general surpasses the number of cities considered in most studies. This calls for a more rigorous method to estimate the exponent, as has been long pointed out by Shalizi (2011) and more recently by Leitão et al. (2016). In the next section, we look into the limitations that these methods have indicated.

Before outlining other methodologies, let us look at other examples and implications resulting from these disparities. In Cottineau et al. (2018), we highlight the effect of the population cut-off on the resulting scaling exponent for France over many variables from the census. In particular, we illustrate the case for the number of individuals employed in manufacturing and in education, see Fig. 6 therein. For urban cores (Unités Urbaines), if no population cut-off is applied, both variables scale superlinearly, while after applying a cut-off of 50,000 people, both scale proportionally to the population size. If instead of urban cores one considers metropolitan areas (Aires Urbaines), a different picture emerges. For no population cut-off, the number of people employed in manufacturing scales sublinearly, indicating that there is no

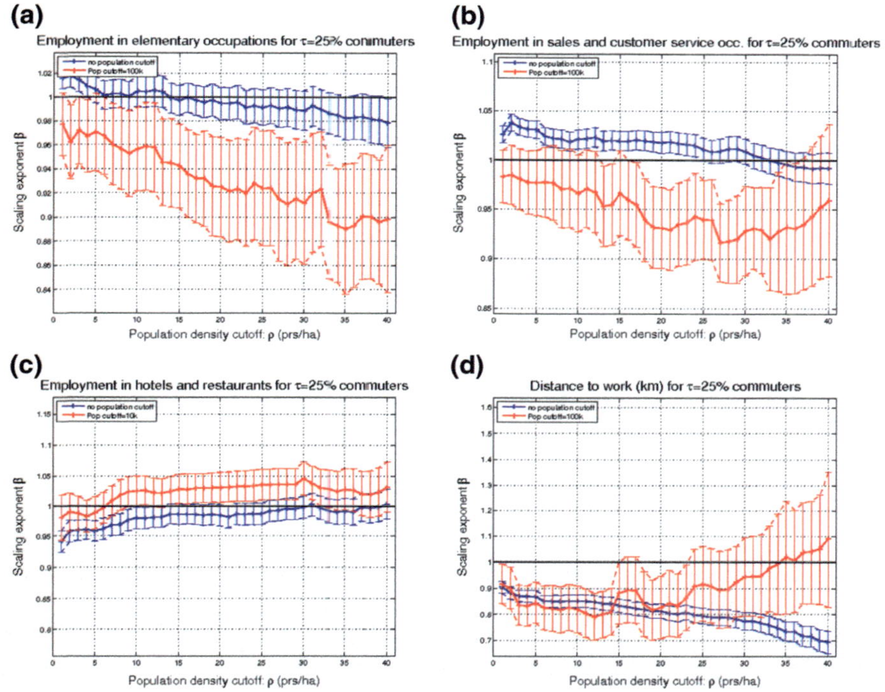

Fig. 4.8 Variation of the scaling exponent following different population cutoffs according to the definitions given above for different systems of cities in England and Wales for the 2001 census. In blue: value of the exponent and confidence intervals according to the definition of cities at different density thresholds (x axis) including areas from which 25% of people commute; in red: value of the exponent if only cities higher than a population threshold are considered (10k = 10,000, 100k = 100,000). A change in regime is observed for certain definitions (recall that at 14 prs/ha cities are well defined)

concentration of people employed in this specific job in large cities, in opposition to the previous finding. After a population cut-off, the exponent becomes linear. There is no universal expected exponent for this variable that would enable us to assess which is the result to bring forward, since the number of people employed in a specific sector depends on the maturity of the sector itself as has been pointed out by Pumain on various occasions (Lane et al. 2009; Pumain 2004; Pumain et al. 2006, 2009). In addition, Finance and Swerts indicate in chapter 5 of this book, that the manufacturing industry has different sectors with different degrees of maturity that cannot be encompassed under one umbrella. The discrepancies found between the different results might be due to these intricacies, but in any case, they indicate that the maturity of the sector cannot be assessed through the regime in which the scaling exponent is found using the methodology outlined so far.

There is a compelling need to disentangle the origin of these inconsistencies, since these can lead to contradictory policies regarding pressing societal challenges.

The puzzle of whether larger cities are more emissions efficient or not, for example, impacts on policies regarding global warming mitigation. Within the literature (Fragkias et al. 2013; Oliveira et al. 2014; Louf and Barthelemy 2014; Rybski et al. 2017), authors have observed that different results arise depending on how the system of cities is defined, in addition to the type of CO_2 emissions under consideration, i.e. originating from transport, industries, or all.

In the next section, we will see that indeed, in the light of a more rigorous approach, many of the results presented so far need to be reviewed, hopefully shedding some light into finding substantive outcomes that can be adopted for policy.

4.4 To Scale or Not to Scale?

The previous sections have illustrated that measuring exponents for power laws or power law relationships using least-squares fitting on a logarithmic scale is not a statistically robust method, and it can lead to inconsistent results. Furthermore, contrary to its common use, the coefficient of determination R^2 does not provide a measure of accuracy of the exponent, since this can be very high for divergent results when sub-setting the number of data points.

The problem of fitting a power law to data, is a well-known open problem, since power-laws are ubiquitous in many complex systems. For more than a century these have been observed in natural and man-made systems. They are particularly important, since in such distributions, a few large events can have an extraordinary impact on the system, that can never be reached through the very many small ones. One example of this is earthquakes (Gutenberg and Richter 1944). There are very few that are very big, but these are the ones bringing devastating consequences as opposed to the regular small ones, that in many cases go unnoticed. In man-made systems, a similar effect is unfortunately observed for wars, where the number of deaths represents the size of a conflict (Roberts and Turcotte 1998). Another example of man-made systems, is the distribution of income (Pareto 1906), where very few hold most of the wealth, and this is observed at a global scale. Such a distribution is also observed for the size of firms (Axtell 2001; Gabaix 1999), where key players are identified as driving the market in specific sectors. Less gloomy examples are found within the usage of language (Zipf 1949), and the number of species of plants and animals (Willis and Yule 1922) to name a few.

Given their ubiquity and relevance, many methods have been developed throughout the years in the search for a better statistical test and fit (Adamic 2011; Cancho and Solé 2001, 2002; Clauset et al. 2009; Deluca and Corral 2013; Goldstein et al. 2004; Huberman and Adamic 1999; Mitzenmacher 2004; Stumpf and Porter 2012). In Clauset et al. (2009), the authors propose a method based on the cumulative distribution and maximum likelihood, which provides a goodness-of-fit between the power law and the data, and a minimum size $x_{\min}$ from which such a distribution holds, recalling that it only holds for the tail of the distribution. They looked at several datasets described as power-law distributions in the literature, and they found that in several occasions, the hypothesis for a power-law distribution was rejected. Such a methodology could help settle the issue raised in Levy (2009), Cottineau (2017) with respect to Zipf's law. In particular, it is important to note that the method in Clauset et al. (2009) returns the same exponent if a population cut-off p_0 is applied such that $p_0 < x_{\min}$. Hence, sub-setting the data before the minimum point at which the distribution holds will not make a difference, while this is nevertheless not the case any longer if $p_0 > x_{\min}$. In such a case, the value of the exponent will be different.

In the light of the many inconsistencies arising from taking different definitions of cities, sceptics questioned whether the hypothesis for urban scaling was correct. In particular, Marc Barthélémy on the one hand, and the group of Eduardo Altmann (Leitão et al. 2016) on the other, started investigating this from the perspective of testing whether such inconsistencies arose from an incompatibility of the data with a model of scaling relationships. In particular, although everybody fitting power-laws knows this, people chose to ignore the fact that when evaluating scaling laws ($y \sim x^{\beta}$) using the linearised equation to estimate the exponent β, i.e. $\ln(y) \sim \beta \ln(x) + c$, there is an underlying assumption that the data points are independent and that the fluctuations around the mean of $\ln(y)$ are Gaussian. This is nevertheless never verified, and unfortunately, in most occasions it does not hold, giving rise to an "inflated" R^2, which although high, it is unreliable and cannot be used as goodness of fit. In Leitão et al. (2016), the authors show that contradictory outcomes are obtained for the exponent β, depending on the choice of underlying model under consideration. An interesting discussion emerges, whether models should describe most of the population or most of the cities. In detail, the authors propose a thorough methodology through the following steps:

1. Estimate β through maximum-likelihood.
2. Estimate β's confidence interval using bootstrapping with replacement.
3. Test whether the data is compatible with the model. This returns a p-value, such that if $p > 0.05$ then the model is not rejected.
4. Test whether there is evidence for non-linear scaling, i.e. $\beta \neq 1$.
5. Test for evidence of non-trivial scaling in the fluctuations (Taylor's law).
6. Check which model describes best the data using Bayesian information criterion.

In their paper, they applied their statistical framework to 15 datasets treated as cases of scaling in the literature and found that there was only 1 case where non-linear scaling was a better model compared to a simple linear fit $\beta = 1$, and where the power law hypothesis could not be rejected. This was the case for GDP for OECD data.

Before these more robust methods were devised, Shalizi has already pointed out the existing flaws in looking at scaling laws from the perspective of non-interacting cities. In Shalizi (2011), the author argues that the observed good fit is just an artefact of using extensive quantities instead of intensive ones, i.e. regressing the total at city level instead of the per-capita contribution. In addition, he argues that agglomeration effects, well-known in economic geography, are indeed a result of the hierarchical structure and interactions within the urban system, where cities exchange and complement production and distribution processes. Although pointed out by Lösch

(1954) and Christaller (1933) for a very long time, unfortunately, these factors are not being considered when looking at increasing returns through the framework of scaling laws, where only city size is taken into account.

4.5 Regional Effects and Local Heterogeneities

We have seen that there are many limitations in trying to fully describe emergent urban quantities using size as the only explanatory variable. In addition, as pointed out in Shalizi (2011) and Leitão et al. (2016), when considering extensive quantities, the interactions and contributions at the local level are overlooked. In this respect, an aggregated measure at city level does not tell us anything with respect to the local heterogeneities. The same scaling exponent can be found for variables that are uniformly or skewly distributed, and hence could not be used to inform on the level of inequality that a city sustains for example. This is further exacerbated when defining the "performance" of cities according to the residuals in the scatter plot of the scaling fit. For example, a city that has managed to attract enough wealth, such that the aggregated value exceeds the expectations for its size, will be noted as outperforming. Nevertheless, such a city will also most likely be highly unequal. In this context, the notion of performance defined through the residuals is a poor measure of what is going on at the city level, since it does not contain information on how the urban variable, e.g. wealth, is distributed among the population and the neighbourhoods of the city.

Another shortcoming of this approach, is the isolation of cities from their regional embeddedness. This can be illustrated considering income, which is known to be commonly Pareto distributed. Figure 4.9 shows the heterogeneities and regional effects for income at the household level for the 2001 census. In addition, it illustrates the North-South divide that we introduced in the previous sections.

Regional effects cannot be understood without considering the impact of the connectivity, or lack of, between cities, which gives rise to a hierarchical structure, see Lösch (1954), Christaller (1933). In Pumain (2004), the author shows how the spatial organisation of urban systems is very much dictated by its connectivity represented through the speed of transportation, which tunes the intensity and likelihood of the interactions. In this sense, city size is no longer the main variable driving the system, but the intensity of the connectivity between the different urban elements, defining an organisation beyond the administrative demarcations. An example of this, is the indisputable spatial organisation resulting from trade between cities and regions.

The hierarchical organisation refers to the way a process will take place in a system. In this sense, it is important to note that there isn't a unique way to determine such an organisation, since it directly refers to the process, and hence to the way the network of interactions is defined. In Arcaute et al. (2016), we explored deriving such a hierarchical structure from the connectivity of the urban space given by infrastructure. We considered the road network, since it naturally encompasses the oldest means of trading and communicating, although as noted by Pumain, in

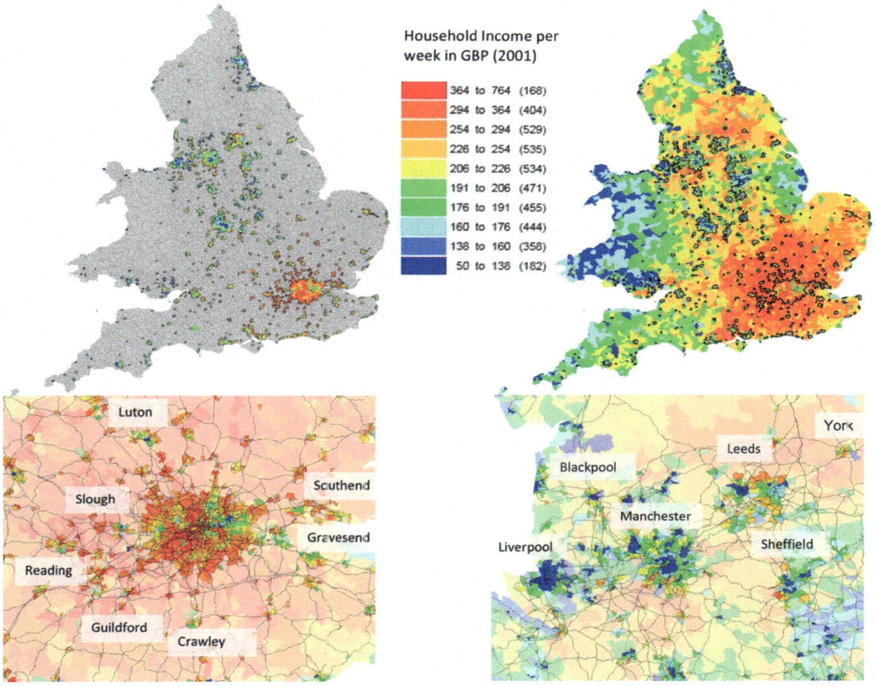

Fig. 4.9 Household income per week at ward level for the 2001 census. The top figures show the heterogeneities within cities and regions, and the bottom figures zoom into the London area, left, and the Manchester and Liverpool area, right. Maps created by Peter Ferguson

order to properly look at the likelihood of interactions, different modes of transport should be incorporated so that speed instead of distance, is used as proximity. Nevertheless, even within such a broad approximation, we find that the emergent hierarchical structure also contains a strong socio-economic syntax. The polarisation of wealth expressed through the North-South divide, is also observed within the different regional divisions obtained through this methodology, see Fig. 4.10. The regional effects can be better appreciated through the representation of the process in the form of a tree. At 1400m, a strong socio-political division is observed: the country is split in two, and Scotland becomes a separate cluster. And at 920m, another critical separation takes place, which mimics the North-South divide. We observe that cities belonging to the blue cluster are part of the cities where wealth is concentrated. The tree gives an indication of the interdependencies at a multi-scalar level, revealing strategic linkages between cities which can be used to inform policies or interventions, such as those concerned with knowledge or economic spillovers, or with changes in transport infrastructure to name a few. In this sense, heterogeneities can be analysed beyond localised properties of cities, and studied within the context

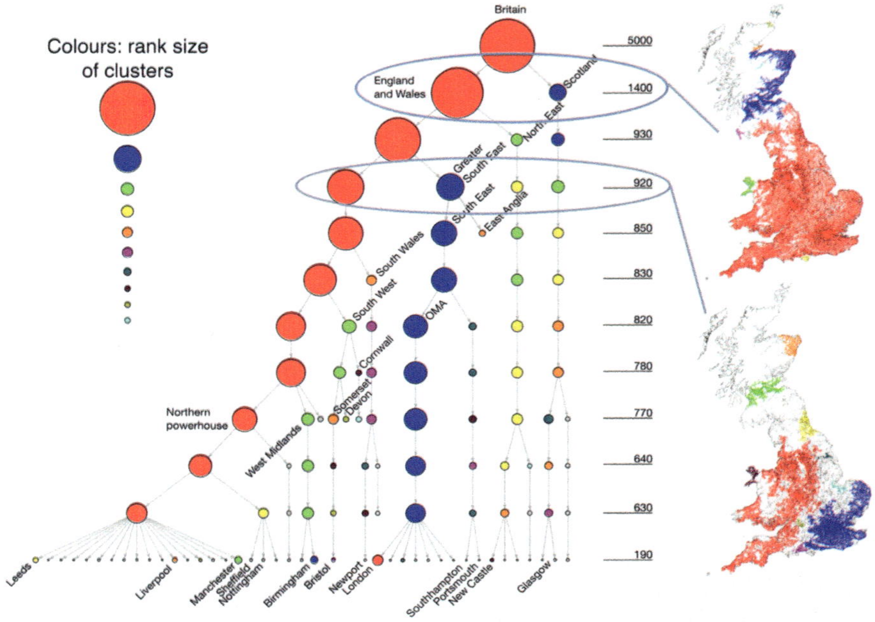

Fig. 4.10 Hierarchical organisation of Britain according to the connectivity given by the road infrastructure. Images taken from Arcaute et al. (2016)

of their region, encoded in the mother cluster, and the proximity to other regions, settlements and city-regions.

Self-similarities of measures in urban systems exist beyond this aggregated approach and have been well-studied through the fractal properties of cities as pointed up earlier. Although initially conceived as mono-fractals, measures in cities can be approximated to multifractals (Murcio et al. 2015; Salat et al. 2017). One important characteristic of multifractals is that the probability of growth is not homogeneous. This immediately suggests a solution to characterising heterogeneities through different self-similarities. In Salat et al. (2018), the analysis of the multifractal spectrum of house prices proved to be a powerful tool to look at inequality, since in addition to looking at the statistical dispersion of the variable, as is generally the approach to inequality through for example the Gini coefficient, it also captures the spatial dispersion, which is normally used for measures of segregation. In this respect, multifractal analysis captures different self-similarities of the system, indicating the level of heterogeneities of the variable itself and of its spatial pattern.

4.6 Conclusions

The search for a science of cities through mechanisms giving rise to universal behaviours has been widely advanced through the complexity sciences, in which the emergent phenomenon is considered to be the outcome of the local interactions which feedbacks into the system, evolving its current status. Cities are composed of a myriad of complex systems themselves, such as systems relating to transport, housing, economics, and education to name a few. The ubiquitous presence of power laws in nature within self-organising principles (Bak 1996), has made it very attractive to explain urban phenomena through this framework. We are still far from an overarching science of cities, but the surge in interest with respect to looking at urban characteristics through scaling laws has helped advance this goal. Self-similarities can be observed in many different ways, beyond the described allometries, and looking for these is certainly still an open path that could bring new insights.

References

Adamic, L.: Complex systems: unzipping Zipf's law. Nature **474**, 164–165 (2011). https://doi.org/10.1038/474164a

Arcaute, E., Hatna, E., Ferguson, P., Youn, H., Johansson, A., Batty, M.: Constructing cities, deconstructing scaling laws. J. R. Soc. Interface **12**, 20140745 (2015). https://doi.org/10.1098/rsif.2014.0745

Arcaute, E., Molinero, C., Hatna, E., Murcio, R., Vargas-Ruiz, C., Masucci, A.P., Batty, M.: Cities and regions in Britain through hierarchical percolation. R. Soc. Open Sci. **3**, 150691 (2016). https://doi.org/10.1098/rsos.150691

Arcaute, E and Hatna, E (eds.): Urban scaling, World Scientific, forthcoming (2020)

Auerbach, F.: Das Gesetz der Bevölkerungskonzentration. Petermanns Geogr. Mitteilungen **59**, 73–76 (1913)

Axtell, R.L.: Zipf distribution of U.S. firm sizes. Science **293**, 1818–1820 (2001). https://doi.org/10.1126/science.1062081

Bak, P.: How nature works: the science of self-organized criticality. Copernicus, Copernicus (1996)

Batty, M.: The New Science of Cities. MIT Press, Cambridge, MA (2013)

Batty, M., Longley, P.: Fractal cities: a geometry of form and function. Academic Press, CA (1994)

Batty, M., Longley, P., Fotheringham, S.: Urban growth and form: scaling, fractal geometry, and diffusion-limited aggregation. Environ. Plan. A **21**, 1447–1472 (1989). https://doi.org/10.1068/a211447

Bettencourt, L.M.A., Lobo, J., Helbing, D., Kühnert, C., West, G.B.: Growth, innovation, scaling, and the pace of life in cities. Proc. Natl. Acad. Sci. U.S.A. **104**, 7301–7306 (2007). https://doi.org/10.1073/pnas.0610172104

Bettencourt, L.M.A., Lobo, J., Strumsky, D., West, G.B.: Urban scaling and its deviations: revealing the structure of wealth, innovation and crime across cities. PLoS ONE **5**, e13541 (2010). https://doi.org/10.1371/journal.pone.0013541

Bretagnolle, A., Paulus, F., Pumain, D.: Time and space scales for measuring urban growth. Cybergeo Eur. J. Geogr. (2002). https://doi.org/10.4000/cybergeo.3790

Cancho, R.F. i, Solé, R.V.: Zipf's law and random texts. Advs. Complex Syst. **5**, 1–6 (2002). https://doi.org/10.1142/S0219525902000468

Cancho, R.F. i, Solé, R.V.: Two regimes in the frequency of words and the origins of complex lexicons: Zipf's law revisited*. J. Quant. Linguist. **8**, 165–173 (2001). https://doi.org/10.1076/jqul.8.3.165.4101
Christaller, W.: Central places in Southern Germany. Englewood Cliffs, NJ: Prentice-Hall, 1966. Original work published in 1933 as "Die Zentrale Orte in Süddeutschland", Jena, Germany: Gustav Fisher (1933)
Clauset, A., Shalizi, C., Newman, M.: Power-law distributions in empirical data. SIAM Rev. **51**, 661–703 (2009). https://doi.org/10.1137/070710111
Cottineau, C.: MetaZipf. A dynamic meta-analysis of city size distributions. PLOS ONE **12**, e0183919 (2017). https://doi.org/10.1371/journal.pone.0183919
Cottineau, C., Finance, O., Hatna, E., Arcaute, E., Batty, M.: Defining urban clusters to detect agglomeration economies. Environ. Plann. B Urban Anal. City Sci. **46**, 1611–1626 (2018). https://doi.org/10.1177/2399808318755146
Cristelli, M., Batty, M., Pietronero, L.: There is more than a power law in Zipf. Sci. Rep. **2**, 812 (2012)
Deluca, A., Corral, Á.: Fitting and goodness-of-fit test of non-truncated and truncated power-law distributions. Acta Geophys. **61**, 1351–1394 (2013). https://doi.org/10.2478/s11600-013-0154-9
Eeckhout, J.: Gibrat's law for (All) cities. Am. Econ. Rev. **94**, 1429–1451 (2004). https://doi.org/10.1257/0002828043052303
Fluschnik, T., Kriewald, S., García Cantú Ros, A., Zhou, B., Reusser, D.E., Kropp, J.P., Rybski, D.: The size distribution, scaling properties and spatial organization of urban clusters: a global and regional percolation perspective. ISPRS Int. J. Geo-Inf. **5**, 110 (2016). https://doi.org/10.3390/ijgi5070110
Fragkias, M., Lobo, J., Strumsky, D., Seto, K.C.: Does size matter? Scaling of CO_2 emissions and U.S. urban areas. PLOS ONE **8**, e64727 (2013). https://doi.org/10.1371/journal.pone.0064727
Frankhauser, P.: The fractal approach. A new tool for the spatial analysis of urban agglomerations. Popul. Eng. Sel. **10**(1), 205–240 (1998)
Gabaix, X.: Zipf's law for cities: an explanation. Q. J. Econ. **114**, 739–767 (1999). https://doi.org/10.1162/003355399556133
Goldstein, M.L., Morris, S.A., Yen, G.G.: Problems with fitting to the power-law distribution. Eur. Phys. J. B **41**, 255–258 (2004). https://doi.org/10.1140/epjb/e2004-00316-5
Gutenberg, B., Richter, C.F.: Frequency of earthquakes in California. Bull. Seismol. Soc. Am. **34**, 185–188 (1944)
Huberman, B.A., Adamic, L.A.: Internet: growth dynamics of the world-wide web. Nature **401**, 131 (1999). https://doi.org/10.1038/43604
Jiang, B., Jia, T.: Zipf's law for all the natural cities in the United States: a geospatial perspective. Int. J. Geogr. Inf. Sci. **25**, 1269–1281 (2011)
Kleiber, M.: Body size and metabolic rate. Physiol. Rev. **27**, 511–541 (1947)
Lane, D., van der Leeuw, S., Pumain, D., West, G.: Complexity Perspectives in Innovation and Social Change Introduction. Springer, Heidelberg and Berlin (2009)
Leitão, J.C., Miotto, J.M., Gerlach, M., Altmann, E.G.: Is this scaling nonlinear? R. Soc. Open Sci. **3**, 150649 (2016). https://doi.org/10.1098/rsos.150649
Levy, M.: Gibrat's law for (All) cities: comment. Am. Econ. Rev. **99**, 1672–1675 (2009). https://doi.org/10.1257/aer.99.4.1672
Lösch, A.: The Economics of Location. New Haven (1954)
Louf, R., Barthelemy, M.: How congestion shapes cities: from mobility patterns to scaling. Sci. Rep. **4**, 5561 (2014). https://doi.org/10.1038/srep05561
Malevergne, Y., Pisarenko, V., Sornette, D.: Testing the Pareto against the lognormal distributions with the uniformly most powerful unbiased test applied to the distribution of cities. Phys. Rev. E **83**, 036111 (2011). https://doi.org/10.1103/PhysRevE.83.036111
Mandelbrot, B.B.: The Fractal Geometry of Nature. Freeman and Co., New York, W.H (1983)
Marshall, A.: Principles of Economics. Macmillan and Co., New York (1890)

Masucci, A.P., Arcaute, E., Hatna, E., Stanilov, K., Batty, M.: On the problem of boundaries and scaling for urban street networks. J. R. Soc. Interface **12**(111), 20150763 (2015)

Mitzenmacher, M.: A brief history of generative models for power law and lognormal distributions. Internet Math. **1**, 1385 (2004). https://doi.org/10.1080/15427951.2004.10129088

Molinero, C., Murcio, R., Arcaute, E.: The angular nature of road networks. Sci. Rep. **7**, 4312 (2017). https://doi.org/10.1038/s41598-017-04477-z

Murcio, R., Masucci, A.P., Arcaute, E., Batty, M.: Multifractal to monofractal evolution of the London's street network. Phys. Rev. E **92**, 062130 (2015). https://doi.org/10.1103/PhysRevE.92.062130

Newman, M.E.J.: Power laws, Pareto distributions and Zipf's law. Contemp. Phys. **46**, 323–351 (2005). https://doi.org/10.1080/00107510500052444

Nordbeck, S.: Urban allometric growth. Geogr. Ann. Ser. B Hum. Geogr. **53**, 54–67 (1971). https://doi.org/10.1080/04353684.1971.11879355

Oliveira, E.A., Andrade Jr., J.S., Makse, H.A.: Large cities are less green. Sci. Rep. **4**, 4235 (2014). https://doi.org/10.1038/srep04235

Pareto, V.: Manuale de economia politica. Societa Editrice (1906)

Pumain, D.: Scaling Laws and Urban Systems. SFI Working Paper 2004-02-002 (2004)

Pumain, D.: Une approche de la complexité en géographie. Géocarrefour **78**, 25–31 (2003). https://doi.org/10.4000/geocarrefour.75

Pumain, D.: Pour une théorie évolutive des villes. L'Espace géographique **26**, 119–134 (1997)

Pumain, D.: Les systèmes de villes. Encyclopédie de Géographie 20 (1992)

Pumain, D., Paulus, F., Vacchiani-Marcuzzo, C.: Innovation cycles and urban dynamics. In: Lane, D., Pumain, D., VanderLeeuw, S., West, G. (eds.) Complexity Perspectives in Innovation and Social Change. Springer, Berlin (2009)

Pumain, D., Paulus, F., Vacchiani-Marcuzzo, C., Lobo, J.: An evolutionary theory for interpreting urban scaling laws. Cybergeo (2006). https://doi.org/10.4000/cybergeo.2519

Pumain, D., Rozenblat, C.: Two metropolisation gradients in the European system of cities revealed by scaling laws. Environ. Plan. B Urban Anal. City Sci. 2399808318785633 (2018). https://doi.org/10.1177/2399808318785633

Roberts, D.C., Turcotte, D.L.: Fractality and self-organized criticality of wars. Fractals **06**, 351–357 (1998). https://doi.org/10.1142/S0218348X98000407

Rybski, D., Ros, A.G.C., Kropp, J.P.: Distance-weighted city growth. Phys. Rev. E **87**, 042114 (2013). https://doi.org/10.1103/PhysRevE.87.042114

Rybski, D., Reusser, D.E., Winz, A.-L., Fichtner, C., Sterzel, T., Kropp, J.P.: Cities as nuclei of sustainability? Environ. Plan. B Urban Anal. City Sci. **44**, 425–440 (2017). https://doi.org/10.1177/0265813516638340

Salat, H., Murcio, R., Arcaute, E.: Multifractal methodology. Physica A **473**, 467–487 (2017). https://doi.org/10.1016/j.physa.2017.01.041

Salat, H., Murcio, R., Yano, K., Arcaute, E.: Uncovering inequality through multifractality of land prices: 1912 and contemporary Kyoto. PLoS ONE **13**, e0196737 (2018). https://doi.org/10.1371/journal.pone.0196737

Sémécurbe, F., Tannier, C., Roux, S.G.: Applying two fractal methods to characterise the local and global deviations from scale invariance of built patterns throughout mainland France. J. Geogr. Syst. (2018). https://doi.org/10.1007/s10109-018-0286-1

Shalizi, C.R.: Scaling and Hierarchy in Urban Economies (2011). arXiv:1102.4101 [physics, stat]

Smith, A.: An Inquiry Into the Nature and Causes of the Wealth of Nations. W. Strahan and T. Cadell (1778)

Strano, E., Nicosia, V., Latora, V., Porta, S., Barthélemy, M.: Elementary processes governing the evolution of road networks. Sci. Rep. **2**, 296 (2012). https://doi.org/10.1038/srep00296

Stumpf, M.P.H., Porter, M.A.: Critical truths about power laws. Science **335**, 665–666 (2012). https://doi.org/10.1126/science.1216142

West, G.: Scale: the universal laws of life and death in organisms. Cities and Companies, Hachette UK (2017)

West, G.B., Brown, J.H., Enquist, B.J.: A general model for the origin of allometric scaling laws in biology. Science **276**, 122–126 (1997)
Willis, J.C., Yule, G.U.: Some statistics of evolution and geographical distribution in plants and animals, and their significance. Nature **109**, 177–179 (1922). https://doi.org/10.1038/109177a0
Woldenberg, M.J.: An allometric analysis of urban land use in the United States. Ekistics **36**, 282–290 (1973)
Zipf, G.K.: Human behavior and the principle of least effort. Addison-Wesley, Cambridge, MA (1949)

Chapter 5
Scaling Laws in Urban Geography. Linkages with Urban Theories, Challenges and Limitations

Olivier Finance and Elfie Swerts

Abstract Scaling laws are simple, easily usable and proven relevant models used in geography for validating various urban theories. These non-linear relationships may reveal physical constraints on the structure and evolution of complex systems, and underline the relationship between urban functions, size of cities and innovation cycles. In this contribution, we examine to what extent scaling laws are transferable towards urban theories and in which specific fields of urban geography these models may be relevant. We thus focus on the accuracy of scaling laws when exploring structures and processes of systems of cities, the diffusion of innovation, metropolization and intra-urban dynamics. We therefore use several examples taken in different regions of the world, embedded in various historical, political and economic contexts. However, in some cases, care must be taken not to over-interpret the results obtained from scaling laws and not to give scaling laws more explanatory power than they can describe. We illustrate this point by providing recommendations relying for instance on the sensitivity of measurements to the delineation of each object of the system under study and to the definition of the system itself. These recommendations can help to get robust results in order to understand the generic evolutionary mechanisms in urban systems.

O. Finance (✉)
UMR 7362 Laboratoire Image Ville Environnement, Université de Strasbourg, Strasbourg, France
e-mail: olivier.finance@live-cnrs.unistra.fr

E. Swerts
FORS—Swiss Centre of Expertise in the Social Sciences, Université de Lausanne, Lausanne, Switzerland
e-mail: elfie.swerts@fors.unil.ch

O. Finance · E. Swerts
UMR 8504 Géographie-cités, Paris, France

D. Pumain (ed.), *Theories and Models of Urbanization*, Lecture Notes in Morphogenesis, https://doi.org/10.1007/978-3-030-36656-8_5

5.1 Introduction

As various social sciences, geography benefits from inter- and multi-disciplinarity through cross-disciplines interrelationships. Geography and spatial analysis feed other disciplines with their specific relation towards space as input of analysis; in the opposite way, quantitative geography often benefits from approaches previously developed in other fields. The search for regularities in geographical systems sometimes revisit models and (statistical) laws firstly developed in other fields, as with the clear parenthood between Newton's law of universal gravitation and Tobler's first law of geography[1] (Tobler 1970). Theoretical implications and interpretations evidently differ in those cases, as well as in the application of scaling laws.

Scaling laws are among these models assimilated and revisited by geographers. These statistical power-law relationships link an attribute distributed among elements under consideration and the size of each of these elements. In other words, they allow observing if there is a regular elevation of a quantity while considering bigger elements in size, not only considering a proportionality assumption (linear shape of the relationship) but also sub-linearly or super-linearly. It is the shape of the relationship which is under study: does the attribute variate "faster" or "slower" than proportionally with the size of the elements when bigger elements are considered?

Scaling laws first emerged in biology, from the systematic observation that the energy expended by living organisms is a function of their size, and that this energy does not vary linearly but according to a law of scale interpreted as economies of scale (West et al. 1997, 1999). Scaling laws are now broadly used in geography especially in urban quantitative geography. They combine and slowly overpass some other typical statistical methodologies previously used in the field as Zipf's law or rank-size rule (Zipf 1949). Scaling laws replace a unidimensional approach (comparing population value and population rank of elements, especially cities) by a two-dimensional approach (based on an attribute distributed among elements, and their population). The rank-size rule is of course still useful for geographers to study the hierarchical structure of an entire set of cities, yet scaling laws now permit to confront diverse urban attributes to this hierarchical structure of the system.

A lot of urban attributes have actually be found as scaling either linearly, sub-linearly or super-linearly with city size. They are as diverse as: morphological indices (for example the length of roads, the built area or the total area of green spaces; Levinson 2012; Bettencourt 2013; Louf and Barthelemy 2014; Arcaute et al. 2015; Rybski et al. 2016; Cottineau et al. 2017); wages and incomes in diverse geographical contexts (Um et al. 2009; Paulus and Pumain 2011; Arcaute et al. 2015; Cottineau et al. 2018; Sarkar et al. 2018); economic figures (as the distribution of some activities among cities of an urban system: for example, the magnitude of research and scientific activities scales super-linearly with city size both in terms of jobs and patents filed; Pumain et al. 2006; Bettencourt et al. 2007b; Arcaute et al. 2015).

---

[1] "Everything is related to everything else, but near things are more related than distant things" (Tobler 1970).

This contribution will focus on several aspects of the use of scaling laws in urban geography. It deals with the relevance and limitations of scaling laws to explore urban theories. In these urban theories, the city is understood in dynamic terms, as a product of interdependence and interactions that take place at different temporal and spatial scales (Pumain 1982; Batty 2013). Scaling laws, like other statistical, mathematical and physical models, make it possible through abstraction, to compare the dynamics of cities and the different systems they form. Thereby, they allow to highlight the regularities of cities and urban systems, as well as some of their specificities. Among these regularities, some are strongly linked to the size of cities and can therefore be studied by scaling laws (Pumain 2004; Batty 2013; Arcaute et al. 2015; Barthelemy 2016). Thus, a remarkable regularity is the way economic activities are spatially distributed in systems of cities and the way this distribution evolves in space and time. This distribution is strongly linked with city size.

In this context, this chapter questions to what extent scaling laws make it possible to test these urban theories, to highlight the invariants and specificities of geographical structures and dynamics by testing theories with empirical case studies. In doing so, they also question the general limitations that have to be kept in mind while using this methodology in geography. Section 5.2 focuses on the process of adoption of scaling laws in geography and on the various theoretical interpretations that can be given to results obtained through this methodology. Through Sect. 5.3, the relevance of scaling laws is highlighted when studying geographical structures, dynamics and theories. Each point is illustrated with chosen examples taken in the literature. Other examples are used in Sect. 5.4 to highlight and discuss some limitations of scaling laws in geography and more specifically key-points which should not be forgotten when using them in urban studies to get robust results. Section 5.5 concludes.

5.2 Origins of Scaling Laws and Their Adoption in Geography

Scaling laws have been transposed from biology to social sciences in the 1990s, as in urban geography. After some words on the use of scaling laws in biology (Sect. 2.1), the specific analytical framework in which scaling laws have been adopted—systems of cities—will be presented (Sect. 2.2). Various and opposing interpretations of these scaling laws have been theorized in this framework, mainly by Pumain and Bettencourt research groups. These various interpretations are here reminded (Sect. 2.3).

5.2.1 *Scaling Laws in Biology: The Metabolic Rate of Organisms*

One of the major application of scaling laws in biology has been the confrontation between energy consumption per unit of time and the body mass of living organisms, i.e. the metabolism versus the size. Results of systematic analyses on several orders of magnitude in terms of size of animals or mammalians (West et al. 1997, 1999; Savage et al. 2004) show that the metabolic rate of an organism of mass m is proportional to m power ~0.75, the minimal rate of energy expenditure per unit of time by endothermic animals at rest (Basal Metabolic Rate) (Fig. 5.1). In other words, bigger animals in terms of body mass expand a lesser quantity of energy than smaller ones compared to the proportionality assumption. This economy of energy is function of a power-law relationship, and this scaling law is obeyed with remarkable precision ($R^2 = 0.99$ in the relationship exposed in Fig. 5.1).

As reminded by Pumain (2012), West takes the example of ants that are able to lift one hundred times their own weight, which a human being would be incapable of. Can we then say that ants are stronger than human beings? "To say so would amount to saying that if an ant could reach the size (or weight) of a man it would be 100 times stronger (…). Yet the change in scale from the ant to the man is not a linear function of weight" (Pumain 2012). In fact, this strength varies according to a power law. "It is only when an appropriate form of relationship is used (…) that it is possible to obtain an evaluation that does not defy intuition: human strength is then more or less equivalent to that of an ant" (ibid.).

The metabolic rate of organisms therefore scales sub-linearly with size with a three quarter exponent, which is explained by West et al. (1997) by "a general model that describes how essential materials are transported through space-filling fractal networks of branching tubes". Three assumptions are emphasized in this general model to explain this sub-linearity. First, a space-filling fractal-like branching pattern is

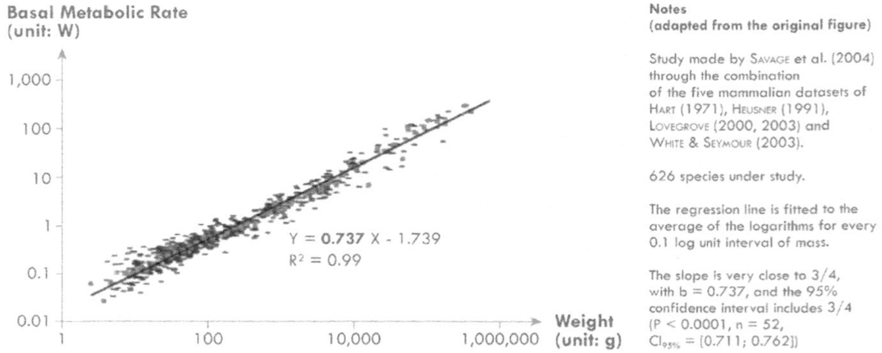

Fig. 5.1 Basal metabolic rate versus size for mammalian species (adapted from Savage et al. 2004)

required in order the biological networks to supply the entire volume of the organism. Moreover, terminal units (the final branches of the network) are size-invariant. The system has at last to be optimized to minimize the energy required to distribute resources (West et al. 1997). The sub-linear scaling regime observed is therefore the consequence of the combination of physical and geometric constraints encompassed in these three assumptions.

As the objects that are observed in geography don't share all the properties of those in these studies in ecology, there are no reasons that the same sub-linearity is found when using scaling laws in urban geography. Nevertheless, urban systems appear to be strongly shaped by scaling laws.

5.2.2 *Urban Systems Shaped by Scaling Laws*

The scaling laws have been transferred in geography within the specific analytical framework of systems of cities (Berry 1964) which allow to understand the city in an evolutionary way, at different temporal and spatial scales. Systems of cities result from the different forms of exchange and interaction that occur between cities at varying distances, resulting from their mutual dependence, complementarity, cooperation and competition. They are carried out through various types of networks such as infrastructure, migration, trade or information (Bretagnolle et al. 2009). Resulting functional regularities are then observed at several levels of observation, from the intra city-level to the systems they form.

Geography is not the only discipline that considers cities as a relevant level of observation, economics being an example. The specificity of geography is to consider different levels of observation of urban entities: cities can be considered as "systems within city systems" (Berry 1964), an interpretation of the city which has even been refined to consider three distinct but interrelated levels of observation. Cities can first be studied through elementary units constituting a first micro-level of observation, which are individuals, companies or institutions that live together in a city. Their actions and exchanges build and shape individual cities, observable at a meso-level. This level of observation is that of the city entity as a whole, defined as a coherent geographical entity, either morphological or functional. A morphological city is defined as a set of continuous urban buildings, whereas a functional one is defined as the whole area over which the city exerts a strong influence and attractiveness, for example by including places in strong interrelationship with an urban core through commuting flows. This second meso-level of observation is covered by a third macro-level, that of the system of cities, "composed of a large number of cities and towns that interact under unified control" (Bretagnolle et al. 2009). These systems are traditionally considered to extend within national borders, but with the increase in long-distance interactions, their actual limits may overpass them (at the European scale for example).

Systems of cities are strongly shaped by a hierarchy of size and functions, either economic, political, administrative or social (Christaller 1933; Berry 1964; Pred

1977; Pumain 1997), due to interactions both at each single level and between micro, meso and macro-levels (Fig. 5.2). Interactions between levels (cooperation or concurrence for example) produce strong interdependencies in the evolution of cities, both in their demographic, social or economic dimensions (Pumain et al. 2009), resulting for example in a strong co-evolution of their socio-economic profiles (Paulus 2004). Some profiles can nevertheless diverge from the general trend through effects of selection at diverse stages of their history, related to faster than expected development of some business sectors in some cities due to their location near a reservoir, a border or metropolises. This selection can lead to a self-reinforcing effect of economic specialization through economics of location, and the marks of this specialization can be observed even after the decline of what has led to this specialization.

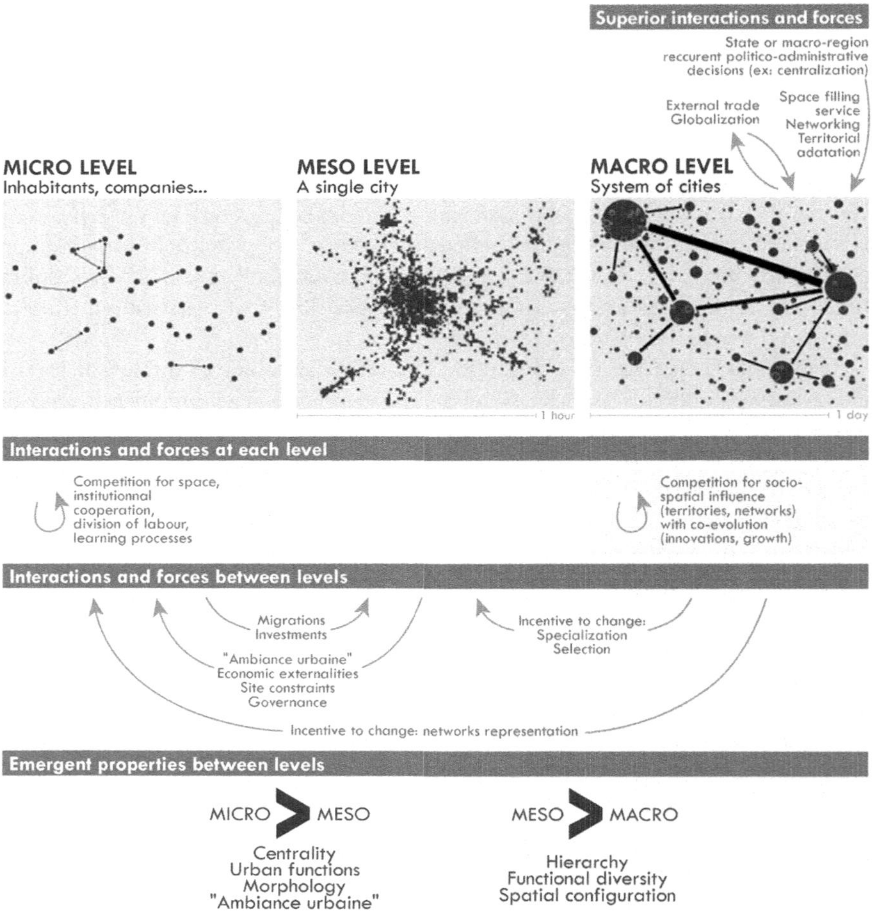

Fig. 5.2 Multi-level organization of systems of cities, interactions and emergences between levels (adapted from Pumain 2006 and Pumain et al. 2009)

As in various other complex multi-level systems, emergent properties link the three levels of observations (bottom part of Fig. 5.2). A city (meso-level) can for example be seen "as a collective entity whose specific properties, although mostly produced by intentional agents at the individual level, cannot be simply explained or predicted from these intentions, nor derived by summing the characteristics of its inhabitants" (Bretagnolle et al. 2009). From the micro to the meso-levels, the emergent properties are as diverse as centrality, morphology or urban functions. Analogously, the hierarchical structure of systems of cities, the diversity of functions among cities or a specific spatial configuration emerge at the macro-level.

Scaling laws in urban geography are useful to detect some of these emergences at the macro level, by helping to catch how some urban attributes are distributed in an urban system by comparing this distribution to the hierarchical distribution of city sizes. Similarly to other disciplines, scaling laws in urban geography are power-law relationships connecting attributes to size according to the following formalization:

$$Y = Y_0 N^{\beta}$$

where Y is an urban attribute, N the size of each city (usually its population), β the power exponent and Y_0 a normalization constant. For convenience, the relationship is usually regressed in its log-transformed form:

$$\log(Y_i) = a_0 + \beta \log(N_i) + \varepsilon_i$$

where a_0 is a general constant and ε_i the residual. β therefore corresponds to the slope of the linear trend line obtained on a bi-logarithmic scatterplot or on log-transformed values of the attribute and the size (hereinafter referred to as "the β parameter" or the "scaling parameter"). β is usually computed through Ordinary Least Squares (OLS) regression; the quality of fitting is quantified by the coefficient of determination R^2 and the 95% confidence interval ($CI_{95\%}$) helps to validate the regime of scaling (Leitão et al. 2016). If $CI_{95\%}$ is totally included in the interval [0.95; 1.05], the scaling regime is considered as linear; the scaling regime is considered either as sub-linear or super-linear when $CI_{95\%}$ is entirely below 0.95 or above 1.05. If there is an overlap, conclusions are mixed or uncertain. Thus, various scaling regimes are observed in urban geography contrary to the previous example in biology. Their interpretation need some geographical-oriented theories that are detailed in the next subsection.

5.2.3 *Diverse Theories to Interpret Urban Scaling Laws*

The transfer of such a theory from experimental to social sciences and especially in urban geography required an adaptation. Consequently, this step is not only a *diffusion* from biology to geography, with an adoption in geography of a model elaborated in another field; it is much more a *transfer* as both the objects studied and the concepts and theories developed to explain the form, dynamics, and long-term evolution of these objects are different. Sub-linearity is still observed for some urban attributes, but linearity and super-linearity can also be observed in urban geography. Various theoretical interpretations are given by physicists and geographers to explain these three scaling regimes.

A first interpretation is given by Bettencourt (2013) and Bettencourt et al. (2007a). An important contribution of this first interpretation is to provide a linkage between the value of scaling law exponents and growth processes via a mathematical model (Bettencourt et al. 2009). Their interpretation of sub-linear regimes is linked with economies of scale: urban attributes that are liable to achieve scale economies would show scaling exponents smaller than one. They expect these sub-linear regimes to represent infrastructure variables, such as the number of gas stations (Bettencourt et al. 2007a). Sub-linearity evidences constraints on development, which translate into a restriction of growth that takes on a logistic function. The existence of linear regimes is linked with individual basic needs, independent from city size, such as water consumption. In the case of linear regimes, unlimited exponential growth can be observed. Third, super-linear regimes of scaling are understood by increasing returns to network interactions. Socio-economic variables such as Gross Domestic Product are expected to scale super-linearly with city size (Bettencourt et al. 2007a). In this case of super-linearity, "constraint then tends to produce development that is all the more marked where the system is already large" (Pumain 2012), that is what economists call *agglomeration economies* or *increasing returns* (Feldman and Florida 1994).

This general interpretation is criticized by Pumain (2012) as "physicists in this case conclude to a "singularity in finite time" of the growth curve of cities, a quantitative explosion which then translates into an abrupt decline in growth if there is no innovation to provide new resources and alter the energy patterns in the system". This interpretation "concluding that there is an increase in the "pace of life" in relation to city size, seems to us to miss all the social organization that builds up through a city's history". Moreover, it assumes ergodicity; yet "this hypothesis is not consistent with an evolutionary theory of urban systems integrating the spatial distribution of labour and the hierarchical diffusion of innovation". This interpretation is rejected as it has been shown that on the very long term, the growth of cities is more or less exponential. Each city cannot reach any possible state in the system, as each kind of city (in terms of size) is not able to catch any kind of innovation at random times.

This concept of innovation is at the core of another interpretation produced by Pumain et al. (2006): an evolutionary theory for interpreting urban scaling laws, linking together innovation, distribution of activities and the hierarchical properties of urban systems. They clearly oppose to Bettencourt's group assumptions considering that "on average different cities are scaled up versions of each other" (Bettencourt et al. 2008). Empirically based on the distribution of economic activities among cities of a given system, this theory is evolutionary as the observations about the way cities co-evolve are considered both taking into account the spatial (hierarchical) structures of urban systems and the course of time. It provides a "linkage between the concepts of urban functions, city size and innovation cycles" (Pumain et al. 2006). Three types of exponents of the power function linking the size of cities to the variable applied to the activities they host are then identified. When the exponent is close to 1, activities are distributed proportionally to the city size (population) in the system. When the exponent is greater than 1, activities are relatively concentrated in the upper-part of the urban hierarchy, and conversely among the smaller units when β is below 1. In this proposal, super-linear regimes characterize innovative business sectors emerging at the top of the urban hierarchy. In parallel with the emergence of new innovation cycles, the process of hierarchical diffusion leads to the propagation of innovations towards smaller towns in the system, leading the scaling regime for a given activity to become first linear (and the activity to become common), then sub-linear when other innovation waves occur (and the activity mature) (Fig. 5.3).

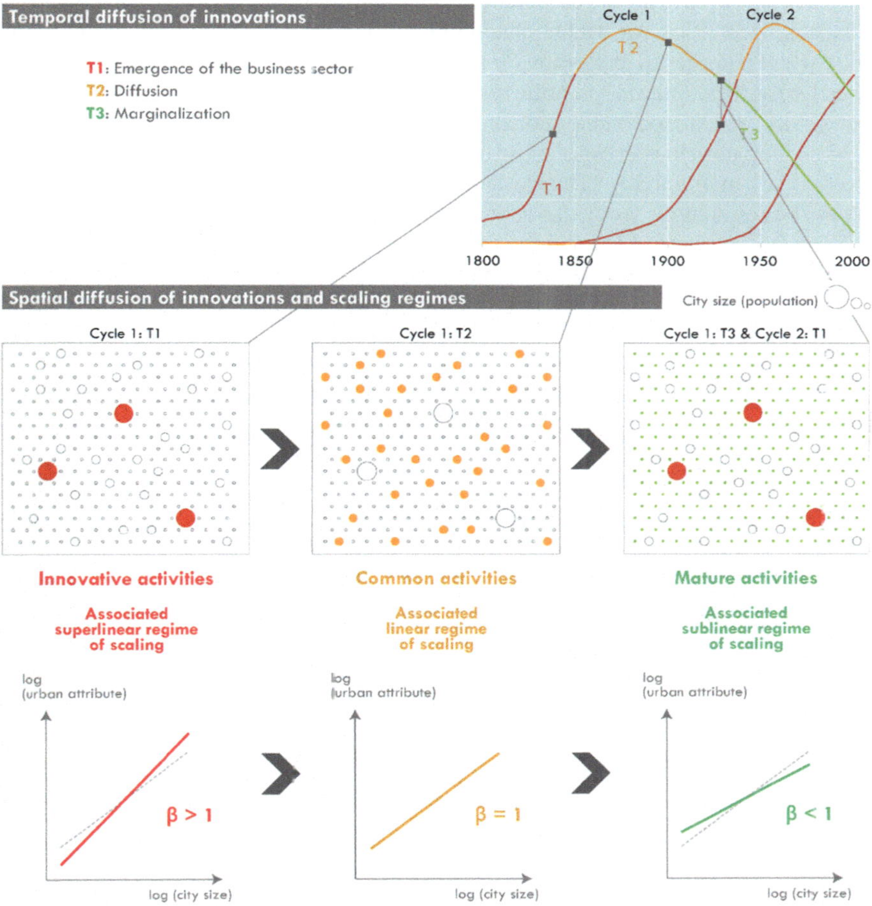

Fig. 5.3 Diffusion of innovations in space and time and scaling laws (adapted from Pumain et al. 2006)

This theory reinterprets the principle of hierarchical diffusion of innovations as formulated by Hägerstrand (1968). It is strongly linked to the evolution cycle of products (and services): initially when the product/service emerges, it is innovative and is first captured by large cities that have the capacity to develop it and benefit from it despite the high costs of production. In a second stage, that of trivialization, the product or service (then called "adapted") spreads to smaller cities because of the lower production and acquisition costs. Finally, if a product or service is replaced by a new one, or in case of obsolescence, its diffusion continues and attached employment reaches and concentrates in a few small specialized cities. This process may also depend on the choice of cities to create an innovative product or to adopt an innovative sector of activity, because of an environment, intrinsic to the city or cities of the same region, conducive to the development of these innovations. This leads to the formation of specialized cities, for which location factors can be more independent of city size (e.g. when it comes to exploiting "resource deposits" for extractive activities, scenic sites for tourism, or even scientific research when it is concentrated in relatively small but highly specialized cities, such as in Europe Oxford, Cambridge or Heidelberg).

The hierarchical processes of diffusion of innovations tend to strengthen the pre-existing urban hierarchy. They explain the persistence of the relative weight of large cities, which benefit from the advantages of capturing the benefits generated by innovations, a relative diversification of their economies and a more complex organization resulting from their previous adaptations. Conversely, cities that specialize in an innovative sector of activity initially experience an increase of their weight in the system as a result of the spillover effects of innovation. However, they may subsequently present difficulties in adapting to new innovations due to their excessive specialization. They are thus weakened by their dependence on the evolution of product cycles (this has been the case in France, for example, in the manufacturing cities of northern France).

From a dynamic point of view, business sectors with a scaling exponent larger than 1 therefore correspond to innovative activities, which at the stage of their emergence are captured by large cities. As they become more commonplace, they spread throughout the system and the value of their scaling exponent tends to 1. Their distribution is then approximately proportional to city size. Activities at the end of a cycle are then concentrated in a few small specialized cities, and the exponents of the scaling law of these sectors are then less than 1. We mainly rely on this second theory to interpret scaling exponents computed in various geographical contexts and about various urban attributes in the next section, showing how scaling laws can be fundamental in geography.

5.3 Scaling Laws and Geographical Theories, Structures and Dynamics

Scaling laws are useful at the macro level to describe the distribution of an urban attribute among cities of a system while taking into account its hierarchical structure. Focusing on the level of systems of cities, we recall how scaling laws can be used as a concentration index of urban attributes among cities and their global added-values compared to other indexes (Sect. 3.1). We show that scaling laws are much more than simple concentration indexes by testing the evolutionary theory for interpreting scaling laws (Pumain et al. 2006) detailed in the previous subsection. The relationship between distribution of activities and the hierarchical position of cities makes it possible to question the processes of diffusion and concentration of such activities within urban systems as metropolization (Sect. 3.2). Once these regularities explored, it makes it possible to highlight outliers, i.e. cities that deviate from the expected pattern, to further enlighten the factors that generate such singularities (Sect. 3.3). Yet results also allow to test both the center-periphery gradient emerging in each city at the meso-level, when various definitions of the city are used, and more generally, the impact of the variation in city definitions on the interpretation of the results of scaling laws (Sect. 3.4).

5.3.1 Urban Hierarchy and Concentration

While analysing the distribution of an urban attribute among cities of an urban system, scaling laws allow to measure its degree of concentration along the urban hierarchy. Several types of measurements can be used to assess the concentration of a variable in an urban system, such as the Gini index or the rank-size rule. Their comparison remind the added-value of scaling laws compared to these other methods (Sect. 3.1.1). Yet once included in an evolutionary theory of systems of cities, scaling laws are much more than a concentration index (Sect. 3.1.2).

5.3.1.1 Added-Value Compared to Gini Index and Rank-Size Rule

Scaling laws are more efficient than some other methods allowing measuring the distribution of an attribute in a system of cities. Scaling laws are compared to the

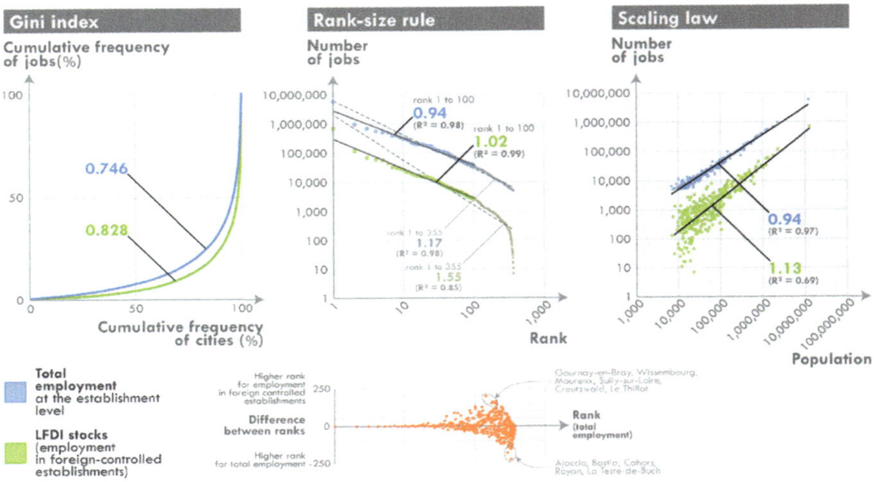

Fig. 5.4 Comparison of the Gini index, the rank-size rule and scaling laws to compare the distribution of total employment and LFDI stocks in the French system of cities

Gini index[2] and the rank-size rule[3] in Fig. 5.4 to compare the distribution of Localized Foreign Direct Investment stocks (LFDI stocks, i.e. employment in every foreign-controlled economic establishment in 2008, aggregated at the functional city level—the 'aires urbaines') in the French system of cities, to the one of total employment at the establishment level.

With a higher Gini index, LFDI stocks appear to be generally more concentrated than total employment. But there is no evidence on which cities have been specifically selected by foreign investors, either metropolises, small cities or even at random positions in the system. With a steeper slope while considering a rank-size rule, LFDI stocks appear once again to be more concentrated than total employment; but as assessed by the bottom part of Fig. 5.4, as cities are separately ordered according to their rank for each attribute, a given rank on one attribute may not necessarily represent the same city as the same given rank of the other attribute. Some cities

[2]The Gini index is based on the Lorenz curve which summarizes the distribution of an attribute among elements (the cumulative frequency of the elements is plotted on the x-axis and the cumulative frequency of the attribute is plotted on the y-axis). The value of the Gini index corresponds to the area between the line of perfect equality (dotted line on Fig. 4) and the Lorenz curve computed for a given attribute. The index varies from 0, a situation of perfect homogeneity, to 1, the maximal inequality or heterogeneity of distribution.

[3]Zipf and some of his predecessors (Auerbach 1913; Zipf 1949) have formulated an empirical law, the rank-size rule or Zipf's law, used in urban geography to illustrate the general hierarchical regularity of city size found in each system of cities in the world. This regularity is expressed as an inverse geometric progression between the population P_i of a city and its rank R_i, as $P_i = K/R^{\alpha}$, where K and α are constants, α being the slope of the trend line on a bi-logarithmic graph. α is found to not pull away strongly from 1 (see meta-analysis by Cottineau 2017; see also an application of Zipf's law on diverse urban systems by Pumain et al. 2015).

find themselves dozens of ranks above or below in terms of LFDI stocks compared to total employment, but the rank-size rule cannot reveal it by itself. Moreover, the rank-size rule, when computed on the 355 'aires urbaines' on the LFDI stocks, is skewed by the shape of the tail. Therefore two values of the slope α are computed (from rank 1 to rank 100 or to rank 355) for each attribute; in both cases, LFDI stocks are more concentrated than total employment, even if the comparison can be impacted by the diverging position of a given city on each curve. Scaling laws overcome this difficulty (right part of Fig. 5.4) by considering another variable on the x-axis: the size of each city (in this case its population). Then, two cities of same size or twice the same city on two scatterplots have exactly the same position on the x-axis. The comparison between two distributions (in this case LFDI stocks and total employment) is consequently much more rigorous while using scaling laws than the rank-size rule.

Using scaling laws, conclusions towards a higher concentration of LFDI stocks are verified. Contrary to the two other methodologies, scaling laws clearly allow to assess a higher concentration of LFDI stocks in the highest part of the urban hierarchy (the scaling regimes are either linear or sub-linear for total employment, significantly super-linear for LFDI stocks, which means that biggest cities concentrate even more LFDI stocks than expected under a proportionality assumption). The high variability in the relationship when the smallest cities are considered (note that it is related to the curved shape of the rank-size rule) implies some other factors than the urban hierarchy to explain the distribution of LFDI stocks in the French system of cities (specialization, proximity to investors or to market potential, etc.; Finance 2016).

The concentration of transnational investment in selective places had already been stated; the use of scaling laws here demonstrates it more rigorously than with other methods and with more control than studies on World Cities (Taylor 2004). Transnational investors don't select the sole metropolises but tend to have a higher presence in the upper-part of the urban hierarchy according to a scaling law (Finance 2016).

5.3.1.2 More Than a Simple Concentration Index

As detailed in the previous section, scaling laws provide a better concentration index than Gini or the rank-size rule. When the scaling regime is sub-linear or super-linear, it can be concluded that the attribute under study is, first, concentrated, and that this more-than-proportional concentration is located in respectively the lowest or the upper-part of the urban hierarchy. But this method is much more than a concentration index when innovation cycles and the location of activities are considered. As assessed by Pumain et al. (2006) in their evolutionary theory, there is a link between innovations and the scaling parameter values when the distribution of economic activities is considered in systems of cities. This assumption can be tested in two ways: first by classifying economic activities into innovative, common and mature sectors according to the values of the scaling parameters, second by computing the scaling parameter values on economic activities beforehand classified in innovative/less innovative clusters.

As stated above, the analysis of the distribution of a business sector in an urban system through scaling laws can lead to the classification of the sector as innovative when $\beta > 1$, common when $\beta \approx 1$ or mature when $\beta < 1$. Applied to the urban systems of France, the USA, South Africa and China—on functional city definitions proper to each geographical context—it confirms that activities developed in large cities are the most advanced and complex ones, according to the geographical, historical, economic context of each system (Table 5.1). The FIRE business sectors (financial activities, insurance and real estate) rank as innovative activities in each of the four systems, in relation with both globalization and metropolization processes. This is also notably the case in China, where the banking sector—initially built on the model of the Soviet monobank system—is developing and becoming more complex. This is largely in relation with the arrival of foreign banking institutions and the emergence of peripheral financial institutions, which have long been limited in the country (Svejnar 2007). Yet other innovative sectors are context-dependent: manufacturing is innovative in so-called emerging countries (South Africa and China), and retail trade in China, whereas these activities are common or mature in formerly industrialized countries.

On the other hand, economic activities can be clustered from the most innovative to the less innovative ones before computing the scaling parameters, to verify the concordance between scaling laws results and the economic nomenclatures. In Fig. 5.5, this clustering is done according to the OECD nomenclature in knowledge-intensive and less-knowledge intensive services as well as high-tech and low-tech manufacturing industries, and scaling parameters are then computed in the French case. Services scale super-linearly with city size whereas manufacturing industries linearly or sub-linearly ($CI_{95\%}$ prevents excluding linearity). Most of the innovative (according to the nomenclature) services and manufacturing industries show

Table 5.1 Scaling parameters of diverse economic activities in four systems of cities (combining Paulus and Pumain 2007; Vacchiani-Marcuzzo and Paulus 2008; Pumain et al. 2009; Swerts 2013)

	France	USA	South Africa	China
Innovative sectors $\beta > 1$	FIRE (financial activities, insurance, real estate)			
	Research & Development Business services, consultancy		Manufacturing	
				Retail trade
Common sectors $\beta \approx 1$	Hotels and Restaurants Community, social, personal services		Retail trade Social services	Public administration
		Manufacturing	Utilities	
Mature Sectors $\beta < 1$	Manufacturing	Retail trade Utilities	Private households	Social services

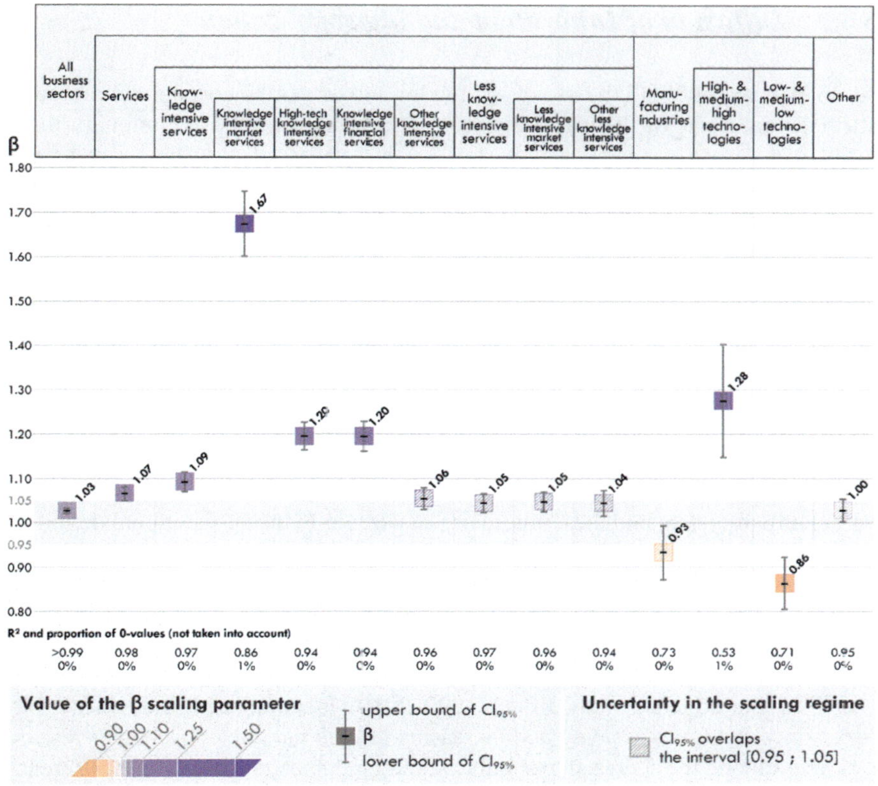

Fig. 5.5 Values of scaling parameters for activities located in the French system of cities after clustering of services and manufacturing industries of similar innovation degrees

clear super-linear regimes (up to 1.67 for knowledge-intensive market services such as activities of head offices, air transport, legal and accounting activities, advertising, etc.), whereas less innovative services show values of β closer to 1 and less innovative manufacturing industries clear sublinear regimes. These results are in total concordance with the previously developed theory. Services are more innovative than manufacturing; innovative services and manufacturing industries are much more metropolitan; above all, innovative manufacturing industries are much more concentrated in the upper-part of the urban hierarchy than less innovative services. Therefore caution is needed when the manufacturing sectors are all put together: there is a strong variability of profiles under this label, and innovation do not only concentrate in globalized services.

5.3.2 Diffusion of Innovation and Metropolization

As an extension of both the use of scaling laws to detect the concentration of urban attributes in a part of the urban hierarchy (Sect. 3.1) and the evolutionary theory to interpret urban scaling laws (Sect. 2.3), this statistical modelling can be used to detect the stage in the diffusion process of diverse innovations. By comparing scaling parameters obtained for many different periods, it is possible to consider the evolution of the concentration of the attribute in the urban system. Following the simple confrontation of an urban attribute and the hierarchical structure of a system of cities, and in close relationship with the geographical interpretation of scaling laws exponents, scaling laws are useful to validate theories of hierarchical diffusion of innovations and the process of metropolization in systems of cities. Therefore, the same methodology can be applied in the same geographical context but at different periods in order to study the evolution of β exponents over time and test these two hypotheses.

Scaling parameters are computed over a selection of business sectors both in the French urban system and the Chinese one in Fig. 5.6 by using functional definitions of cities in both cases. Many manufacturing industries exhibit decreasing scaling parameters over time in France (1962–1999): these sectors are diffusing into the system of cities. At the opposite, research and development is becoming more and more concentrated in the upper-part of the urban hierarchy, being one of the most innovative business sector in France. The Chinese system of cities is quite different even if the period under study differs (1990–2010): manufacturing industries exhibit super-linear regime of scaling and increasing scaling parameters, whereas research classifies among the decreasing or stable scaling parameters. Other slight differences oppose these urban systems: retail trade exhibits a linear regime in France whereas the scaling parameter is now above 1 in China; education stays linear in France while sub-linearity is detected in the Chinese case. Therefore it has to be reminded that the results of each computation of scaling laws may be better understood in its geographical context (a given business sector is not as innovative in two distinct urban systems) and in its temporal context (as the theory for interpreting scaling laws is evolutionary).

5.3.3 Scaling Laws as a Filter to Detect Outliers

As one can expect from some wide confidence intervals detected previously (knowledge-intensive market services or high-tech manufacturing industries), some cities deviate from the statistical model. In urban geography, rare are the attributes (as the distribution of jobs of a given business sector in an urban system) scaling as perfectly with city size as the metabolic rate of organisms. Urban scaling laws can reveal robust relationships between urban attributes and the urban hierarchy, but some cities may deviate significantly from the value expected by the scaling

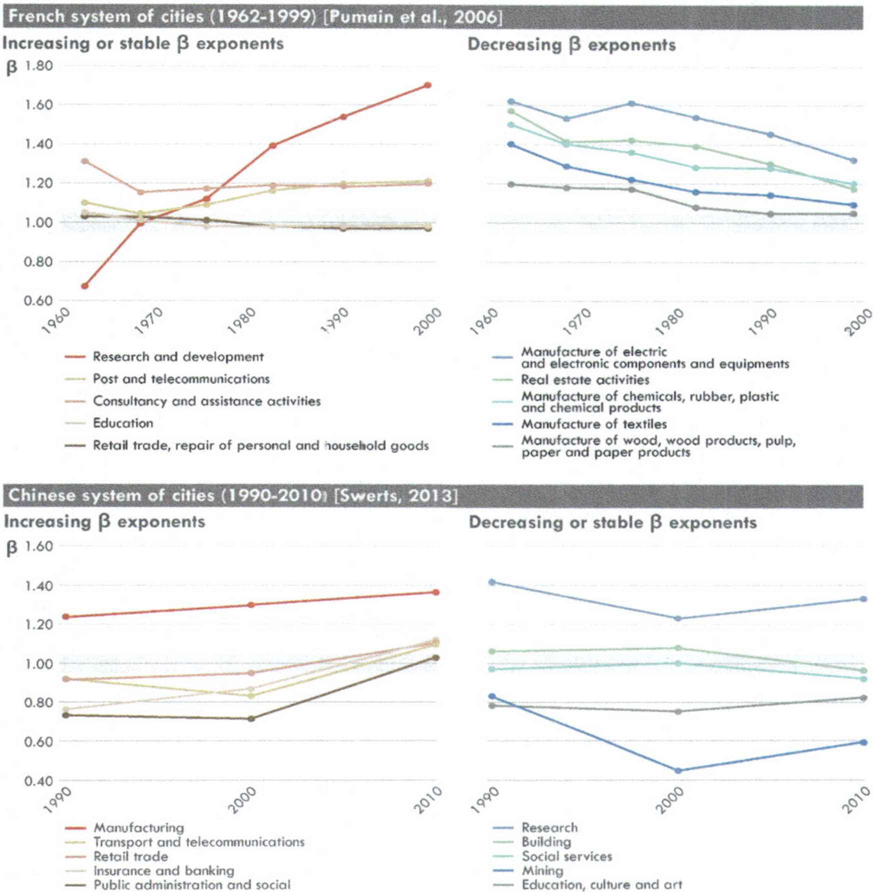

Fig. 5.6 Temporal evolution of scaling parameters computed over a selection of business sectors in the French and Chinese systems of cities

relationship. Therefore, deviations to scaling laws can be computed to observe local specificities that are a higher or lower concentration of the urban attribute under study than expected from city size and the bivariate relationship.

Two different examples of computation of deviations to scaling laws are shown in Fig. 5.7, about patents in cities of the United States and LFDI stocks in the French ones (functional definitions in both cases). Cities in blue are below the computed trend line of a scaling law and the ones in red above. In the first case, a correction has been applied to make the deviations independent to city size, but the semiology makes the confrontation between the urban hierarchy and the deviations difficult to read as the size of the circles represents the intensity of the deviation rather than the population. In both cases, deviations are strongly affected by regional effects. Patenting appears

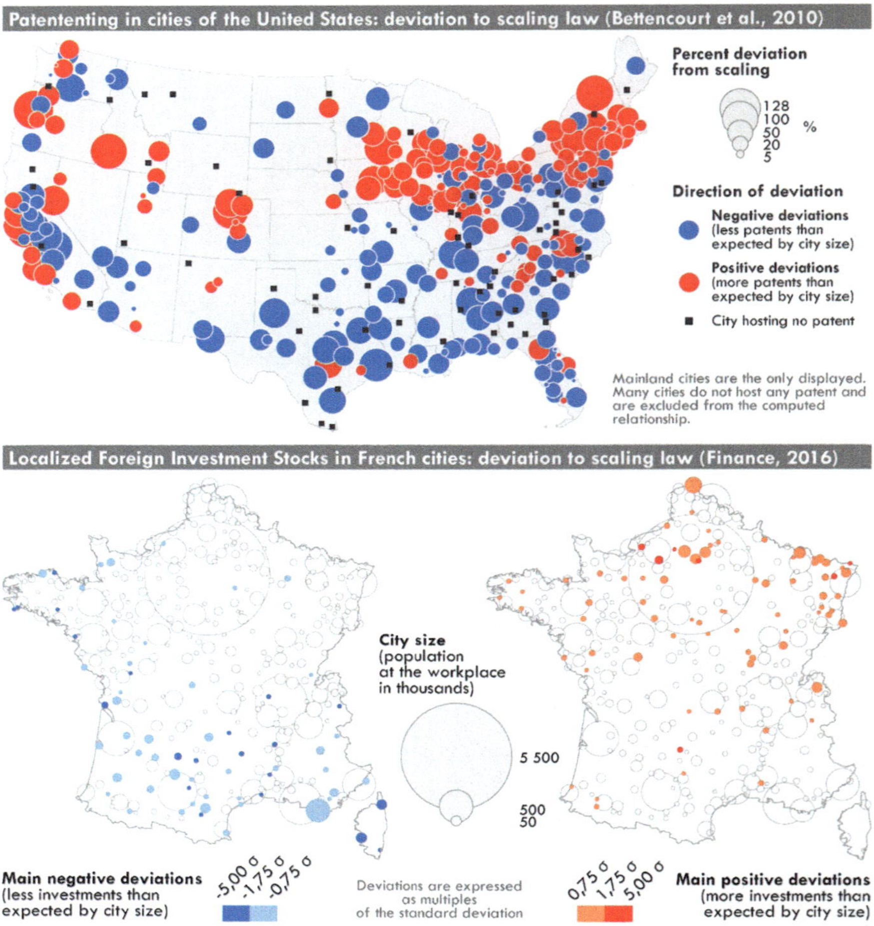

Fig. 5.7 Deviations to scaling law in the cases of patents in the United States and the LFDI (Localized Foreign Direct Investment) stocks in France

to be more intense in the Great Lakes region and on the West coast, in relation with the economic profiles of these cities (for example more manufacturing oriented in the Great Lakes region) (Bettencourt et al. 2010). LFDI stocks are also much more concentrated in the northern and eastern parts of France (cities surrounding the metropolitan area of Paris included), while some cities in southern and western France are marginalized in foreign transnational investment networks (Finance 2016). This is also partly linked with other attributes than city size: economic profiles, proximity to investors, proximity to a high market potential, etc. Therefore, simple bivariate relationships as scaling laws can help to detect outliers in order to try to understand why some cities have a distinct profile compared to the one predicted by scaling laws.

Further analyses are then needed to understand these deviations that supplement the strong hierarchical effect.

5.3.4 Scaling Laws and the Center/Periphery Gradient in Cities

Another regularity that can be observed is the center-periphery structure of cities that emerges at the meso-level. Scaling laws can help to learn more about this gradient.

A first way to consider this gradient is to work directly on its morphological shape. Studies by Lemoy and Caruso (2018) and Delloye et al. (2019) have shown empirical evidences across 300 European cities "of the simple geometrical scaling of cities as seen from the radial evolution of their land use and population density" (Lemoy and Caruso 2018). Radial profiles of cities (considered as monocentric) are very well captured and are homothetic to each other by rescaling them by the square root for land-use and by the cube root for the population density.

Another way to consider this gradient is to consider various definitions of cities in a system and to compare scaling parameters obtained on given urban attributes. Cities can be defined in either administrative, morphological or functional approaches, which do not describe the same territories. Cities defined by official censuses correspond to an administrative vision of territorial management and population counting. These definitions are a key to understand valuable territories in that they reflect the political, cultural and even historical consideration of cities. However, these administrative boundaries of cities do not always reflect their morphological and functional extent, are rarely compatible from one country to another. Finally, their designation is often political, which does not make it possible to capture all forms of urbanization. Harmonization of the city definitions over time and as independently as possible of their official definitions and administrative boundaries makes it possible to overcome this difficulty. Morphological and functional approaches are much more appropriate to measure the evolution of cities over long periods of time in a consistent way, to compare the size and evolution of cities defined according to the same reference and to integrate "unofficial" cities into the databases.

A morphological city is defined by delineating its dense and continuous urban structure, to which a statistical threshold of resident population is added. In the e-Geopolis project (whose objective is to describe and compare cities over time), a distance threshold of maximum 200 m and a minimal population threshold of 10,000 inhabitants are simultaneously considered (Moriconi-Ébrard 1994). A functional city includes the area of influence of a central city (without the necessity of a spatial continuity of the built-up area), detected through the intensity of relations between a central city and its periphery (usually through commuting flows). Comparing a morphological definition to a functional one can lead to apparently contradictory results, that can be explained by the differences in the nature of what is a dense urban core and what are the peripheries or the suburbs included in functional definitions. The area of greenspace in cities of England and Wales, which scales linearly to sub-linearly when defined according to a density threshold, becomes clearly super-linear when a functional dimension (30% of commuters) is added (Arcaute et al. 2015). Manufacturing, vacant dwellings, basic services and education are some examples of diverging conclusions of scaling laws in the French urban systems when morphological ('unités urbaines') and functional ('aires urbaines') cities are compared (Cottineau et al. 2017).

We take the example of the Chinese urban system to further the analysis of the influence of the definition of cities on scaling results. Three definitions are considered and compared. The first definition is the official one, which is an administrative and juridical approach. The perimeter of cities does not delimit a "strictly urban" area, but includes rural areas as well as cities at lower administrative levels (Gipouloux 2006). Indeed, Chinese cities are political-administrative entities at all levels of Chinese administration. It encompasses and administrates rural areas that are all the more extensive when the administrative level is high. Chinese cities have the particularity of being nested: cities in the lower administrative ranks are included in the administrative boundaries of higher ranking cities. Despite the high relevance of the administrative definition of Chinese cities, the definition of Chinese cities as morphological agglomerations makes it possible to capture all urban entities regardless of their administrative status, and to compare their temporal and spatial evolution, including within China itself. The delineation of Chinese cities as morphological entities is based on the definitional criteria used in the aforementioned e-Geopolis project. As Chinese rural areas are very densely populated and built, this definition may nevertheless include dense rural settlements. To overcome this potential limitation, an economic criterion has been added to exclude each settlement whose assets were mainly engaged in the primary sector through a clustering analysis (i.e. an agglomerative hierarchical clustering on the agglomerations' active population). Data describing each district being contained in the agglomeration are aggregated (either classified as urban and rural districts; *Qu*, *Xianjishi* and *Xian*); when agglomeration only partly includes a district, only the data describing the villages (*Zhen*) were aggregated with the rest of the agglomeration. The database set up (named ChinaCities; Swerts 2018) thus includes 1664 agglomerations built from districts, from metropolises to district level cities (ChinaCities V1), and 9476 agglomerations from metropolises to towns, built from districts but also from towns (ChinaCities V2). The morphological agglomerations thus constructed include a much smaller dense urban area than the official cities, which include a large area of agricultural land and several cities of different sizes.

The application of scaling laws to these databases of different entities shows the very different results of the measures depending on how the cities to which they are applied are defined. These differences make it possible to capture that the distribution of activities in Chinese cities are strongly linked to the country's stage of development and history, but also submitted to the influence of the political and administrative system. The example developed in Table 5.2 shows the distribution of four activities in the Chinese urban system: Manufacturing, Scientific Research & Geological

Table 5.2 Scaling laws applied to Manufacturing, Scientific Research & Geological Prospecting, Education and Public Administration and Social Organizations sectors on three different definitions of the city in China (2010)

	657 official city centers and urban districts (including large parts of rural areas)		1664 agglomerations from metropolises to district level cities (ChinaCities V1 database)		9476 agglomerations from metropolises to towns (ChinaCities V2 database)	
	β $CI_{95\,\%}$	R^2	β $CI_{95\,\%}$	R^2	β $CI_{95\,\%}$	R^2
Manufacturing	1.28 [1.25; 1.31]	0.82	1.37 [1.35; 1.39]	0.84	1.03 [1.02; 1.04]	0.76
Scientific Research & Polytechnic Services and Geological Prospecting Industry	1.37 [1.33; 1.44]	0.63	1.33 [1.30; 1.36]	0.66	0.94 [0.90; 0.98]	0.77
Education	1.00 [0.97; 1.03]	0.92	0.82 [0.80; 0.84]	0.84	0.91 [0.89; 0.93]	0.87
Public Administration and Social Organizations	0.93 [0.90; 0.96]	0.83	1.03 [1.01; 1.05]	0.87	0.93 [0.90; 0.96]	0.80

Prospecting, Education and Public Administration and Social Organizations. The scaling exponents seem strongly dependent to the city delineation.

For the Manufacturing and Scientific Research & Geological Prospecting sectors, the variation of the β exponent shows that the distribution of activities in the urban hierarchy is marked by the development stage of these two activities. Thus, manufacturing activities are over-represented in large cities if we consider the official database and even more so the ChinaCities V1 database. This shows that in China, as in South Africa, manufacturing is an innovative activity developed in large cities. At the same time, it is over-represented in smaller entities when scaling is applied to the ChinaCities V2 database, which means that when the towns ("*Zhen*") agglomerations are included, manufacturing follows a linear regime of scaling. This can be explained by the fact that several stages of development of manufacturing activity overlap in China. The ancient plants from the Maoist period are still located in small towns, while the new waves of manufacturing occur above all in largest cities. The fact that the β exponent is higher with the ChinaCities V1 database than with the official one suggest that the manufacturing activities are located in the center of the Chinese cities rather than in the surrounding countryside; indeed, we recall that the official database cover a city perimeter encompassing a larger territory than ChinaCities V1 agglomerations.

Similar trends are observed for the Scientific Research & Geological Prospecting sector. The scaling regime is super-linear when the official cities or the ChinaCities V1 agglomerations are considered. Including the Zhen agglomerations (ChinaCities V2), the relationship between scientific research and Geological Prospecting becomes sub-linear. These variations could be explained by the fact that the Scientific Research is likely an innovation located in largest cities and the Geological Prospecting Industry is located in smaller towns, close to mineral resources. Available nomenclatures don't allow to distinguish these activities, slightly different by nature.

Activities of Education and Public Administration & Social Organizations show the importance of the administrative status of cities in China in the hierarchical distribution of activities. Education activities are proportionally distributed within the cities of the Chinese system when the official base is considered. They are slightly over-represented in small cities while considering the ChinaCities V2 database, even more strongly when the towns are excluded (ChinaCities V1). The global sublinear/linear relationship between education and city size (contrary to USA, France and South Africa) is a Chinese specificity due to the political system and the decentralization policies, which lead to educational activities being present throughout the territory, including in small cities. However, the fact that there is a less overrepresentation in small towns when Zhen are included show the effect of the administrative status of cities and the limitation of decentralization policy. Finally, Public Administration and Social Organizations exhibits a linear regime of scaling while considering ChinaCities V1 database excluding towns, but sublinear with the two other definitions. This shows that the Public Administration reflects the Chinese political system and is linked to the cities' status.

We conclude this section by highlighting a last example on how city delineation may strongly affect scaling parameters. Sensitivity analyses of scaling parameters have been computed by Arcaute et al. (2015) and Cottineau et al. (2017, 2018) to detect variations of scaling regimes when thousands of concurrent definitions of the city are considered, in England and Wales and in France (these definitions are based on the systematic combination of three definitional criteria: density, commuting flows and population cutoffs). Figure 5.8 shows through heat maps some results about the sensitivity of scaling parameters to city definitions for two urban indicators: the total length of roads and the total wages, both in the French urban system. For the most realistic definitions of the city, it appears that "larger cities appear either richer or as rich as smaller cities, but never poorer on average" (Cottineau et al. 2017). Moreover, "with respect to the relationship between total wages and total residents, we see that larger cities, when they are defined as sprawling metropolises (bottom-right), do seem richer than their smaller counterparts. This is not true when we look at city cores only (top-left)" (ibid.). This mainly reflects the difference in spatial distribution of jobs and residents in cities, with a much stronger concentration of jobs in the central parts of cities. The sensitivity analysis is even more striking while considering the total length of roads as "not only do the scaling exponent values vary: the scaling regime (sub- or superlinear) depends on the combination of density, commuting flows and population cutoffs". When the city is delineated as its dense urban core, the scaling regime is sub-linear (among realistic definitions); yet "when one considers cities along with their functional peripheries (…), then we find the opposite result: the largest cities become relatively more consumptive of infrastructure per capita" (Cottineau et al. 2017).

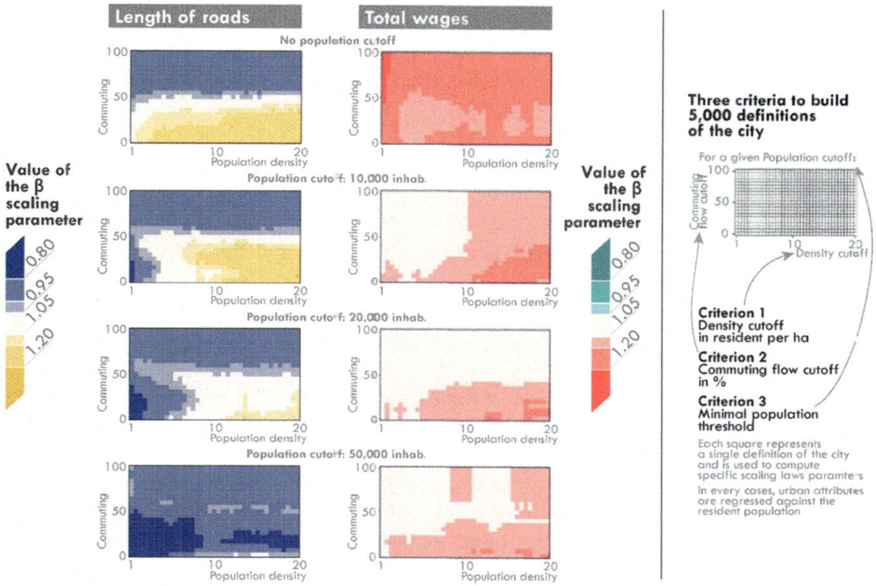

Fig. 5.8 Scaling parameters computed on the length of roads and total wages in French cities based on 5000 different definitions of the city (adapted from Cottineau et al. 2017, 2018)

Scaling laws are relevant to describe the relationships between urban attributes and city size. They make it possible to highlight generic and regular mechanisms (such as the hierarchical diffusion of innovations), and by studying the gaps in regularity, the specific characteristics of some cities. All these examples show how each definition criterion affects results that are expected from scaling laws computations, and supports once again the need of relying on appropriate delineations of cities in urban scaling. Scaling laws are very sensitive to city delineation and city systems on which they are applied. When the definition of the objects studied is mastered, this is a richness in terms of interpretation. On the other hand, it can also render obsolete the interpretation of scaling laws. We thus develop some recommendation for a proper use of scaling laws in the fourth section.

5.4 Some Guidelines for Proper Use of Scaling Laws in Urban Geography

As underlined in the previous sections, scaling laws are relevant for validating urban theories. However, this method should be used with many precautions. Indeed, the exponents of the scaling laws are very sensitive to the definition of the entities on which the measurements are made. This extreme responsiveness, without totally questioning the use of scaling laws, makes the interpretation of the results at least tricky, even meaningless or false. Due to the number of limitations partly illustrated before in studies relying on scaling laws in urban geography, this final section offers some recommendations for further studies.

Recommendation 1: ***always consider proper definition of the city***. The main danger when handling scaling laws is first of all to build conclusions on data aggregated at erroneous or inconsistent definitions of the city. As we have seen in the previous section, scaling laws exponents are extremely sensitive to cities' delineation. Choosing one or another definition consistent definition of the city can lead to different or event opposite results; building the computation on cities that are not well defined could lead to doubtful or erroneous conclusions.

It is therefore dangerous to work directly with official databases if the detailed knowledge of how entities are calculated is not known, and if they include entities with different definitions as communes, places, counties, municipalities. To give meaning to the results obtained, it is therefore necessary to have a coherent and homogeneous definition of a city. A reliable solution is to work with databases whose definition of the city is harmonized, either morphological or functional. It has the advantage of allowing the results computed through scaling laws to be compared over time and space.

In a recent paper focusing on Brazil, some authors wonder about the surprising deviation of some infrastructural and individual basic services variables from the scaling regimes they were expecting. The number of hospital beds appear to scale linearly with the size of elements under consideration; they "propose that these deviations are a product of top-down decisions/policies" specific to the Brazilian context (Meirelles et al. 2018). Yet elements under study are the Brazilian municipalities; functional cities as Rio de Janeiro or São Paulo are consequently split in various elements, each of them becoming an individual element. It is probable that the super-linearity that has been detected at the municipality level is not only linked with the unequal distribution of hospital beds among functional cities in Brazil, but also within functional cities. It has not been demonstrated by the authors if the aggregation of municipalities into functional urban areas would have produced less surprising scaling parameters. Results comparing scaling parameters computed in various geographical contexts without using harmonized definitions of what is a city have to be considered carefully.

Recommendation 2: ***avoid working on too small samples of cities***. As with all statistical or regression models, the number of cities on which scaling laws are applied influences the value of the scaling exponent but also and above all the validity and the significance of the results. As aforementioned, considering or excluding small towns of the set of Chinese cities affects both the results and the conclusions that can be made through scaling laws. In the same vein, introducing population minimal threshold can make the results variate slightly.

The distribution of activities within a system of cities only makes sense if the system under consideration includes a significant number of cities, and not a sample of metropolises (if this case, one should call it a system of metropolises!). Even if they don't miss recommendation 1, Bettencourt and Lobo (2016) miss this second one while comparing the distribution of GDP, urbanized area, employment and patents in various national urban systems of Europe (namely France, UK, Spain, Italy and Germany). The set of functional cities considered in each case varies from 8 (Spain) to 24 (Germany) only. This makes the results interpretable only for the upper-part of the complete urban hierarchies. Therefore the delineation not only of each city but of the system of cities has to be judicious to enable conclusions on the whole system of cities. Beyond that the sample size determines the significance of the results. That lead to the third recommendation:

Recommendation 3: ***don't forget to validate the scaling parameters through the confidence intervals***. For the interpretation of the results to make sense, over-interpretation of results should be avoided. Values of β give a valuable indication to validate urban theories, but must be significant. Therefore the interval [0.95; 1.05] is commonly used as the range of linear regimes of scaling; if there is any overlay of this range by confidence intervals of apparently super-linear or sub-linear regimes of scaling, linearity cannot be excluded.

Once again, Bettencourt and Lobo's results on some national European urban systems (2016) have to be interpreted carefully. Values of β seem to agree with their expectations, but confidence intervals are extremely broad (partly due to the small samples considered). Consequently, GDP cannot be considered as scaling purely super-linearly with city size in the cases of UK, Spain or Italy; the distribution of patents cannot be considered as super-linear in any case, even not only super-linear or linear regarding the extremely wide confidence intervals.

Recommendation 4: ***never forget that a minimal size threshold can affect the results***. In defining the cities and the system of cities on which the scaling laws are applied, the question of the size threshold of the cities considered also has a strong impact on the results of the scaling and the interpretation that can be given to them. Some urban attributes are quite ubiquitous in the upper-part of the urban hierarchy, and even if distributed in every city of a system, show a more selective propensity among the smallest cities. The previous example of LFDI stocks in France is one of them: reducing progressively the system of cities from 355 to 150 cities by excluding the smallest city at each step produces a slow but slight increase of β (Finance 2016; Finance and Cottineau 2018).

Recommendation 5: ***don't always trust OLS.*** Ordinary Least Square regression is the usual way to compute scaling laws parameters; but as exposed by Leitão et al. (2016), other ways of estimating non-linearity in systems of cities are available. The way the computations are performed can be another source of variability in the results. Alternative methods proves in some cases a better robustness.

Rather than considering linearity as the range of values into the interval [0.95; 1.05], the authors compare diverse models of scaling with different shapes of the noise (lognormal, Gaussian, etc.) in the Maximum Likelihood Estimation (MLE) approach. In each case, a first instantiation is computed where β is allowed to vary, another one where β has a fixed value of 1 (the linear model). Then, constrained and unconstrained models are compared through a Bayesian Information Criterion, an index of performance of the model in terms of fit and parsimony (Finance and Cottineau 2018). As an example, when scaling parameters are computed on total employment in French cities on various definitions of the system by reducing progressively the system of cities from 355 to 150 cities (see recommendation 4), constrained models are often more effective in terms of fit and parsimony and proved linearity with higher reliability.

Recommendation 6: ***never forget that zero values are not taken into account***. Another point reminded by Leitão et al. (2016) and further detailed by Finance and Cottineau (2018) is the influence of the presence of zero-values in the dataset.

As scaling laws are usually computed through OLS regressions on log-transformed data, an artificial filtering is introduced: it requires removing the data points with $y_i = 0$ to calculate the scaling exponent, as log(0) is not defined. "The problem with the zero count for cities where these attributes are absent is that the technical necessities of usual estimation procedures make the analysis ignore them altogether even when they represent some valid information" (Finance and Cottineau 2018): the fact that no patent is registered in some cities (as in the example in Fig. 5.7 based on Bettencourt et al. 2010), or that a given business sector is absent in some cities (see proportion of zero-values in Fig. 5.5), or that some cities don't host any transnational greenfield investment during a given period (see Finance and Cottineau 2018) is as interesting as the effective intensity of concentration of these attributes in cities hosting these attributes. Therefore, β computed through OLS on these cases does not represent the relationship between the quantity under study and city size in the whole system, but only in the subset of cities which concentrate a positive part of this quantity.

When zero-values are mostly registered in the smallest cities, a judicious minimal population threshold can be introduced to reduce them (Arcaute et al. 2015; Finance and Cottineau 2018). But some alternative methods, some of them taking specifically into account zero-values like Hurdle models, are able to deal specifically with zero-values without filtering. These methods being much more computationally demanding, a first recommendation would be to clearly enumerate the proportion of zero-values in the dataset considered when scaling laws are used.

Recommendation 7: ***be creative!*** **And recommendation 8:** ***don't hesitate to formulate new recommendations to the community***.

5.5 Conclusion

Scaling laws offer the advantage of being simple, easily usable and proven relevant models for validating the theories of cities dynamic. The simplicity of scaling laws can certainly be a questionable argument as to the relevance of the formalizations they allow. In particular, the low number of explanatory parameters considered, unlike econometrics, where a whole set of explanatory variables are integrated into the models, could appear to be an element that gives scaling laws limited explanatory potential.

However, it is in this apparent weakness that a strength of scaling laws lies. Their simplicity allows them to be used as a filter to highlight urban attributes whose distribution within cities of a system is related to the size of cities. They thus make it possible to highlight hierarchical regularities of city systems, but also to detect cities that deviate from these regularities. This then makes it possible to investigate the factors that influence major deviations from expected values. In addition, one of the added values of scaling laws is that they offer a wide field of application in urban geography, in particular because they make it possible to study a whole set of urban attributes, from the distribution of economic activities to transport networks. Scaling laws allow to observe and explain these distributions and regularities over time, if data are available, and especially to make international comparisons.

Scaling laws have thus allowed new advances in geography as well as the confirmation of theories, structures and dynamics already verified with other methods previously. In particular, they made it possible to validate urban theories relating to the hierarchical diffusion of innovations and those relating to metropolization. The validation of these theories can also be enriched by the contributions of other disciplines in the sense that these simple models used in other disciplines make it possible to build an interdisciplinary dialogue, with physicists in particular, or economists, for example on questions relating to elasticity. As with any statistical or mathematical model, however, care must be taken not to over-interpret the results, and not to give scaling laws more explanatory power than they can describe. Particular attention must also be paid to the objects to which they are applied, otherwise interpretations of the results may be false or even impossible.

To conclude, scaling laws that result of a methodological transfer from biology, are a good example of the relevance of transferring tools from other disciplines in the Social Sciences and more particularly here in urban geography. Used with all the necessary precautions, these tools allow to test and validate formalizations and theorization of city dynamics, in order to better understand their generic evolutionary mechanisms and thus to better understand what makes each city unique.

References

Arcaute, E., Hatna, E., Ferguson, P., Youn, H., Johansson, A., Batty, M.: Constructing cities, deconstructing scaling laws. J. R. Soc. Interface **12**(102), 1–8 (2015). https://doi.org/10.1098/rsif.2014.0745

Auerbach, F.: Das Gesetz der Bevölkerungskonzentration. Petermanns Geogr. Mitt. **59**, 74–76 (1913)

Barthelemy, M.: The Structure and Dynamics of Cities. Cambridge University Press, 278p (2016). https://doi.org/10.1017/9781316271377

Batty, M.: The new science of cities. The MIT Press, Cambridge, MA (2013)

Berry, B.J.: Cities as systems within systems of cities. Pap. Reg. Sci. **13**(1), 147–163 (1964). https://doi.org/10.1111/j.1435-5597.1964.tb01283.x

Bettencourt, L.M., Lobo, J., Helbing, D., Kuhnert, C., West, G.B.: Growth, innovation, scaling, and the pace of life in cities. Proc. Nat. Acad.Sci. **104**(17), 7301–7306 (2007a). https://doi.org/10.1073/pnas.0610172104

Bettencourt, L.M., Lobo, J., Strumsky, D.: Invention in the city: increasing returns to patenting as a scaling function of metropolitan size. Res. Policy **36**(1), 107–120 (2007b). https://doi.org/10.1016/j.respol.2006.09.026

Bettencourt, L.M., Lobo, J., West, G.B.: The self-similarity of human social organization and dynamics in cities. In: Lane, D., Pumain, D., van der Leeuw, S., West GB. (eds.) Complexity Perspective in Innovation and Social Change, Methodos Series. Springer, Dordrecht, pp. 221–236 (2009). https://doi.org/10.1007/978-1-4020-9663-1_8

Bettencourt, L.M.: The origins of scaling in cities. Science **340**(6139), 1438–1441 (2013). https://doi.org/10.1126/science.1235823

Bettencourt, L.M., Lobo J.: Urban scaling in Europe. J. R. Soc. Interface. **13**(116), 20160005 (2016). https://doi.org/10.1098/rsif.2016.0005

Bettencourt, L.M., Lobo, J., West, G.B.: Why are large cities faster? Universal scaling and self-similarity in urban organization and dynamics. Eur Phys J B **63**(3), 285–293 (2008). https://doi.org/10.1140/epjb/e2008-00250-6

Bettencourt, L.M., Lobo, J., Strumsky, D., West, G.B.: Urban scaling and its deviations: revealing the structure of wealth, innovation and crime across cities. PLoS ONE **5**(11), 1–9 (2010). https://doi.org/10.1371/journal.pone.0013541

Bretagnolle, A., Pumain, D., Vacchiani-Marcuzzo, C.: The organization of urban systems. In: Lane, D., Pumain, D., van der Leeuw, S., West G.B. (eds.) Complexity Perspective in Innovation and Social Change, Methodos Series. Springer, Dordrecht, pp. 197–220 (2009). https://doi.org/10.1007/978-1-4020-9663-1_7

Christaller, W.: Die zentralen Orte in Süddeutschland: eine ökonomisch-geographische Untersuchung über die Gesetzmässigkeit der Verbreitung und Entwicklung der Siedlungen mit städtischen Funktionen. University Microfilms (1933)

Cottineau, C.: MetaZipf. A dynamic meta-analysis of city size distributions. PLoS ONE **12**(8) (2017). https://doi.org/10.1371/journal.pone.0183919

Cottineau, C., Hatna, E., Arcaute, E., Batty, M.: Diverse cities or the systematic paradox of urban scaling laws. Comput. Environ. Urban Syst. **63**, 80–94 (2017). https://doi.org/10.1016/j.compenvurbsys.2016.04.006

Cottineau, C., Finance, O., Hatna, E., Arcaute, E., Batty, M.: Defining urban clusters to detect agglomeration economies. Environ. Plan. B Urban Anal. City Sci. **46**(9), 1611–1626 (2018). https://doi.org/10.1177/2399808318755146

Delloye, J., Lemoy, R., Caruso, G.: Alonso and the scaling of urban profiles. In: Geographical Analysis (2019)

Feldman, M.P., Florida, R.: The geographic sources of innovation: technological infrastructure and product innovation in the United States. Ann. Assoc. Am. Geogr. **84**(2), 210–229 (1994). https://doi.org/10.1111/j.1467-8306.1994.tb01735.x

Finance, O.: Les villes françaises investies par les firmes transnationales étrangères: des réseaux d'entreprises aux établissements localisés. Ph.D. thesis, Université Paris I Panthéon-Sorbonne, France (2016)

Finance, O., Cottineau, C.: Are the absent always wrong? Dealing with zero values in urban scaling. Environ. Plan. B Urban Anal. City Sci. OnlineFirst (2018). https://doi.org/10.1177/2399808318785634

Gipouloux, F.: Attractivité, concurrence et complémentarité: la place ambiguë des villes côtières chinoises dans la dynamique économique du corridor maritime de l'Asie de l'Est. Outre-Terre **2**, 149–160 (2006)

Hägerstrand, T.: Innovation Diffusion as a Spatial Process (1968)

Leitão, J.C., Miotto, J.M., Gerlach, M., Altmann, E.G.: Is this scaling nonlinear? R. Soc. Open Sci. **3**(7), 1–13 (2016). https://doi.org/10.1098/rsos.150649

Lemoy R., Caruso G.: Evidence for the homothetic scaling of urban forms. Environ. Plan. B Urban Anal. City Sci. (Online first) (2018). https://doi.org/10.1177/2399808318810532

Levinson, D.: Network structure and city size. PLoS ONE **7**(1), 1–11 (2012). https://doi.org/10.1371/journal.pone.0029721

Louf, R., Barthelemy, M.: How congestion shapes cities: from mobility patterns to scaling. Sci. Rep. **4**(5561), 1–9 (2014). https://doi.org/10.1038/srep05561

Meirelles, J., Neto, C.R., Ferreira, F.F., Ribeiro, F.L., Binder, C.R.: Evolution of urban scaling: evidence from Brazil. PLoS ONE **13**(10), 1–15 (2018). https://doi.org/10.1371/journal.pone.0204574

Moriconi-Ébrard, F.: Geopolis: pour comparer les villes du monde. Paris: Anthropos, coll. “Villes”, pp. 246. ISBN: 2-7178-2721-8 (1994)

Paulus, F.: Coévolution dans les systèmes de villes: croissance et spécialisation des aires urbaines françaises de 1950 à 2000. Ph.D. thesis, Université Paris I Panthéon-Sorbonne, France (2004)

Paulus, F., & Pumain, D.: Lois d’échelle et activités urbaines: une comparaison France-États-Unis. In: Mattéi, MF., & Pumain, D. (eds.) Données urbaines 5. Economica-Anthorpos, Paris, 315–323 (2007)

Paulus, F., Pumain, D.: Salaire et hiérarchie urbaine. In: Pumain, D., Mattéi, MF. (eds.) Données urbaines 6. Economica-Anthropos. Paris, 205–216 (2011)

Pred, A.R.: City Systems in Advanced Economies: Past Growth, Present Processes, and Future Development Options. Wiley (1977)

Pumain, D.: La dynamique des villes. Economica, Paris (1982)

Pumain, D.: Pour une théorie évolutive des villes. L’Espace géographique **26**(2), 119–134 (1997). https://doi.org/10.3406/spgeo.1997.1063

Pumain, D.: Scaling laws in urban systems. Santa Fe Institute Working Papers, n°04-02-002, 26p (2004)

Pumain, D.: Alternative explanations of hierarchical differentiation in urban systems. In: Pumain, D. (ed.) Hierarchy in Natural and Social Sciences, Methodos Series. Springer, Dordrecht, pp. 169–222 (2006)

Pumain, D.: Urban systems dynamics, urban growth and scaling laws: the question of ergodicity. In: Portugali, J., Meyer, H., Stolk, E., Tan E. (eds.) Complexity Theories of Cities Have Come of Age. Springer, Berlin (2012). https://doi.org/10.1007/978-3-642-24544-2_6

Pumain, D., Paulus, F., Vacchiani-Marcuzzo, C., Lobo, J.: An evolutionary theory for interpreting urban scaling laws. Cybergeo Eur. J. Geogr. 343 (2006). https://doi.org/10.4000/cybergeo.2519

Pumain, D., Paulus, F., Vacchiani-Marcuzzo, C.: Innovation cycles and urban dynamics. In: Lane, D., Pumain, D., van der Leeuw, S., West GB. (eds.) Complexity Perspective in Innovation and Social Change, Methodos Series. Springer, Dordrecht, pp. 237–260 (2009). https://doi.org/10.1007/978-1-4020-9663-1_9

Pumain, D., Swerts, E., Cottineau, C., Vacchiani-Marcuzzo, C., Ignazzi, C.A., Bretagnolle, A., Delisle, F., Cura, R., Lizzi, L., Baffi, S.: Multilevel comparison of large urban systems. Cybergeo Eur. J. Geogr. document 706 (2015). https://doi.org/10.4000/cybergeo.26730

Rybski, D., Reusser, D.E., Winz, A.-L., Fichtner, C., Sterzel, T., Kropp, J.P.: Cities as nuclei of sustainability? Environ. Plan. B Urban Anal. City Sci. **44**(3), 425–440 (2016). https://doi.org/10.1177/0265813516638340

Sarkar, S., Phibbs, P., Simpson, R., Wasnik, S.: The scaling of income distribution in Australia: possible relationships between urban allometry, city size, and economic inequality. Environ. Plan. B Urban Anal. City Sci. **45**(4), 603–622 (2018). https://doi.org/10.1177/0265813516676488

Savage, V.M., Gillooly, J.F., Woodruff, W.H., West, G.B., Allen, A., Enquist, B.J., Brown, J.H.: The predominance of quarter-power scaling in biology. Funct. Ecol. **18**, 257–282 (2004). https://doi.org/10.1111/j.0269-8463.2004.00856.x

Svejnar, J.: China in light of the performance of Central and East European economies. Working Paper: 41p (2007)

Swerts, E.: Les systèmes de villes en Inde et en Chine. Ph.D. thesis, Université Paris I Panthéon-Sorbonne, France (2013)

Swerts, E.: A data base on Chinese urbanization: ChinaCities. Cybergeo Eur. J. Geogr. 830 (2018). https://doi.org/10.4000/cybergeo.28554

Taylor, P.J.: World City Network: A Global Urban Analysis. Routledge, London (2004)

Tobler, W.R.: A computer movie simulating urban growth in the detroit region. Econ. Geogr. **46**(sup1), 234–240 (1970). https://doi.org/10.2307/143141

Um, J., Son, S.W., Lee, S.I., Jeong, H., Kim, B.J.: Scaling laws between population and facility densities. PNAS Proc. Natl. Acad. Sci. USA **106**(34), 14236–14240 (2009). https://doi.org/10.1073/pnas.0901898106

Vacchiani-Marcuzzo, C., Paulus, F.: Scalings laws and economic specialization: an experiment United States–France–South Africa. AAG Ann. Meet., Boston (2008)

West, G.B., Brown, J.H., Enquist, B.J.: A general model for the origin of allometric scaling laws in biology. Science **276**(5309), 122–126 (1997). https://doi.org/10.1126/science.276.5309.122

West, G.B., Brown, J.H., Enquist, B.J.: The fourth dimension of life: fractal geometry and allometric scaling of organisms. Science **284**(5420), 1677–1679 (1999). https://doi.org/10.1126/science.284.5420.1677

Zipf, G.K.: Human behaviour and the principle of least-effort. Addison-Wesley, Reading, Cambridge, MA (1949)

Chapter 6
The Topology of Communicating Across Cities of Increasing Sizes, or the Complex Task of "Reaching Out" in Larger Cities

Horacio Samaniego, Mauricio Franco-Cisterna and Boris Sotomayor-Gómez

Abstract Cities have been compared to social reactors constrained by the communication and coordination possibilities offered by an urban environment that has only grown since the advent of the industrial age. We attempt to provide a first description of human interactions in the urban environment using Call Detailed Records (CDR) of the major mobile phone communication network operator in Chile. We build communication networks for 145 Chilean cities to describe and characterize the communication behavior of urban dwellers. We center our analysis in observed indicators of social activity, such as the number of contacts, number of calls and total communication time in each city and evaluate their scaling relationship with the number of mobile phones assigned to each city as an approximation of city size. Interestingly, the values of scaling exponents closely match recent explanations proposed in the literature. The topologies of voice-call networks among cities of increasing sizes are slightly assortative, albeit assortativeness decreases with size. Additionally, they show small average path length relative to their sizes, a typical feature of small-world networks. However, they decrease instead of growing when size is taken into account, unlike other complex networks. Different transitivity indices show mixed results. Average Watts-Strogatz clustering coefficient increases in larger cities much more than expected by pure chance as it has been shown in other social networks. On the other hand, the fact that classic transitivity index decreases seem to exhibit a regime change with a decreasing relation with size and an unexpected growth in larger cities. Both transitivity indices, as a whole, could describe among those who are making new interactions as the city grows. All these results indicate that while tightly knit human communities seem to lose cohesion as they grow, such community properties may progressively disappear among the three to four largest urban centers in Chile where the coordination of complex functions requires each city dweller to reach out to a larger network of people and speak for longer periods

---

H. Samaniego (✉) · M. Franco-Cisterna · B. Sotomayor-Gómez
Laboratorio de Ecoinformática, Instituto de Conservación, Biodiversidad y Territorio, Universidad Austral de Chile, Campus Isla Teja s/n, Valdivia, Los Ríos, Chile
e-mail: horacio@ecoinformatica.cl

H. Samaniego
Instituto de Ecología y Biodiversidad (IEB), Santiago, Chile

Instituto de Sistemas Complejos de Valparaíso (ISCV), Valparaíso, Chile

D. Pumain (ed.), *Theories and Models of Urbanization*, Lecture Notes in Morphogenesis, https://doi.org/10.1007/978-3-030-36656-8_6

of time as compared to smaller cities. Finally, although these results are valid for all networks, there is a division into two regimes when networks reach a critical size of ~10,000 nodes, which raises the possibility of an empirical definition of city for Chile.

6.1 Introduction

The massive amount of data that is usually recorded and stored by telephone companies has high scientific value to unveil how humans interact among each other and with the urban infrastructure. Mobile phone records have been instrumental to revisit studies addressing the intensity and shape of social behavior (Onnela et al. 2007; Candia et al. 2008; Hidalgo and Rodriguez-Sickert 2008; González et al. 2008; Iqbal et al. 2014; Lenormand et al. 2014; Louail et al. 2014; Barbosa et al. 2018), and in particular, social interactivity (Kovanen 2009; Reades et al. 2009; Ratti et al. 2010; Palchykov et al. 2014; Schläpfer et al. 2014). The deep prevalence of mobile phones and smart cards allows to generate contextually detailed descriptions of the networks describing social interactions and their dependence on the structural environment in which they are embedded (González et al. 2008; Hidalgo and Rodriguez-Sickert 2008; Steenbruggen et al. 2014).

While the study of networks is now an active interdisciplinary field, it has historically been linked to the social sciences, where it pioneered the development of metrics and standards to describe the structure of social networks (Jackson 2008; Scott and Carrington 2011). Among the most important metrics used for such purpose are: centrality, a mean to identify the influence of an actor in the network; clustering and transitivity metrics seeking to unveil the presence of microstructures grouping members of the network as a function of various characteristics; and distance or accessibility which seeks to characterize the likelihood of members of the network to find each other (Watts and Strogatz 1998). Later contributions from the field of physics expanded considerably the tools and algorithms used to explore larger datasets while placing particular emphasis on the dynamic patterns responsible for particular topologies.

Important efforts seek to associate topology to function, in order to distinguish empirical networks from random networks with known generative processes that shape their structure (Ben-Naim et al. 2004; Clauset et al. 2009; Newman 2010). The contributions of these research lines to the study of social networks has granted its labeling as a field of its own often referred to as Social Physics (Scott 2011) or Network Science (Barabási and Pósfai 2016; Latora et al. 2017). While this labeling remains a controversial issue, it has sparked an explosion of scientific publications providing important and fresh descriptions of the shape of dynamics, ranging from cellular biology and metabolic chain reactions to particular description of social organization across urban systems (Schläpfer et al. 2014; Zhong et al. 2015).

The large success of the study of networks comes from its ability to provide deep insights based on the description of the internal dynamics of highly interactive systems. Moreover, current data availability and the emergence of new tools have made obvious that any thorough understanding of complex systems is rather incomplete without the effective description of the set of networks linking their individual components (Albert and Barabási 2002; Freeman 2011). The study of networks has thus opened the possibility to explore the dynamic structure of highly interactive systems. It has allowed the generation of specific predictions based on the patterns unveiled by the synoptic descriptions of their topological structure in ecology, epidemiology, economics, transportation, and urban sciences, among others (Estrada 2011).

Various studies suggest that some properties of cities vary systematically with size and may be described by a scaling pattern (Pumain 2004; Bettencourt et al. 2007; West 2017). Empirical analysis show that $Y = Y_0 P^{\beta}$, describes the ubiquitous relation of urban indicators, Y, with P, the size of the city (i.e. urban population), Y_0 a normalization constant and β the scaling exponents, also termed elasticity of Y with respect to P in the economics literature. These regularities, integrate important theoretical developments in biology (Marquet et al. 2005) and statistical mechanics (Barenblatt 1996). The premise is that the interaction among components of complex systems like cities, that emanates from the flows of information, matter, and energy are somehow responsible for the emergence of scaling relationships (Batty 2013; Bettencourt 2013a). However, this approach tends to envision the city as a homogenous system of interacting individuals, which may not be adequate if the city is made of loosely interacting subsystems whose agents (e.g. urban dwellers) take almost independent livelihood decisions. In this context, the integration of urban scaling into more traditional approaches in economics and geography, like those focused on the spatial behavior of individual agents (Thisse 2014), will be key to unveil specific aspects of the underlying dynamics. Moreover, an integration of both views should substantially contribute to develop better theories and tools for a sustainable management of complex urban adaptive systems. For instance, several studies have proposed different models on how interpersonal interactions are structured within a city and how scaling relationships may emerge from them (Onnela et al. 2007; Pan et al. 2013; Martínez 2016; Ribeiro et al. 2017).

While we do not attempt to provide any alternative explanation to scaling, or lack of thereof, we here attempt to contribute to the discussion by providing a thorough description and analysis of the networks of social interaction across cities. We reconstruct and analyze the network of voice call interactions among urban dwellers in Chile. We characterize the topological structure of the network produced from the voice communication between individuals by abstracting the relationships between caller and called and seek to describe changes in the social network topology across urban systems of different sizes. The notion is that of the urban environment impacts on the topological structure of the voice call networks in a country with fairly homogeneous constrains in its socio-demographic and economic composition. This task requires to extend the subject of analysis beyond the particular characteristics of individual cities to, hopefully, unveil macroscopical patterns regarding how social interactions may in fact change with the size of the urban system in which they are embedded.

6.2 Methods

6.2.1 Data

Call Detail Records (CDR) for Chile were provided by Telefonica's research and development center in Chile. Telefonica is among the largest mobile phone providers and holds just under a 40% of the market share across the country. It has a fairly uniform base of users among cities making this dataset well suited for interurban comparisons across the gradient of urban sizes in Chile (Fig. 6.1). We analyzed four weeks of CDR collected during 2015 between March 15th and November 26th summing a total of 5.4×10^8 s of voice conversations (i.e. 171 yrs) among 2,831,085 unique users. This dataset has no information, whatsoever, on the identity of users, nor the content of voice conversations as CDRs were anonymized by the company

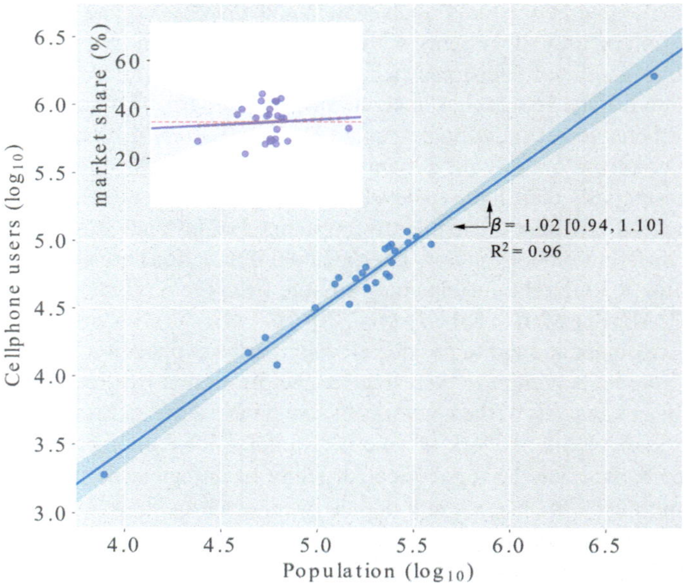

Fig. 6.1 Cell phone users in major urban areas reported by mobile phone provider Telefonica Chile. Market share is calculated as the users in the dataset as a fraction of all users across the different phone companies reported to the governmental telecommunication agency in Chile in 2016 (see: https://www.subtel.gob.cl/estudios-y-estadisticas/telefonia/)

before our analysis. Users ID were available to us as a unique encrypted 32 digit long number. Each record in the CDR contains an identifier for both, the caller and called of a voice interaction. Additionally, the dataset includes the duration, and timestamps for the initiation and termination of the call and the geographic coordinates of the closest antenna where the call was initiated. Since no information on the location of the receiver is available, we used the contemporary Data Detail Record (XDR) dataset to determine the receiver location. XDR are analogous to the CDR but for digital data. A record is produced every time a mobile phone connects to a given antenna to send, or request, a data package. While the number of individual users in the XDR dataset is similar to the CDR, the spatial and temporal density of such records is roughly seven times the size of voice call records and allows to precisely identify the location (i.e. antenna) of the receiver of the call in each CDR by finding the location of the receiver in the XDR right after the call was initiated.

6.2.2 Cities

We used the official definition of cities in Chile to define the geographic extent of urban voice-call interactions. This definition, provided by the Ministry of Housing and Urban Planning,[1] states that "Urbanized areas are bounded by an officially established urban limit with a population larger than 5,000 inhabitants. In the event that a set of spaces with an independent urban boundary are linked together by frequent public transport systems, they are considered as a single city" (Infante Fabres and Pino Castillo 2005). While the definition of urban boundaries may be an important source of heterogeneity in urban scaling analyses (Masucci et al. 2015; Arcaute et al. 2015; Cottineau et al. 2016; Humeres and Samaniego 2017), we use this standard definition as a starting point to evaluate how the communication among urban dwellers change with the size of interaction networks in Chilean cities. Some cities were considered as outliers and removed from the analysis given their erratic behavior leaving the analysis with a total of 145 urban areas in continental Chile. For example, the city of Gorbea (−39.1 lat, −72.7 lon) was removed due to the exceptionally low number of mobile phones ($n = 40$).

6.2.3 Record Selection and Data Pre-processing

To assign call records to urban areas, we approximated the spatial coverage of antennas using Voronoï cells, as done elsewhere (González et al. 2008; Louail et al. 2014). Every call record was assigned to a city if the Voronoï cell of the caller intersected the urban area. Even so, some cells failed to intersect the city in spite of being surrounded

[1] Also: http://observatoriodoc.colabora.minvu.cl/Documentos%20compartidos/OBSERVATORIO%20URBANO%20Listado%20Ciudades_%20Noviembre%202007.xls

by other intersected cells. Such cases where manually assigned to the nearest urban areas. Antennas with no urban assignation were discarded. In order to depict the social interactions within the urban system, we only considered calls in which both, sender and called were located in the same city. This left us with a total of 90,140,299 call records.

6.2.4 Network Construction and Analysis

Undirected networks of voice-call interactions between urban dwellers were built for each city using nodes as users and links between nodes when a voice-call was performed between them. The graph-tool library (ver. 2.19-1) for Python was used both for the construction of networks and its subsequent analysis (Peixoto 2014).

As we were interested in providing a description of how interaction among urban dwellers change across urban systems of different sizes. We first looked at general topological descriptions of networks and later focused on higher order descriptors of group formation among networks for each city (Caldarelli and Vespignani 2007; Newman 2014; Latora et al. 2017). The network aspect described by these indices and the motivation for their usage are described in what follows (see also Table 6.1 for equations).

Ordinary Least Square methods on linearized power equations were used to evaluate scaling relationships between urban network size and network indices. Identification of best models for higher-order topology indices were performed using piecewise regressions. Piecewise, or segmented, regression are used to evaluate the existence of potential thresholds maximizing the fit in a dataset by partitioning the relationship between two or more sections along the x-axis (Hawkins 1976). Akaike Information Criterion (ΔAIC) and the Bayesian Information Criterion (ΔBIC) were used to assess model fit in two segments for all topological attributes (Burnham and Anderson 2003), except for transitivity and density.

6.2.5 Size

The size of networks were computed counting the number of nodes. The size of each urban network is used as a proxy of city size, as the number of urban dwellers is strongly correlated to the number of nodes (i.e. mobile phones users) (Schläpfer et al. 2014).

Table 6.1 Network indices calculated for 147 urban voice-call networks in Chile

Measure	Feature	Notation
Network size	Number of mobile phones	N_i
Number of links	Connections between mobile phones	L_i
Density	Ratio of actual connections	$d_i = \frac{L_i}{N_i(N_i-1)}$
Accumulated degree	Total contacts in city i	$K_i = \sum_n k_n$
Weighted links (calls)	Number of calls ($m \leftrightarrow n$)	$W_i^{(C)} = \sum_m \sum_{n>m} w_{nm}^{(C)}$
Weighted links (time)	Length of calls ($m \leftrightarrow n$)	$W_i^{(T)} = \sum_m \sum_{n>m} w_{nm}^{(T)}$
Number of components	Network structure/fragmentation	c_i
Nodes in largest component	Network structure/fragmentation	S_i
Transitivity index	Probability of common neighbours	$C_i = \frac{3 \cdot \Delta_i}{l(2)_i}$
Mean clustering coefficient	Probability of common neighbours	$\langle C_i' \rangle = \frac{1}{N_i} \sum_n C_{n,i}'$, where $C_{n,i}' = \frac{L^{(n)}}{k_n(k_n-1)}$
Degree assortativity	Similitude between interactings	$r = \frac{\sum jk(e_{jk}-q_j q_k)}{\sigma_q^2}$
Average path length	Social accessibility between non-interactings	$\langle l \rangle = \sum \sum l_{ij}$

6.2.6 Degree

The number of links associated to a node was computed as the number of unique voice-calls performed, or received, to other users. This is the fundamental characteristics of any network as the degree of a node (user) is a first indicator of how connected a user is to others in the network. The frequency distribution often serves as a defining characteristic of the different network types (Battiston et al. 2002; Barabási and Pósfai 2016).

6.2.7 Cumulative Degree

Computed by counting the number of calls between unique pairs of users and represents the number of voice interactions among different urban dweller per city. Note that the number of links in a network is directly related to the total number of contacts and was measured through the cumulative degree of the network, which is simply twice the number of links (Newman 2014).

6.2.8 *Density*

Defined as the proportion of links to the total number of potential connections available in a fully connected city network. It usually represents the average connectivity among urban dwellers for a given city.

6.2.9 *Component*

A component is a group of nodes linked to any other node in the network. Formally, it can be defined using the notion of *path:* a sequence of connected nodes linking any two nodes. Hence, a component is the maximum subset of nodes reachable if we were to follow a particular path. Consequently, the largest component of a network is the component hosting the largest number of available links and is usually called the *major component* (Newman 2014).

6.2.10 *Number of Components*

The number of components provides a measure of the level of the network's fragmentation. Such fragmentation may be due to the very nature of the network or by the quality of the data gathered to build it. In our case, it provides a mean to identify isolated/segregated groups of users within cities of different sizes.

6.2.11 *Fraction of Nodes in the Major Component*

With the *number of components*, the fraction of nodes in the major component allows to quantify the fraction of users integrated into the major group of interacting users within the city.

6.2.12 *Distance*

The distance between two nodes of the network is the minimum number of nodes required to join two nodes along a path. This measure is also known as the "degrees of separation" (Watts and Strogatz 1998).

6.2.13 Average Path Length

That index is the average of the shortest distances between users in the network. It can be interpreted as a measure of the social accessibility among urban dwellers across the city network. We only compute this particular index for the largest component and exclude Santiago, the largest city, due to computational constrains for his calculation (Albert et al. 1999).

6.2.14 Cliques

Cliques are considered as the most basic units of higher order grouping between actors of a network and represents subsets of three interacting nodes formed in a given network. These cliques, or triple bonded nodes, effectively create a triangle of interaction.

6.2.15 Transitivity

Transitivity is commonly used in Social Network Analysis and seeks to quantify a specific characteristic of social groups. In particular, whether *friends of my friends are also my friends*. In formal terms, it represents the transitive relationship between two actors having a common acquaintance in the network (Newman 2014). Usually, partial transitivity in social networks is an indicator that any two interacting individuals have also a social world's similarity (Borgatti and Halgin 2011).

We use two indices to characterize such property, the *Global Clustering Coefficient* initially conceived by Watts and Strogatz (1998) to distinguish empirical networks from random and small world counterparts. This index in defined locally as the fraction of links between neighbors of node n with respect to the total number of links possible to be established. Because this is then averaged over all nodes in the networks (Table 6.1), it suffers from being highly influenced by low-degree nodes (Estrada 2011). Such issue prompts us to consider a second *Transitivity* index defined globally as the rate of cliques to the total number of paths of length two (Newman 2014). Hence, a larger transitivity is indicative of the tendency to belong to a similar social group.

6.2.16 Assortativity

Also seeks to characterize groups within the network, although through a common characteristic. Hence, it represents the preference of actors to be associate themselves to other actors that have a common characteristic such as the income, or the number of acquaintances (Battiston et al. 2002). The most common characteristic used is the degree, for instance. Hence, actors of networks with high degree assortativity prefer to be linked to other actors matching their own degree, meaning that individuals having a large number of contacts will usually communicate with other individuals also having a large number of contacts (Newman 2003).

6.3 Results

The distribution of users among the voice-call networks built for Chilean cities is left skewed with an average of 23,983 users and a median of 4,022. While a large number of components are observed for most urban areas (Fig. 6.2a), the inset of Fig. 6.2b shows that the largest component largely dominates de networks of at least ¾ of the cities. In fact, the fraction of the major component (S) rapidly grows to nearly 90% for networks having more than 10,000 users (Fig. 6.2b). At the lower end of the spectrum, we find a small sized mining town, El Salvador, with only 120 (4.9%) isolated subnetworks, and on the upper end, the largest capital city, Santiago, with the largest number of isolated groups of interacting dwellers distributed among 43,459 isolated subnetworks representing only a 4.3% of interacting urban dwellers (Fig. 6.2b).

The frequency distribution of users' degree provides a first view at the shape of voice-call interactions in these urban networks. A comparison of empirical networks with an Erdos-Renyi random network shows that most users exhibit local interactions with a small number of pairs, while others still have a large number of interactions (Fig. 6.2). This produces the large difference of the Erdos-Renyi model at low degrees and the typical heavy tail when using linear binning. However, a closer analysis fitting

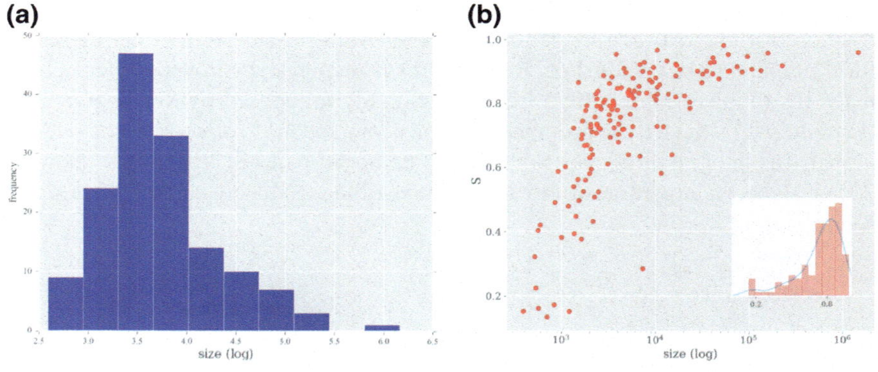

Fig. 6.2 **a** Frequency distribution of the size of voice-call networks in 147 cities in Chile. **b** Relationship between the fraction of nodes in the major components (S) and the size of the voice-call networks among cities in Chile. The inset shows the probability distribution of S

several probability distributions using the method of Clauset et al. (2009) (see also Alstott et al. 2014), suggests that most networks have a scale-free behavior, at least at the tail of their distribution (Fig. 6.3). In fact, fitting lognormal, power law, and a truncated power law distributions favors the latter two for most cities.

Interestingly, no significant relationship seem to exist between the size of the network and the power law exponent α (Pearson correlation: -0.01, p-value: 0.92). However, a larger variation of α is apparent for smaller sized cities compared to larger ones (Fig. 6.4).

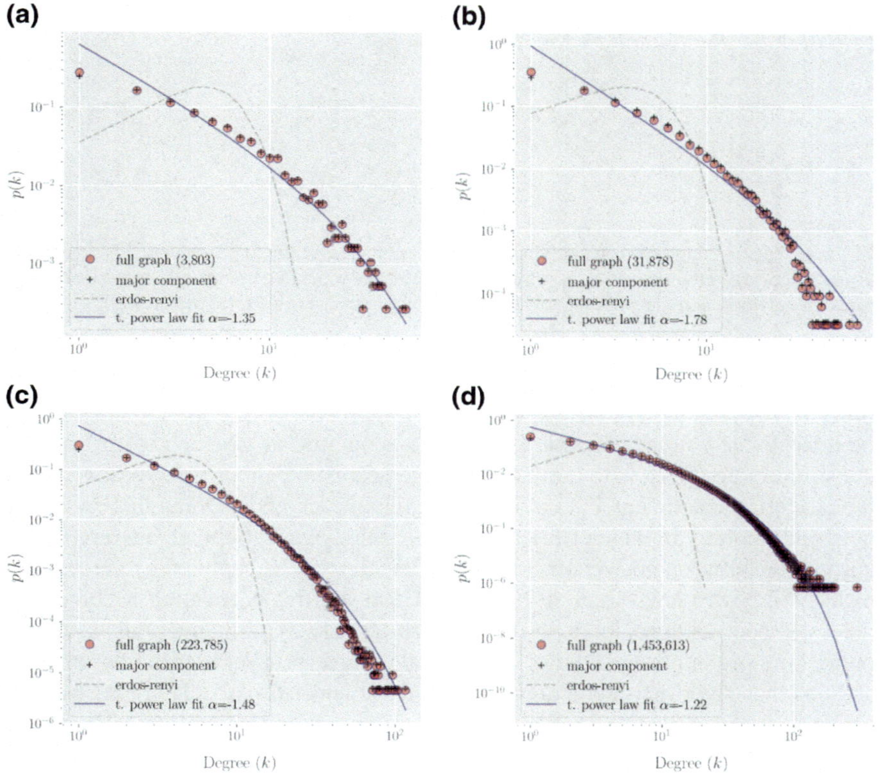

Fig. 6.3 Probability degree distribution for a sample of four cities analyzed in Chile. Cities range from small to large. **a** El Salvador a small mining town of nearly 10,000 people. **b** Valdivia is a medium-to-small university town of nearly 200,000 inhabitants. **c** Gran Valparaíso is the second largest conurbation and includes the city of Viña del Mar, and; **d** Santiago, the largest city in Chile with nearly 5 million urban dwellers. The empirical probability degree distribution is shown for the full graph, the size, i.e. nodes in the network, are in parenthesis. The major component, and a random Erdos-Renyi graph built preserving the original full graph degree distribution. The fit and the exponent of the power law section a truncated power law model is shown in continuous blue line (see text for fitting procedure)

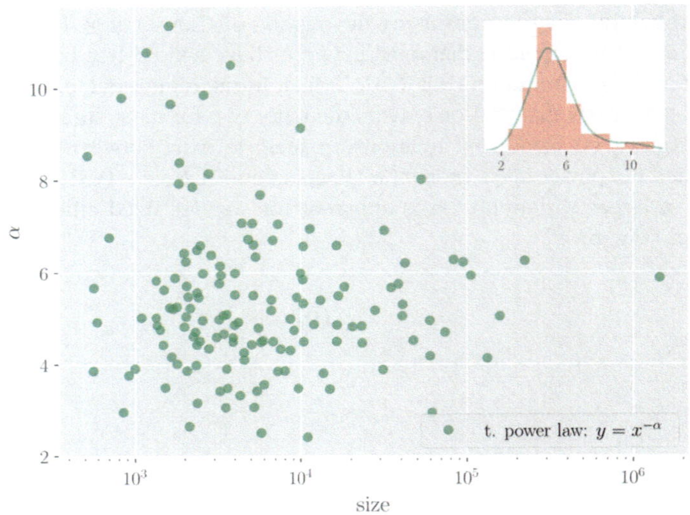

Fig. 6.4 Relationship between size (log) and the exponent α of the fitted truncated power law, for urban voice-call networks in Chile. A truncated power law was fitted to the node degree probability for each network, and the parameter α was then plotted against the size of the network

Further descriptions regarding how urban dwellers establish voice-call interactions may be obtained by inspecting the correlation between nodes. While this may be done for any node characteristics, socio-economic or gender gap for instance, the simplest approach is to correlate average nearest neighbor's degrees, $\overline{k_{nn}}(k)$, or the clustering coefficient, $\overline{c}(k)$, with the degree of each node. Doing this describes a network topology in which Chilean urban system have a slight, albeit very marked correlation between node characteristics at low degrees, k, captured by the exponent of a fitted power law (i.e. α in Fig. 6.5). However, this correlation becomes very heterogeneous as users increase their number of voice-call interactions, k. At such point, the correlation signal fades which, again, suggests the existence of users acting as hubs that favor the interaction with poorly connected pairs. This may be at the origin of the observed scale-free behavior of urban voice-calls as the size of urban networks increases (Fig. 6.5).

The description of topological patterns are complemented by a deeper scrutiny of the interactions among voice-callers in urban Chile. To such end, we looked at an aggregate measure of k, the cumulative degree, K, and its relation to the size of the system where interactions take place (i.e. network size). Doing so, shows that interactions among urban dwellers grow at a rate faster than the size of the network. While this super linear scaling has previously been reported in Schläpfer et al. (2014), the fraction of the total interactions occurring in the network associated to an average user, or density, decreases as the size of cities get smaller (Fig. 6.6). This strongly suggests that, while the size of urban voice-call networks increases, the fraction of interactions per individual gets actually smaller. Hence, while urban

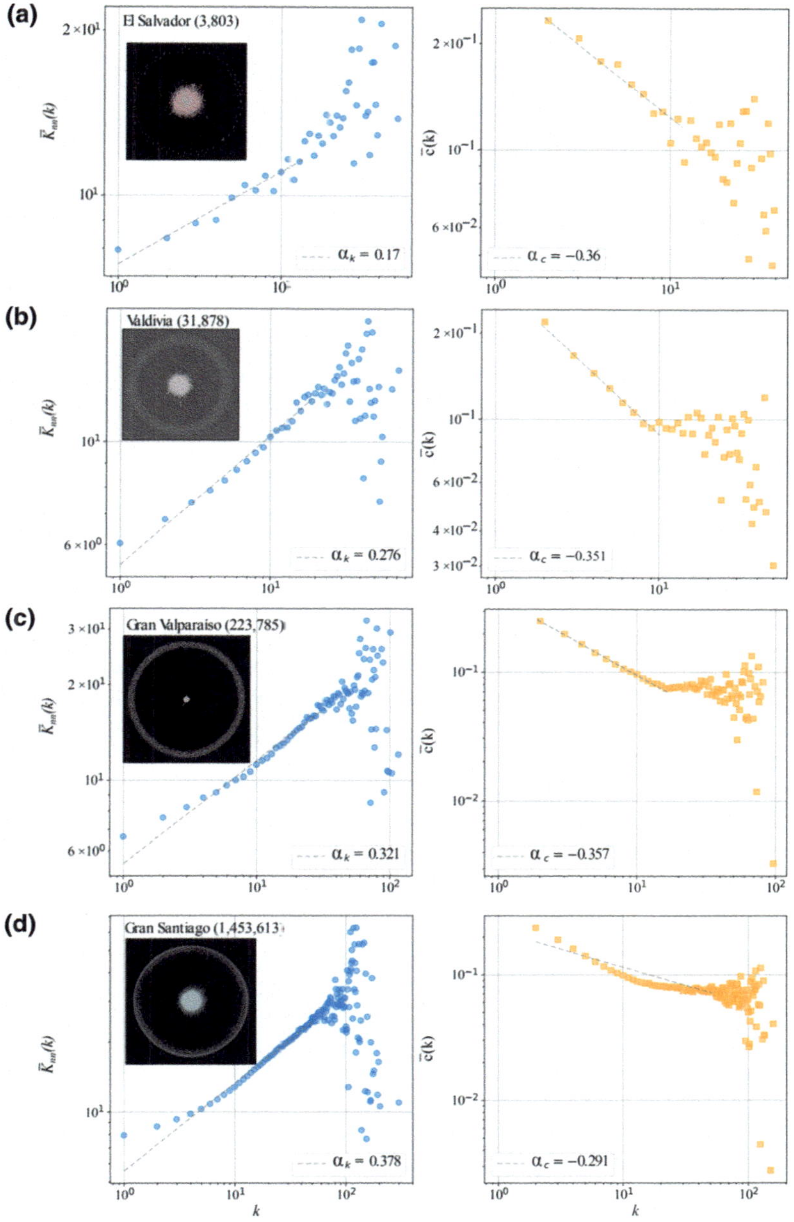

Fig. 6.5 Node degree correlations among four urban voice-call networks (inset). **a** El Salvador (−26.25 lat, −69.63 lon); **b** Valdivia (−39.80 lat, −73.26 lon); **c** Gran Valparaiso (−33.04 lat, −71.62 lon), and **d** Santiago (−33.44 lat, −70.65 lon). The left column shows the correlation between the average degree of node's nearest-neighbors, $\overline{K_{nn}}(k)$, against the degree of nodes, k. The right columns show the correlation between average node clustering, $\overline{c}(k)$ and degree, k. Dotted line is a power law fit with exponent α

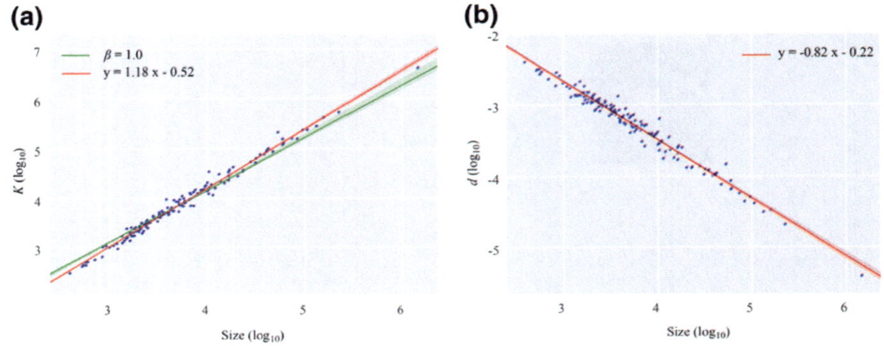

Fig. 6.6 Rates of changes of networks' topological descriptors. **a** There is a superlinear relationship between the cumulative degree and size of the urban voice-call network, meaning that urban dwellers in larger cities tend to have a larger number of voice-call interactions compared to smaller urban areas (linear model fitted using OLS, $R^2 = 0.98$). **b** The rate of change of node density across networks of different sizes suggests that while urban voice-call networks increases, the fraction of interactions per individual decreases, so that the size of the network increases proportionally more that the number of interactions involving an average individual among the Chilean urban system ($R^2 = 0.97$)

dwellers have on average, a larger pool of possible interactions available in larger cities, the number of pairs they actually interact with is a smaller fraction of the whole network. Additionally, the time spent on voice-call interactions and the number of calls (T and W, respectively) also shows a steady, and superlinear, increase with the size of networks (Fig. 6.7). While this might be a mere effect of the size of the network increasing faster than the number of interactions involving an average individual among the Chilean urban system, it also speaks eloquently about how local interaction patterns change among larger cities. Deeper analyses of the robustness of

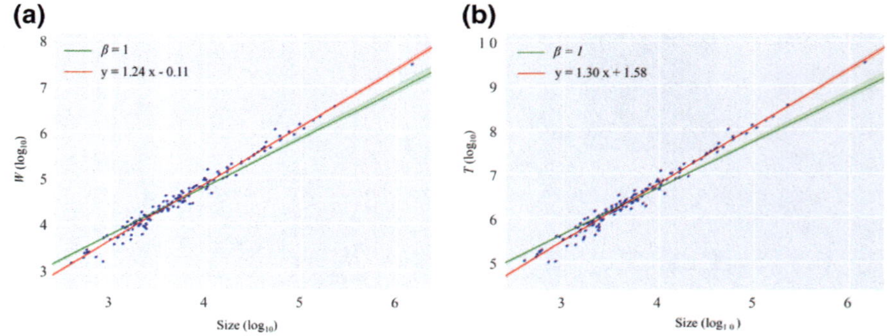

Fig. 6.7 Time and number of communication events across the size of voice-call networks in urban systems in Chile. **a** Shows W, the rate of change in the number of callers as the urban voice-call network change in size ($R^2 = 0.98$). **b** The actual time, T, spent on voice-calls as the size of the urban system gets larger ($R^2 = 0.98$). Both relationship show a superlinear increase

Table 6.2 Estimation of scaling exponent, β, in Fig. 6.6a and 6.7 using a probabilistic framework with explicit error modeling as proposed in Leitão et al. (2016)

	Log-normal error		Normal error		Pearson model
	$\delta = 2$	$\delta \in [1,3]$	$\delta = 1$	$\delta \in [1,2]$	
Cumulative grade (*K*)	**1.17** ± 0.04	1.17 ± 0.06	1.11 ± 0.09	1.17 ± 0.05	1.10 ± 0.05
Calls (*W*)	1.22 ± 0.04	**1.23** ± 0.03	1.19 ± 0.08	1.23 ± 0.03	1.18 ± 0.06
Time (*T*)	1.28 ± 0.04	**1.29** ± 0.02	1.26 ± 0.10	1.29 ± 0.03	1.25 ± 0.06

OLS fits in log-scale is $\delta = 2$. Errors were computed using bootstraps at 95% CI for each model, all models showed *p*-values <0.05. Grey backgrounds indicate best model for each error types modeling (i.e. log normal, Gaussian, or Pearson). Bold fonts indicate the best overall model. Model selection is based on lowest ΔBIC variation. All estimated parameters show significant differences with a fixed $\beta = 1$. Further details on error modeling procedures are in Leitão et al. (2016)

OLS estimations for such super linearity using various error modeling approaches proposed by Leitão et al. (2016) shows two interesting features. The first is that log-linear regression is the best model for estimating parameters of the scaling of *K* and size. It also shows that common error modeling using OLS may not hold for *W* and *T*, which paradoxically shows no significant differences in the parameter estimation (Table 6.2).

Second order descriptors of urban voice-call networks show how social interactions change with the size of the network. In general terms, transitive properties shown in Fig. 6.8a and b indicates that while networks increase with the size of the city, users have the tendency to interact with pairs having a similar number of interactions, or acquaintances. Now, the joint inspections of both transitivity indices in Fig. 6.8a and b, indicates that the likelihood to be linked to a similarly connected pair increases inversely with the number of links available to users. This is partly due to the fact that the average clustering coefficient (Fig. 6.8a) tends to be biased by low degree nodes compared to the transitivity index proposed by Newman (2003) (Fig. 6.8b) (Estrada 2011). However, the increase in transitivity seems to be more likely associated to a narrowing of the index's variance towards higher transitivity values as the network size increases, reaching a maximum at ~0.12 for all networks (Fig. 6.8a). This is a slightly lower value compared to previously reported values (e.g. in Schläpfer et al. 2014), however, C becomes essentially a constant for urban areas having networks larger than ~10^4 (i.e. green fit in Fig. 6.8b).

While, individuals are more likely to contact other individuals having similar number of contacts (*K*) for all networks. The magnitude of assortativity decreases as the size of the city increases (*p*-value <10^{-3}) (Fig. 6.8c).

Finally, urban voice-call networks exhibit low average distances, on the order of *log*(*n*). This readily suggests themselves as classic *small world* types of networks (Travers and Milgram 1969; Watts and Strogatz 1998).

The variation of higher order network indices across cities of different sizes is documented here, however, the large variability observed in Fig. 6.8 prompts to question the robustness of such patterns across the full range of network sizes. While

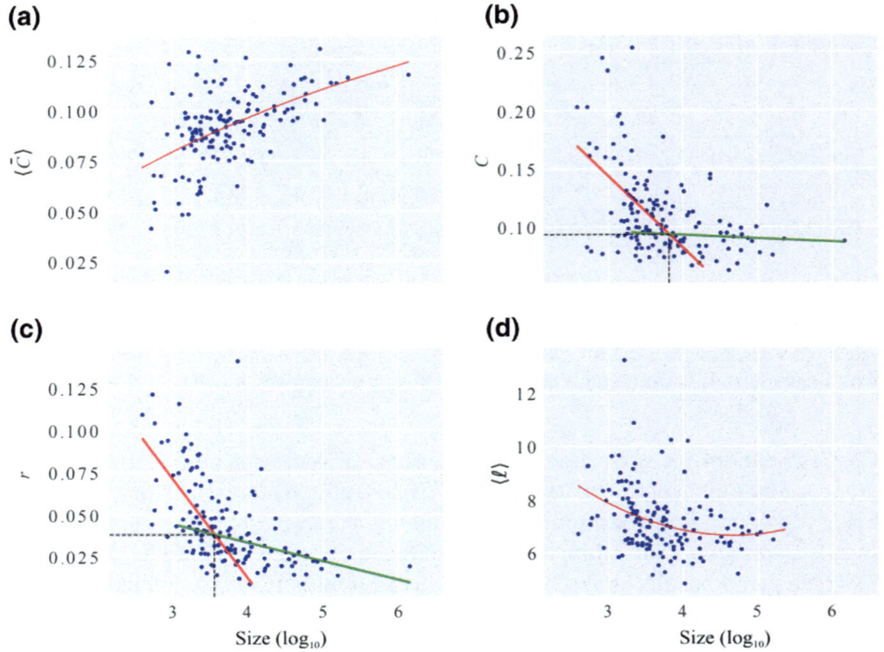

Fig. 6.8 Higher-order topological indices. **a** Average network clustering; **b** transitivity; **c** assortativity; and **d** average path length

it shows that not accounted factors may have important roles in shaping the topology of urban voice-call interaction networks, it may as well points towards the existence of a critical size in which social interactions may operate differently. For instance, a piecewise regression shows the existence of a common threshold when network size reaches ~10^4 (Table 6.1). Thus, our data shows a critical network size that may mark specific shapes in social interactions, however, the large variability observed among small networks could also be a consequence of poor data quality for smaller cities.

6.4 Discussion

While our analysis of urban voice-call networks shows that these networks conforms to topological patterns described for social networks elsewhere (McPherson et al. 2001; Battiston et al. 2002; Newman 2002; Vespignani 2010; Goyal 2011), it is interesting to note that the number of links scale with sizes with a robust statistical regularity (Table 6.2). Hence, while the number of average acquaintances of an urban dweller increases with the size of the city, so does the density. The scaling of the number of links with network size (Fig. 6.6a) shows that sociability increases more

Table 6.3 Parameter estimation of the linear regression of calculated urban voice-call network indices on network sizes

Measure	Linear fit	Range	Confidence interval for β	R^2
Fraction of nodes in largest component	$S = 0.73 \log_{10}(n) - 1.74$	$\log_{10}(n) \leq 3.45$	[0.58, 0.90]	0.62
	$S = 0.09 \log_{10}(n) + 0.44$	$\log_{10}(n) > 3.45$	[0.06, 0.13]	0.22
Transitivity	$C = -0.06 \log_{10}(n) + 0.33$	$\log_{10}(n) \leq 3.81$	[− 0.09, −0.04]	0.19
	$C = 0.00 \log_{10}(n) + 0.11$	$\log_{10}(n) > 3.81$	[− 0.01, 0.01]	0.01
Assortativity	$r = -0.06 \log_{10}(n) + 0.25$	$\log_{10}(n) \leq 3.55$	[− 0.08, −0.03]	0.27
	$r = -0.01 \log_{10}(n) + 0.08$	$\log_{10}(n) > 3.55$	[− 0.02, 0.00]	0.10

For each index, two models were evaluated within the full range of sizes. The best model selected had the lowest variation in Akaike Information Criterion (ΔAIC) and the Bayesian Information Criterion (ΔBIC)

than proportionally, requiring urban dwellers to reach-out to a wider number of pairs as the city increases in size (Bettencourt 2013b; Bettencourt et al. 2014). This pattern has previously been described by Schläpfer et al. (2014), who shows a similar scaling of the cumulative degree (K) with population size, albeit with an exponent, β between 1.05 and 1.13, and a constant transitivity as the size of the network increases. While we show a decrease of transitivity as network size increase, our piecewise regression analysis also show a constant transitivity when using Newman (2003) global index for cities larger than ~$10^{3.8}$ users (Table 6.3), suggesting that interaction dynamics may shape differently for very small networks.

The seemingly opposed trends shown by the transitivity metrics employed may be indicative of how this property changes with the network size. The fact that average clustering coefficient is strongly influenced by low-degree nodes (Newman 2003), points to the existence of two phenomena as the size of the voice-call network increases. On one hand, there seem to be an increase in the proportion of poorly connected users forming clusters of low-degree individuals; and, on the other hand, as the size of networks increases, the number of highly connected individuals tend to interact with poorly connected pairs, forming *hubs* acting as bridges that connects communities of low-degree individuals. Such heterogeneity apparently typical of large social networks was extensively studied in social physics (Borgatti and Everett 2000) and similar patterns were recently described by Onnela et al. (2007). Recently various mechanisms have been proposed to explain the emergence of this type of networks such as the *preferential attachment* model in which new users fill prefer to interact with highly connected pairs (Barabási and Albert 1999) or the *rich-club effect* model describing how highly connected users will mostly establish connections among each other (Opsahl et al. 2008; Leo et al. 2016).

The drop of assortativity with cities size reflects that contact preferences between individuals of similar social activity is reduced as the size of the urban environment grows. This may be associated to the well described constraints imposed by larger, more heterogeneous, and complex organization of cities through congestion (Louf and Barthelemy 2014; Barthelemy 2016), or structural constraints of urban infrastructure on urban life (Samaniego and Moses 2008; Louf and Barthelemy 2014). In fact, while social and economic diversity has largely been linked to more productive urban systems (Lobo et al. 2013; Bettencourt et al. 2014; Youn et al. 2016), heterogeneity among urban dwellers is syndicated as a major determinant of network structure by sociologists (i.e. regular equivalence index; Hanneman and Riddle 2011). Such observation points towards portraying larger cities as effectively more unequal or concentrated in terms of their social relations, as they become more diverse in terms of the structures of each individual's social networks.

6.5 Conclusion

We here present a topological analysis of social networks derived from telephone call records (CDR). Using standard metrics, we have described the properties of social networks in cities, and observed how these evolve with the size of the city. The majority of our findings coincide to a large extent with empirical observations and theoretical models related to the greater inequality and diversity of social interactions present in larger cities. Even when larger cities have less social accessibility compared to small cities, the results show that in general Chilean cities continue to have *small world* behavior.

This work, and others, clearly suggest that cellphone connections may be a useful proxy of real social interaction (Coscia and Hausmann 2015). This is embedded in the well-known discussion seeking to understand the relationship between social interaction and urban infrastructure (Picornell et al. 2015; Alessandretti et al. 2018; Carrasco et al. 2018). As Carrasco et al. (2008) clearly puts it, this dependence is clearly justified by recognizing that a large part of individual's motivation relates to "*…with whom they interact rather than where they go*".

We hope that our interpretations and conclusions will contribute to the understanding of human interactions in the urban environment. Particularly to foster new approaches and usage of the vast amount of digital traces generated to understand urban segregation and inequality.

Acknowledgements We thank Pablo García and the team at Telefónica I+D Chile for the provision of anonymized X/CDR data for Chile.

Funding Funding for this research was provided by the Chilean research council CONICYT under the grant program FONDECYT #1161280 to Horacio Samaniego.

References

Albert, R., Barabási, A.-L.: Statistical mechanics of complex networks. Rev. Mod. Phys. **74** (2002)
Albert, R., Jeong, H., Barabási, A.-L.: Diameter of the world-wide web. Nature **401**, 130–131 (1999). https://doi.org/10.1038/43601
Alessandretti, L., Sapiezynski, P., Sekara, V., Lehmann, S., Baronchelli, A.: Evidence for a conserved quantity in human mobility. Nat. Hum. Behav. (2018). https://doi.org/10.1038/s41562-018-0364-x
Alstott, J., Bullmore, E., Plenz, D.: Powerlaw: a python package for analysis of heavy-tailed distributions. PLoS ONE **9** (2014). https://doi.org/10.1371/journal.pone.0085777
Arcaute, E., Hatna, E., Ferguson, P., Youn, H., Johansson, A., Batty, M.: Constructing cities, deconstructing scaling laws. J. Roy. Soc. Interface **12**, 20140745 (2015). https://doi.org/10.1098/rsif.2014.0745
Barabási, A.-L., Albert, R.: Emergence of scaling in random networks. Science **286**, 11 (1999). https://doi.org/10.1126/science.286.5439.509
Barabási, A.-L., Pósfai, M.: Network Science. Cambridge University Press, Cambridge (2016)
Barbosa, H., Barthelemy, M., Ghoshal, G., James, C.R., Lenormand, M., Louail, T., Menezes, R., Ramasco, J.J., Simini, F., Tomasini, M.: Human mobility: models and applications. Phys. Rep. **734**, 1–74 (2018). https://doi.org/10.1016/j.physrep.2018.01.001
Barenblatt, G.I.: Scaling, Self-similarity, and Intermediate Asymptotics. Cambridge University Press, Cambridge (1996). https://doi.org/10.1007/978-1-4419-7646-8
Barthelemy, M.: A global take on congestion in urban areas. Environ. Plan. B: Plan. Des. **43**, 800–804 (2016)
Battiston, S., Catanzaro, M., Caldarelli, G.: Social and Financial Networks. In: Caldarelli, G., Vespignani, A. (eds.) Large Scale Structure and Dynamics of Complex Networks: From Information Technology to Finance and Natural Science, pp. 205–233. World Scientific Publishing, River Edge, NJ, USA (2002)
Batty, M.: Sociology. A theory of city size. Science **340**, 1418–1419 (2013). https://doi.org/10.1126/science.1239870
Ben-Naim, E., Frauenfelder, H., Toroczkai, Z.: Complex Networks. Springer, Berlin, Heidelberg (2004)
Bettencourt, L.M.A.: The origins of scaling in cities. Science **340**, 1438–1441 (2013). https://doi.org/10.1126/science.1235823
Bettencourt, L.: The kind of problem a city is. Cities, Scaling and Sustainability Papers, pp. 1–6 (2013b). https://doi.org/10.1016/j.respol.2006.09.026
Bettencourt, L.M.A., Lobo, J., Helbing, D., Kühnert, C., West, G.B.: Growth, innovation, scaling, and the pace of life in cities. Proc. Natl. Acad. Sci. U.S.A. **104**, 7301–7306 (2007). https://doi.org/10.1073/pnas.0610172104
Bettencourt, L.M.A., Samaniego, H., Youn, H.: Professional diversity and the productivity of cities. Sci. Rep. **4**, 5393 (2014). https://doi.org/10.1038/srep05393
Borgatti, S.P., Everett, M.G.: Models of core/periphery structures. Soc. Netw. **21**, 375–395 (2000). https://doi.org/10.1016/s0378-8733(99)00019-2
Borgatti, S.P., Halgin, D.S.: Analyzing affiliation networks. The Sage Handbook of Social Network Analysis, pp. 417–433 (2011)
Burnham, K.P., Anderson, D.R.: Model Selection and Multimodel Inference: A Practical Information-Theoretic Approach. Springer, New York (2003)
Caldarelli, G., Vespignani, A.: Large Scale Structure and Dynamics of Complex Networks: From Information Technology to Finance and Natural Science. World Scientific Publishing, River Edge, NJ, USA (2007)
Candia, J., González, M.C., Wang, P., Schoenharl, T., Madey, G., Barabási, A.L.: Uncovering individual and collective human dynamics from mobile phone records. J. Phys. A: Math. Theor. **41** (2008). https://doi.org/10.1088/1751-8113/41/22/224015

Carrasco, J.A., Hogan, B., Wellman, B., Miller, E.J.: Collecting social network data to study social activity-travel behavior: an egocentric approach. Environ. Plan. B: Plan. Des. **35**, 961–980 (2008). https://doi.org/10.1068/b3317t
Carrasco, J.A., Axhausen, K., Dugundji, E.: Social interactions and travel behaviour. Res. Transp. Econ. **68**, 1 (2018). https://doi.org/10.1016/j.retrec.2018.05.008
Clauset, A., Shalizi, C.R., Newman, M.E.J.: Power-law distributions in empirical data. SIAM Rev. **51**, 661 (2009). https://doi.org/10.1137/070710111
Coscia, M., Hausmann, R.: Evidence that calls-based and mobility networks are isomorphic. PLoS ONE **10**, 1–15 (2015). https://doi.org/10.1371/journal.pone.0145091
Cottineau, C., Hatna, E., Arcaute, E., Batty, M.: Diverse cities or the systematic paradox of Urban Scaling Laws. Comput. Environ. Urban Syst. **59** (2016). https://doi.org/10.1016/j.compenvurbsys.2016.04.006
Estrada, E.: The Structure of Complex Networks: Theory and Applications. Oxford University Press, Oxford (2011)
Freeman, L.C.: The development of social network analysis—with an emphasis on recent events. In: Scott, J., Carrington, P.J. (eds.) The SAGE Handbook of Social Network Analysis, pp. 26–39 (2011). https://doi.org/10.4135/9781446294413
González, M.C., Hidalgo, C.A., Barabási, A.L.: Understanding individual human mobility patterns. Nature **453**, 779–782 (2008). https://doi.org/10.1038/nature06958
Goyal, S.: Social networks in economics. In: Scott, J., Carrington, P.J. (eds.) The SAGE Handbook of Social Network Analysis, pp. 67–79 (2011)
Hanneman, R.A., Riddle, M.: Concepts and measures for basic network analysis. In: Scott, J., Carrington, P. (eds.) The SAGE Handbook of Social Network Analysis, pp. 340–369 (2011). https://doi.org/10.4135/9781446294413
Hawkins, D.M.: Point estimation of the parameters of piecewise regression models. Appl. Stat. **25**, 51 (1976). https://doi.org/10.2307/2346519
Hidalgo, C.A., Rodriguez-Sickert, C.: The dynamics of a mobile phone network. Phys. A: Stat. Mech. Appl. **387**, 3017–3024 (2008). https://doi.org/10.1016/j.physa.2008.01.073
Humeres, F.J., Samaniego, H.: Dissecting the spatial structure of cities from human mobility patterns to define functional urban boundaries. arXiv preprint arXiv:1709.06713
Infante Fabres, N., Pino Castillo, M.: Chile: Ciudades, Pueblos, Aldeas y Caserios. Santiago de Chile (2005)
Iqbal, M.S., Choudhury, C.F., Wang, P., González, M.C.: Development of origin–destination matrices using mobile phone call data. Transp. Res. Part C: Emerg. Technol. **40**, 63–74 (2014)
Jackson, M.O.: Social and Economic Networks. Princeton University Press (2008)
Kovanen, L.: Structure and Dynamics of a Large-Scale Complex Social Network. Helsinki University of Technology (2009)
Latora, V., Nicosia, V., Russo, G.: Complex Networks Principles, Methods and Applications. Cambridge University Press, Cambridge (2017)
Leitão, J.C., Miotto, J.M., Gerlach, M., Altmann, E.G.: Is this scaling nonlinear? Roy. Soc. Open Sci. **3**, 150649 (2016). https://doi.org/10.1098/rsos.150649
Lenormand, M., Picornell, M., Cantú-Ros, O.G., Tugores, A., Louail, T., Herranz, R., Barthélemy, M., Frías-Martínez, E., Ramasco, J.J.: Cross-checking different sources of mobility information. PLoS ONE **9**, e105184 (2014). https://doi.org/10.1371/journal.pone.0105184
Leo, Y., Fleury, E., Alvarez-Hamelin, J.I., Sarraute, C., Karsai, M.: Socioeconomic correlations and stratification in social-communication networks. J. Roy. Soc. Interface **13**, 20160598 (2016). https://doi.org/10.1098/rsif.2016.0598
Lobo, J., Bettencourt, L.M.A., Strumsky, D., West, G.B.: Urban scaling and the production function for cities. PLoS ONE **8**, e58407 (2013). https://doi.org/10.1371/journal.pone.0058407
Louail, T., Lenormand, M., Cantú-Ros, O.G., Picornell, M., Herranz, R., Frias-Martinez, E., Ramasco, J.J., Barthélemy, M.: From mobile phone data to the spatial structure of cities. Sci. Rep. **4**, 5276 (2014). https://doi.org/10.1038/srep05276

Louf, R., Barthelemy, M.: How congestion shapes cities: from mobility patterns to scaling. Sci. Rep. **4**, 1–9 (2014). https://doi.org/10.1038/srep05561

Marquet, P.A., Quiñones, R.A., Abades, S.R., Labra, F., Tognelli, M., Arim, M., Rivadeneira, M.: Scaling and power-laws in ecological systems. J. Exp. Biol. **208**, 1749–1769 (2005). https://doi.org/10.1242/jeb.01588

Martínez, F.: Cities' power laws: the stochastic scaling factor. Environ. Plan. B: Plan. Des. **43**, 257–275 (2016). https://doi.org/10.1177/0265813515604245

Masucci, A.P., Arcaute, E., Hatna, E., Stanilov, K., Batty, M.: On the problem of boundaries and scaling for urban street networks. J. Roy. Soc. Interface **12**, 20150763 (2015). https://doi.org/10.1098/rsif.2015.0763

McPherson, M., Smith-Lovin, L., Cook, J.M.: Birds of a feather: homophily in social networks. Annu. Rev. Sociol. **27**, 415–444 (2001)

Newman, M.E.J.: Assortative mixing in networks. Phys. Rev. Lett. **89**, 1–4 (2002). https://doi.org/10.1103/PhysRevLett.89.208701

Newman, M.E.J.: Mixing patterns in networks. Phys. Rev. E **67**, 026126 (2003). https://doi.org/10.1103/PhysRevE.67.026126

Newman, M.E.J.: Networks: An Introduction. Princeton University Press (2010)

Newman, M.E.J.: Networks: An Introduction, pp. 163–186. Oxford University (2014). https://doi.org/10.1007/978-3-319-03518-5-8

Onnela, J.-P., Saramaki, J., Hyvonen, J., Szabo, G., Lazer, D., Kaski, K., Kertesz, J., Barabasi, A.-L.: Structure and tie strengths in mobile communication networks. Proc. Natl. Acad. Sci. U.S.A. **104**, 7332–7336 (2007). https://doi.org/10.1073/pnas.0610245104

Opsahl, T., Colizza, V., Panzarasa, P., Ramasco, J.J.: Prominence and control: the weighted rich-club effect. Phys. Rev. Lett. **101**, 1–4 (2008). https://doi.org/10.1103/PhysRevLett.101.168702

Palchykov, V., Mitrović, M., Jo, H.-H., Saramäki, J., Pan, R.K.: Inferring human mobility using communication patterns. Sci. Rep. **4**, 1–6 (2014). https://doi.org/10.1038/srep06174

Pan, W., Ghoshal, G., Krumme, C., Cebrian, M., Pentland, A.: Urban characteristics attributable to density-driven tie formation. Nat. Commun. **4**, 1961 (2013). https://doi.org/10.1038/ncomms2961

Peixoto, T.P.: The graph-tool python library. figshare (2014). https://doi.org/10.6084/m9.figshare.1164194

Picornell, M., Ruiz, T., Lenormand, M., Ramasco, J.J., Dubernet, T., Frías-Martínez, E.: Exploring the potential of phone call data to characterize the relationship between social network and travel behavior. Transportation **42**, 647–668 (2015). https://doi.org/10.1007/s11116-015-9594-1

Pumain, D.: Scaling laws and urban systems. Santa Fe Institute, Working Paper No. 04-02-002, pp. 1–26 (2004)

Ratti, C., Sobolevsky, S., Calabrese, F., Andris, C., Reades, J., Martino, M., Claxton, R., Strogatz, S.H.: Redrawing the map of Great Britain from a network of human interactions. PLoS ONE **5**, e14248 (2010)

Reades, J., Calabrese, F., Ratti, C.: Eigenplaces: analysing cities using the space–time structure of the mobile phone network. Environ. Plan. B: Plan. Des. **36**, 824–836 (2009). https://doi.org/10.1068/b34133t

Ribeiro, F.L., Meirelles, J., Ferreira, F.F., Neto, C.R.: A model of urban scaling laws based on distance dependent interactions. Roy. Soc. Open Sci. **4**, 160926 (2017). https://doi.org/10.1098/rsos.160926

Samaniego, H., Moses, M.E.: Cities as organisms: allometric scaling of urban road networks. J. Transp. Land Use **1**, 21–39 (2008)

Schläpfer, M., Bettencourt, L.M.A., Grauwin, S., Raschke, M., Claxton, R., Smoreda, Z., West, G.B., Ratti, C.: The scaling of human interactions with city size. J. Roy. Soc. Interface **11** (2014). https://doi.org/10.1098/rsif.2013.0789

Scott, J.: Social physics and social networks. In: Scott, J., Carrington, P. (eds.) The SAGE Handbook of Social Network Analysis, pp. 55–66 (2011). https://doi.org/10.4135/9781446294413

Scott, J., Carrington, P.J.: The SAGE Handbook of Social Network Analysis. SAGE Publications Ltd. (2011)

Steenbruggen, J., Tranos, E., Nijkamp, P.: Data from mobile phone operators: a tool for smarter cities? Telecommun. Policy **39**, 1–12 (2014). https://doi.org/10.1016/j.telpol.2014.04.001
Thisse, J.-F.: The new science of cities by Michael Batty: the opinion of an economist. J. Econ. Lit. **52**, 805–819 (2014). https://doi.org/10.1257/jel.52.3.805
Travers, J., Milgram, S.: An experimental study of the small world problem. Sociometry **32**, 425 (1969). https://doi.org/10.2307/2786545
Vespignani, A.: Complex networks: the fragility of interdependency. Nature **464**, 984–985 (2010). https://doi.org/10.1038/464984a
Watts, D.J., Strogatz, S.H.: Collective dynamics of 'small-world' networks. Nature **393**, 440–442 (1998). https://doi.org/10.1038/30918
West, G.B.: Scale: The Universal Laws of Growth, Innovation, Sustainability, and the Pace of Life in Organisms, Cities, Economies, and Companies (2017)
Youn, H., Bettencourt, L.M.A., Lobo, J., Strumsky, D., Samaniego, H., West, G.B.: Scaling and universality in urban economic diversification. J. Roy. Soc. Interface **13**, 20150937 (2016). https://doi.org/10.1098/rsif.2015.0937
Zhong, C., Schläpfer, M., Muller Arisona, S., Batty, M., Ratti, C., Schmitt, G.: Revealing centrality in the spatial structure of cities from human activity patterns. Urban Stud. **54**, 437–455 (2015). https://doi.org/10.1177/0042098015601599

Part III
Economy Versus Geography: Theories of Urbanization and Cities Development

Chapter 7
Revisiting Urban Economics for Understanding Urban Data

Marc Barthelemy

Abstract The recent availability of data about cities and urban systems opens the exciting possibility of a 'new Science of Cities'. Urban morphogenesis, activity and residence location choice, mobility, urban sprawl and the evolution of urban networks are just a few of the important processes that can be discussed now from a quantitative point of view. Here, we will discuss how a data-informed approached can elaborate on urban economics models in order to get predictions in agreement with empirical observations. We will illustrate this approach on the polycentric organization of activity in cities and how it evolves when their population grow.

7.1 Introduction

With the availability of large amounts of urban data, we can now hope to bridge the gap between theoretical models and empirical observations. This will help us to provide a quantitative understanding of the phenomenon under study. In the case of a system as complex as a city, the hope is to construct solid, scientific foundations of urban systems (see Batty 2013; Barthelemy 2016). This effort is necessarily interdisciplinary (and this is not always easy, see O'Sullivan and Manson 2015): we have to build up on early studies in urbanism to discuss morphological patterns and their evolution, and on quantitative geography and spatial economics to describe the behavior of individuals, the impact of different transportation modes, and the effect of economic variables (such as the income, the renting market).

We will illustrate this type of approach on some aspect of mobility in cities. The strategy used is in general determined by the following main tasks: First, we extract robust empirical facts and useful information from large amounts of data; second, we identify the relevant mechanisms and parameters describing the behavior of the elementary constituents of the system (in most cases individuals, but also groups, companies or institutions). Then, using these mechanisms and parameters,

---

M. Barthelemy (✉)
Institut de Physique Théorique (IPhT/CEA-CNRS), Gif-sur-Yvette, France
e-mail: marc.barthelemy@cea.fr

Centre d'Analyse et de Mathématique Sociales (CAMS/EHESS), Paris, France

D. Pumain (ed.), *Theories and Models of Urbanization*, Lecture Notes in Morphogenesis, https://doi.org/10.1007/978-3-030-36656-8_7

we construct parsimonious models by combining tools and concepts from statistical physics with ingredients from urban economics and quantitative geography. Finally, we validate the model by data.

We will discuss empirical aspects in Sect. 7.2 with the study of the spatial structure of activities in cities. We will show how we can extract this information from mobile phone data in Sect. 7.2.1. This same dataset can also help us in describing mobility patterns and we will show in Sect. 7.2.2 how we can extract mesoscopic information from these large datasets. In Sect. 7.3, we will focus on theoretical approaches for explaining regularities observed empirically. We will start by discussing how to model the spatial distribution of activities, and then another problem which is the effect of income on commuting. In these theoretical approach we show that classical models fail to predict the empirical observations, therefore calling for the need for different frameworks. These examples highlight the importance of three essential ingredients in the modeling strategy that is used here: (i) extracting useful mesoscopic information from large datasets, (ii) describing complex quantities by random variables and (iii) complex actions by stochastic processes.

7.2 Empirical Studies

7.2.1 Extracting the Spatial Distribution of Activity

Cell phone networks, enable to capture large amounts of human behavioral data but also provide information about the structure of cities and their dynamical properties. We illustrate this point with mobile phone data recorded over two months and for 30 Spanish metropolitan areas (Louail et al. 2014, 2015). We can measure the density of users in certain areas of the city and applying filters such as the frequency of visit of a location and the duration of stay we can infer density of activity during the day in cities. The type of measures that we obtain is shown in Fig. 7.1 in the case of Vitoria and Bilbao.

We observe that there are essentially two types of cities. First we observe that usually smaller cities have a unique activity center (Fig. 7.1, left) and correspond to the classical image of the monocentric city organized around a central business district. Second for larger cities we observe a more complex pattern (Fig. 7.1, right) with more than one activity center.

In order to go further in the quantitative analysis of the spatial organization of activities in cities we have to determine the number of activity centers, or "hotspots". We can see the density of users (or employment for example) as a two dimensional surface and the hotspots are the local maxima of this surface. In order to decide if a location can be considered as a local maximum, we introduced a parameter free method to introduce a threshold allowing to detect hotspots. This method, described in detail in Louail et al. (2014), relies on the Lorentz curve of the density at different points. It is based on the observation that the curvature of the Lorentz curve is

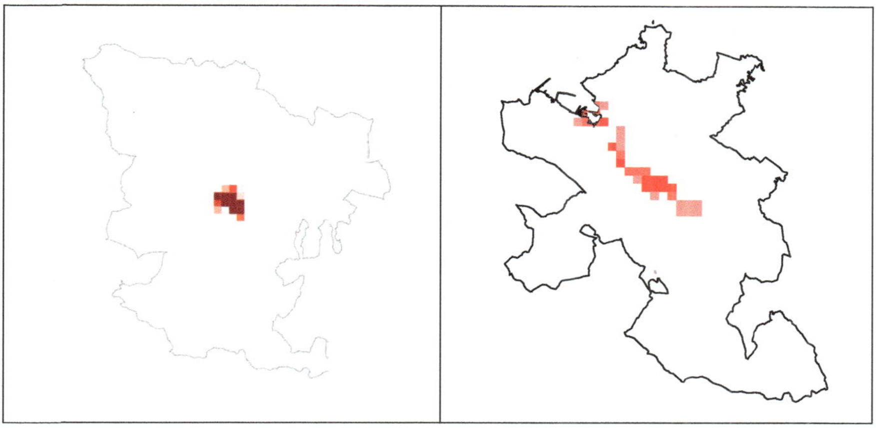

Fig. 7.1 Areas with large density of mobile phone users for (left) Vitoria and (right) Bilbao. The darker the area and the larger the density of users. Figure from Louail et al. (2014)

connected to the heterogeneity of the values of the density (the surface between the diagonal and the Lorentz curve is indeed directly related to the Gini coefficient). This curvature of the Lorentz curve $L(F)$ can be characterized by the slope of the curve at $F = 1$ and it is therefore natural to extract from this slope a threshold (see Fig. 7.2). This threshold (denoted by F_{LouBar} in Fig. 7.2), is naturally larger than the naive one given by the average. Indeed considering that every location with a value larger than the average is a very mild determination of a hotspot and F_{LouBar} would give the strongest constraint.

For a given value of the threshold we can then count the number H of hotspots and any result on this quantity should be robust with respect to small variations of the thresholding procedure. For example, we plot in Fig. 7.3 the number H of hotspots obtained by using the average for the threshold (i.e. each location with a density larger than the average is counted as a hotspot) and by using the quantity F_{LouBar}.

We observe on these results the existence of a robust behavior (confirmed by studies on employment data for 9000 US cities, Louf and Barthelemy 2013)

Fig. 7.2 Lorentz curve used for constructing a parameter-free method for determining hotspots. The intersection of the slope at $F = 1$ with the x-axis gives a threshold naturally related to the heterogeneity of the density distribution. Figure taken from Louail et al. (2014)

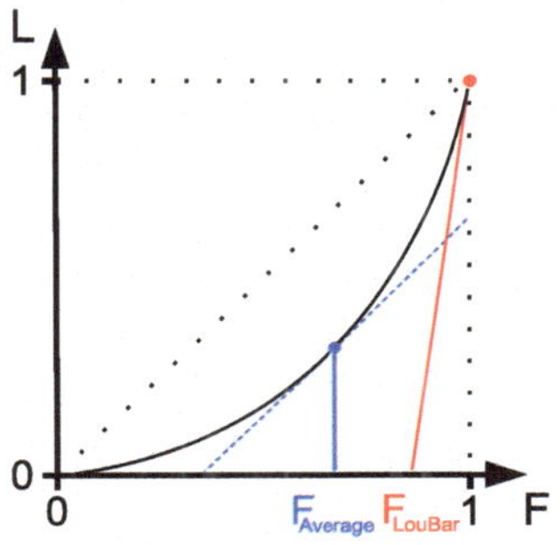

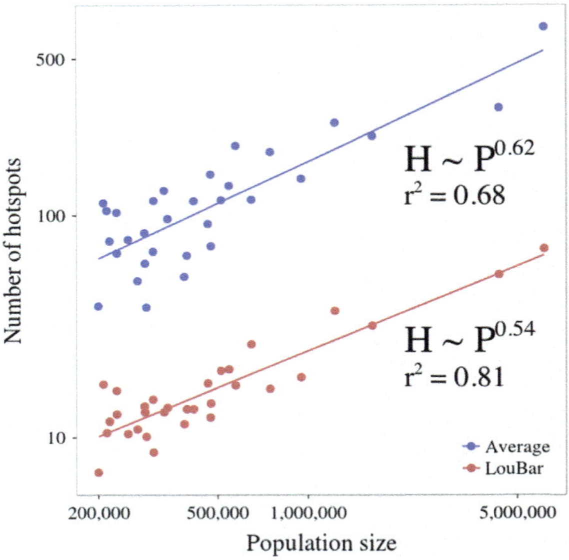

Fig. 7.3 Number of hotspots H versus the population of the city for two different values of the threshold (the average and the one obtained from F_{LouBar}). Both thresholds predict consistently a sublinear behavior. Figure taken from Louail et al. (2014)

$$H \sim P^{\beta}$$

where the exponent β is usually found to be around 0.5–0.6. The number of these hotspots thus scales sublinearly with the population size, a result that will serve as a guide for constructing theoretical models. The spatial structure of these hotspots is also of interest and allows us to distinguish different categories of cities, from monocentric and "segregated" where the spatial distribution is very dependent on land use (residential or activity), to polycentric where the spatial mixing between land uses is much more important. These results point towards the possibility of a new, quantitative classification of cities using high resolution spatio-temporal data.

7.2.2 A Typology of Mobility Patterns

The description of mobility patterns and their statistics is of great interest for understanding the functioning of a city at a large scale and the effect of infrastructures. Surprisingly enough, there were few studies of this problem and the main source of inspiration for many authors is the paper by Bertaud and Malpezzi (2003) who proposed the simple typology of journey-to-work trips shown in Fig. 7.4.

The first class of cities display the type (a) of mobility patterns which correspond to the classical idea of what happens in a monocentric city: all the flows (corresponding here to the journey-to-work commute) are converging to one unique center. In (b), Bertaud and Malpezzi proposed another type of organization where flows appear to be "random". In (c) we have the polycentric—or urban villages—organization with

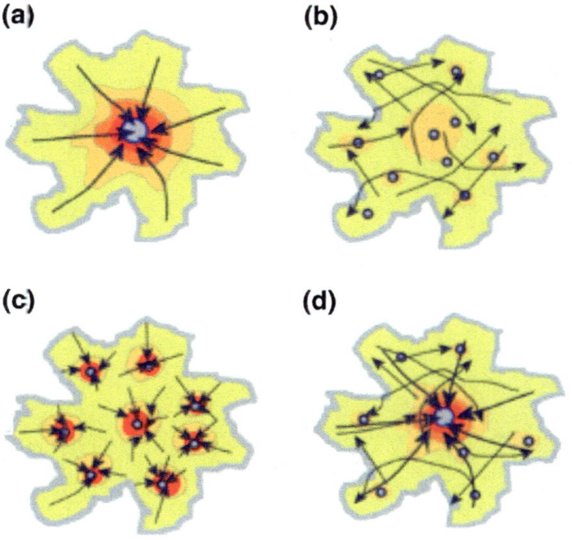

Fig. 7.4 A proposal for a typology of mobility patterns from home to work. The journey-to-work flows are indicated by arrows. Figure taken from Bertaud and Malpezzi (2003)

the existence of many activity centers with their own attraction basin. Finally, they conclude with the possibility of observing a pattern (d) resulting from the superimposition of patterns (a–c). Although this typology proposal is very reasonable it wasn't verified empirically.

The current mobile phone data however allows to test this typology. A more general problem with large datasets such as the one obtained with mobile phone data (or other sources such as GPS or RFIDs), is the extraction of a clear and simple footprint of the structure of large origin-destination matrices which contain the complete information on commuting flows, but are difficult to analyze and compare. We discuss here briefly a versatile method (Louail et al. 2015) which extracts a coarse-grained signature of mobility networks, under the form of a 2×2 matrix that separates the flows into four categories. The main idea is to separate working places in two different categories: either a location is an activity center (i.e. a hotspot) or not. We can also apply the method discussed above to residence places: we then obtain "residential hotspots" for which the population density is far above the average. The commuting flows from home to work can then be aggregated in the four different categories shown in Fig. 7.5.

We then obtain the four numbers I, C, D, R for each city that aggregate the flows. The quantity I represents the flow between residential hotspots and activity centers, while R describes the flows between non-hotspots (both residential and working). The two other quantities C and D represent flows between areas of different types.

We apply this method to origin-destination matrices extracted from mobile phone data for journey to work trips in the 30 largest Spanish urban areas (Louail et al. 2015), and lead to the result shown in Fig. 7.6. We observe that these cities essentially differ

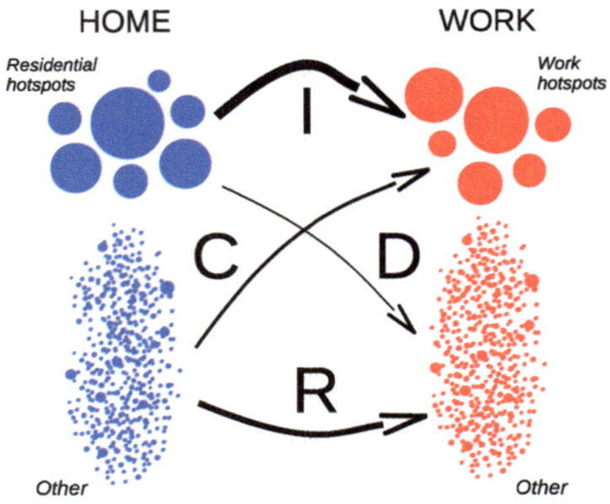

Fig. 7.5 Constructing the four different types of commuting flows: we separate both homes and working locations into hotspots and non-hotspots. Figure taken from Louail et al. (2015)

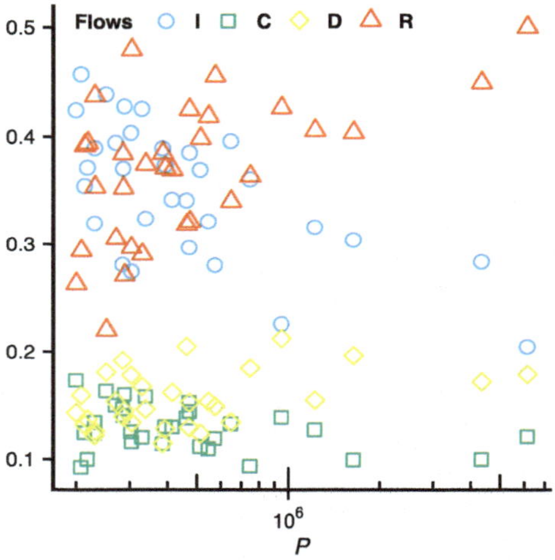

Fig. 7.6 The four different flows I, R, C, D for the 30 largest Spanish urban areas versus their population. We observe that large areas are mainly determined by R and I, the other quantities being negligible. Figure taken from Louail et al. (2015)

by their proportion of two types of flows: integrated (I) between residential and employment hotspots and random flows (R), whose importance increases with city size.

This result is in contrast with the naive expectation of a monocentric city where C flows dominate. For large cities "random" flows described by R are the most important. This might be due to the fact that it is easier to move around in large cities thanks to the various public transportation systems. We see here how the extraction of mesoscopic patterns from large datasets forces us to reconsider our view of cities and more particularly here the spatial organization of mobility patterns.

7.3 Theoretical Approaches: Modelling Strategies

The empirical study of activity centers discussed above shows that their number scales with the population of the city as

$$H \sim P^{\beta}$$

where $\beta \approx 0.5$–0.6. The theoretical question is therefore simple: how can we explain this behavior and can we predict the value of β? (or at least why do we have a sublinear behavior with $\beta < 1$?).

In order to understand the spatial structure of cities and in particular how the number of hotspots varies with population as discussed above, we have to model how an individual chooses her residence and workplace. In the following we will discuss the classical approaches to this problem and we will show that they are unable to predict and understand the scaling of H with P. We then show how we can integrate economical ingredient and make simplifying assumptions typical from statistical physics in order to get a model with predictions in agreement with the observation.

7.3.1 Classical Approaches: Krugman, Fujita and Ogawa

We first follow here Krugman (1996) for a simple model of activity clustering in cities. The distribution of companies is described by the density $\rho(x, t)$ and a fundamental quantity is the market potential given by

$$\Pi(x) = \int dz \rho(z) K(x - z)$$

where $K(x - z)$ is the kernel that describes the impact (or spillover effects) of a company located at z on the attractiveness of location x. We can compute the average market potential $\overline{\Pi}$ over the whole city and Krugman proposes then to describe the evolution of company density by the following equation

$$\frac{\partial \rho}{\partial t} = \gamma \left[\Pi(x, t) - \overline{\Pi} \right]$$

This nonlinear equation states that the density will increase at location where the market potential is larger than the average. Krugman then showed that a uniform distribution of companies $\rho(x, t) = const.$ is unstable where the most unstable mode is given by a quantity k^* that depends on the details of the system such as spillover effects (but is independent from the population size P). This means that the activities will indeed form clusters and this simple mechanism seems to explain the clustering of activities in cities. However, we would like to understand the scaling of the number of hotspots H and this model simply predicts

$$H \sim Ak^{*2}$$

where A is the surface area of the city. This model is therefore unable to predict the evolution of the spatial distribution of cities when the population grows (unless introducing an external assumption about how the area of the city varies with the population). In other words, this model predicts a constant value of H that is independent of the population P.

Another important approach for understand the spatial structure of cities was proposed by the economists Fujita and Ogawa (1982). In this model, agents optimize their utility and companies their profit. Focusing on agents, an agent will choose to live in location x and work in location y such that the quantity (which corresponds to the composite commodity) given by

$$Z(x, y) = w(y) - C_R(x) - C_T(x, y)$$

where $w(y)$ is the wage when working at y, $C_R(x)$ the rent at location x, and $C_T(x, y)$ the transportation to commute from home to work (for companies there is a similar equation for profits). In this model and for a monocentric organization of activities, Fujita and Ogawa choose to take $C_T(x, y) = t|x - y|$ independent of the traffic (i.e. without any congestion effect). They were able to show that this monocentric organization is actually unstable, in particular when transportation cost per unit distance (t) become too large. This formalism however does not allow to predict the number of activity centers when population grows.

We saw on these two examples models that are in agreement with the qualitative organization of the spatial structure of activity in cities, but are so far unable to provide a quantitative prediction. Even if these models are satisfying from an intellectual point of view, as long as their predictions are not in agreement with empirical measures, we can only place a low level of confidence in their ability to describe what is actually happening in cities. Ideally, we would like to have a model with a minimal number of parameters and which is able to predict a large number of unrelated empirical facts. The model described in the following didn't reach this level yet but at least is a proposal for an alternate modeling of cities that is sound from an economical point of view and in agreement with the scaling of the number of hotspots.

7.3.2 Complex Quantities as Random Variables

The problem is to compute the value of the exponent β. In this new way of modelling cities (Louf and Barthelemy 2013, 2014), we integrate ingredients of urban economics, and most importantly we replace an unknown, complex quantity by a random one, a concept introduced in the study of the spectra of heavy atoms (Wigner 1955). More precisely, we assumed that:

(1) At each time step, we add a new individual in the city.
(2) Each individual will optimize its own budget consisting of its wage minus residential and transportation costs.
(3) The wage is described as a random variable.
(4) Transportation costs through congestion integrate interactions between individuals.

With the assumption (1) dynamics is introduced in this system and we do not consider that cities are in equilibrium in contrast with many previous studies such as the Fujita-Ogawa model (1982). With (2), we integrate ingredients coming from urban economics that discussed for a long time the behavior of individuals. The assumption (3) is typical from statistical physics where replacing the wage $w(y)$—a complex quantity that results from a large number of interacting constituents—by a random variable $\eta(y)$ proved to be in some cases accurate (Wigner 1955). We note here that in the original model proposed by Fujita and Ogawa (1982), wages are endogenous variables and are an output of the model. This description leads to complications that however forbids to make clear testable predictions. In (4), the effect of congestion on the time spent to go from one point to another is described in transportation economics and describes effectively interactions between individuals. We use the generalized cost for transportation which is proportional to the time $\tau(x, y)$ needed to go from x to y and which is given by the Bureau of Public function (see for example Branston 1955)

$$\tau(x, y) = \frac{d(x, y)}{v}\left[1 + \left(\frac{T(x, y)}{C}\right)^{\mu}\right]$$

where $T(x, y)$ is the traffic between these two points, C is the capacity and v the average velocity of the road system between x and y, and μ is an exponent that characterizes the sensitivity of the system to the congestion and is generally of order $\mu \approx 2$–5 (see for example Branston 1976). This interaction is at the heart of the non-trivial collective behavior observed in this model and characterized by non-trivial exponent values.

Putting all these ingredients together, the model is now described by the maximization of the following quantity:

$$Z(x, y) = \eta(y) - td(x, y)\left[1 + \left(\frac{T(x, y)}{C}\right)^{\mu}\right]$$

(in this expression, we also note that the wage $\eta(y)$ can be understood as a number encoding the attractiveness of the location y). For this model, using mean-field arguments and numerical simulations, we were able to predict that the monocentric city is stable until a threshold P^* for the population above which another activity center becomes more interesting for individuals (Fig. 7.7).

This spatial splitting of the activity is driven in this model by the congestion: all individuals choose to go to the most attractive center (from the point of view of the

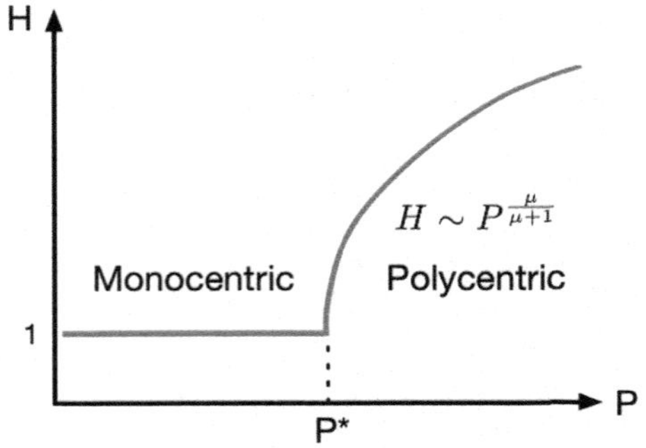

Fig. 7.7 Number of activity centers versus the population. For $P < P^*$, the system is monocentric and over this threshold the activity is dispersed over H different centers

wage), but this increases the transport cost (due to the congestion effect). Another center, less attractive but with a smaller traffic becomes then the most interesting working place. We can estimate this threshold and we can show that increasing the population leads to a larger number of activity centers. More precisely we show that

$$H \sim \left(\frac{P}{P*}\right)^{\frac{\mu}{\mu+1}}$$

and therefore predicts $\beta = \mu/\mu+1$. This result demonstrates that independently from the value of μ, the behavior of H is sublinear with the population, in agreement with empirical observations. In this simple model, non-linear congestion effects imply that cities undergo a 'dynamical' transition from a monocentric to a polycentric structure as their population grow. Congestion is certainly not the only factor that favor the formation of different activity centers, but these results demonstrate that it plays at least a major role for understanding the spatial organization of cities.

Within this model we know the location of residence and work for all individuals and we can therefore estimate other quantities such as the delay spent in traffic jams or the CO_2 emitted by cars. The predictions for these quantities are in excellent agreement with measures for European or OCDE cities (Louf and Barthelemy 2013, 2014). On a more fundamental level, this model predicts that CO_2 emitted by cars or the total spent in cars is not a simple function of urban density. This is in sharp contrast with the celebrated result of Newman and Kenworthy (1989), showing that the gasoline consumption in a city is a decreasing function of the population density. Certainly more theoretical and empirical work is needed here for understanding this issue.

7.4 Discussion

The recent availability of large amounts of data enables us to reveal regularities in the spatial structure and mobility patterns in cities across countries. These regularities suggest that common mechanisms that encompass the differences of cities exist and govern the formation and evolution of these systems. Also, traditional assumptions can now be tested and in some cases a whole new modeling framework is needed in order to understand empirical observations.

Describing several (human) actions by stochastic processes and complex quantities resulting from the interactions of several agents by random variables are the key ideas that are exposed here and that allow to propose new models for understanding the evolution of cities. More generally, we tried to show in this paper, through examples about mobility in cities, how a combination of empirical results, economical ingredients, and statistical physics tools can lead to parsimonious models with predictions in agreement with observations. Obviously, many problems are left. In particular, we have now to integrate within this framework other transportation modes, and socio-economic factors such as the impact of the revenue on the spatial structure of cities. Our hope is that this interdisciplinary theoretical approach, informed by data, will lead to a solid understanding of systems as complex as cities.

References

Batty, M.: The New Science of Cities. The MIT Press, Cambridge, MA, USA (2013)

Barthelemy, M.: The Structure and Dynamics of Cities. Cambridge University Press, Cambridge (2016)

Bertaud, A., Malpezzi, S.: The Spatial Distribution of Population in 48 World Cities: Implications for Economies in Transition. Center for Urban Land Economics Research, University of Wisconsin (2003)

Branston, D.: Link capacity functions: a review. Transp. Res. **10**(4), 223–236 (1976)

Fujita, M., Ogawa, H.: Multiple equilibria and structural transition of non-monocentric urban configurations. Reg. Sci. Urban Econ. **12**(2), 161–196 (1982)

Krugman, P.: The self-organizing economy. No. 338.9 KRU 1996. CIMMYT (1996)

Louail, T., et al.: From mobile phone data to the spatial structure of cities. Sci. Rep. **4** (2014)

Louail, T., et al.: Uncovering the spatial structure of mobility networks. Nat. Commun. **6**, 6007 (2015)

Louf, R., Barthelemy, M.: Modeling the polycentric transition of cities. Phys. Rev. Lett. **111**(19), 198702 (2013)

Louf, R., Barthelemy, M.: How congestion shapes cities: from mobility patterns to scaling. Sci. Rep. **4**, 5561 (2014)

Newman, P.W., Kenworthy, J.R.: Gasoline consumption and cities: a comparison of US cities with a global survey. J. Am. Plan. Assoc. **55**, 2437 (1989)

O'Sullivan, D., Manson, S.M.: Do Physicists Have 'Geography Envy'? And What Can Geographers Learn From It? Series: UC Berkeley Previously Published Works (2015)

Wigner, E.: Characteristic vectors of bordered matrices with infinite dimensions. Ann. Math. **62**(3), 548–564 (1955)

Chapter 8
The French Version of the CAGE Mythology (Competitiveness—Attractiveness—Globalization—Excellence) and Some Ideas on How to Escape from It

Olivier Bouba-Olga and Michel Grossetti

Abstract The recent period is characterized by the emergence of an attractive mythology in the field of economic development: the deepening of globalization would plunge all territories into an imperative of competitiveness, with only a few very big cities able to compete to attract the talents and leaders of tomorrow, which should therefore be supported by concentrating efforts on excellence. We summarize it by the acronym CAGE for Competitiveness, Attractiveness, Globalization and Excellence. A careful analysis of the different components of the CAGE shows that as reassuring as it can be, it tends to lock thought into conceptions that do not stand the test of facts. Despite this, and to a greater or lesser extent influenced by certain researchers and private or public analysis and advisory bodies, it underlies a whole range of public policies; it has even structured part of the debates around the results of elections in different countries. Enclosure in the CAGE causes adverse effects. As public resources are limited, dedicating them strongly to a few actors (start-ups, researchers deemed "excellent" …) or to a few places ("global" cities) leads to reinforcing socio-spatial inequalities. Some elements for reflection on possible alternatives, which seem healthier to us, will be presented in order to help escaping from it.

8.1 Introduction

Since the *Mythologies* of Barthes (1957), it has become common to designate as myths shared beliefs that have become so common that they are no longer even discussed. A mythology is a set of myths that work together. Scientific specialities

O. Bouba-Olga (✉)
Université de Poitiers, Poitiers, France
e-mail: obouba@univ-poitiers.fr

M. Grossetti
CNRS, EHESS, Université de Toulouse, Toulouse, France
e-mail: Michel.Grossetti@univ-tlse2.fr

D. Pumain (ed.), *Theories and Models of Urbanization*, Lecture Notes in Morphogenesis, https://doi.org/10.1007/978-3-030-36656-8_8

thus have their myths, statements that have become stylized and taken up without debate from publication to publication. Their status as a myth does not say anything about their relevance or their agreement with the data—it just means that they are taken for granted, that they form a basis for reasoning more than their objects. In scientific debates, designating statements as myths means that one seeks precisely to remove them from the status of myth, to question them, to re-examine them, to reduce them to the status of statements that can be criticized and compared to data. In social sciences, these myths are, perhaps more systematically than in other sciences, related to political orientations, decisions and public debates.

In the course of our research on the links between economic performance and the size of urban agglomerations, on the geography of innovation and academic scientific activities and on the mobility of skilled workers, we have encountered statements repeated over and over again in texts and debates that appear to work like myths. After discussing them separately, we realized that they often referred to each other and they could be seen as a mythology.

This mythology, which we will call CAGE (Competitiveness, Attractiveness, Globalization and Excellence) is present in many urban and regional studies and it exists simultaneously in public discourse and politics.

In France, the term "global city" is used less than in English-speaking countries and instead the word "metropolis" is used, a term that puts less emphasis on the connection to international socio-economic flows and corresponds more to the idea of an urban hierarchy between large cities with specific "functions" (in terms of economic activity but also with regard to government or public services) and smaller cities that do not have them. If we replace the term "metropolis" or "metropolitanization" with that of "globalization", we obtain a local version of the CAGE mythology[1] which fills the notes of the government agencies and *think tanks*, justifying policies implemented at a national, regional and local level that are concerned with the links between the economy and the geographical area.

The purpose of this text, which we conceived as an essay rather than an academic article, is to give a synthetic presentation of this mythology based on the comments of some of the authors who disseminate it in whole or in part in academic or political debates in France, in order to examine the ingredients, show the limits that each one of them presents and highlight the epistemological and theoretical problems that are common to these ingredients, before concluding with ways to analyse phenomena without resorting to these myths.

[1]In French, CAGE thus becomes CAME which suggests another metaphor. In French slang, "came" means drugs, which suggests that we need to clean up public debates.

8.2 A Golden CAGE

The CAGE mythology can be summed up with a few propositions: the process of globalization is occurring within the framework of the exacerbation of global competition following the emergence of low-cost countries for labour, the reduction of tariff and non-tariff barriers to trade, the development of transportation and communication infrastructure, the financialization of the economy etc., i.e. the deepening of globalization. In developed countries, the only way to create jobs is through innovation. However, innovation requires people who are highly qualified, talented, creative, who know how to create start-ups etc., who have a strong preference for very densely populated areas where they can find the jobs they want, be connected on a global scale, interact with each other face-to-face (thereby increasing the knowledge available) and have access to the best amenities.

To be successful in this global competition, the challenge in terms of public policy is to support these "global" cities by strengthening their visibility and attractiveness in order to achieve high performance ("excellence"). Citizens have taken on board this economic story even in their vote, with the elite of large cities rejoicing in supporting progressive candidates that believe in this vision whilst the citizens of peripheral areas reject it by raising their voices on the fringes.

How, then, is it possible to "compensate" the losers? The essential challenge for the peripheries is to become complementary to global cities. For this purpose, the first course of action is to focus on activities that serve the current population for the benefit of people working in the big cities when they want to go away for the weekend, on holiday or when they reach retirement age. The second course of action is, on the productive side, to question what these peripheral areas can offer to large cities in order to take advantage of their success. The last course of action is to take on board this inescapable trend towards the urban concentration of activity and to promote the mobility of people living outside of large cities, especially young people, so that they can come and settle in these places that are essential for the creation of wealth. By way of compensation, the extra wealth generated by the concentration in global cities makes it viable to make social transfers that benefit the non-mobile inhabitants of the peripheral territories.

This mythology is a continuation of many academic works such as those of Saskia Sassen on "global cities" (Sassen 1991), those of Richard Florida claiming the existence of a "creative class" that is sensitive to urban amenities (*soft factors*), which cities should try to attract to foster innovation and therefore the creation of jobs and wealth (Florida 2002), or those of Richard Baldwin (2017), recommending to "cajole" the elite concentrated in global cities in order to promote economic development.[2]

[2]See also the 2011 book by the economist Edward Glaeser for the nuanced title: "Triumph of the City: How Our Greatest Invention Makes Us Richer, Smarter, Greener, Healthier, and Happier".

If we replace the term metropolis with that of globalization, all of the elements of the mythology are present in the note for the Economic Analysis Council[3] by Askenazy and Martin (2015[4]), in which they state that "*a territorial policy aimed at maximizing a country's growth momentum must therefore facilitate the concentration of activities in metropolitan areas by investing in the fight against the effects of congestion*" (p. 1) in order to "*compete with global metropolises*" (p. 2), making the agglomeration "*more efficient, more productive, and generating more innovation and growth*" (p. 5). As for the other territories, they explain, they would not necessarily be negatively affected because "*the spatial concentration of activities, insofar as it allows productivity gains, indirectly benefits the disadvantaged territories by increasing the solvency of the social transfer system*" (p. 7).

Far from being confined to issues of wealth and job creation, the CAGE mythology has shaped the debates during the last elections. The economist Pierre-Yves Geoffard, the director of the Paris School of Economics, as he states in an article for *Libération*, sees in the election of Trump and the vote for Brexit the revenge of the sedentary population that are imprisoned in their territory, who do not see "*job creation in major cities; they see the local factory shutting down*"; revenge against the nomads "*made up of migrants and refugees but also traders, the bosses of big international companies, artists, academics and high-level athletes who are able to practice their professions in different countries*".[5] He therefore proposes that the elite of these global cities consent to paying a tax to finance "*political programs to better share their gains, thereby improving the situation of others*" (Geoffard 2016). The geographer Jacques Levy, in an article for *Le Monde*, also interprets the results of the American election in the context of this confrontation between two societies, one in the cities and the other outside of cities: "*Urbanity or its rejection, the public space versus the private space resonates with other very strong elements: education, productivity, creativity, globality, openness to otherness, the demand for justice, the presence of the future on the one hand; on the other, contempt for the intellect, economic isolation, a lack of innovation, calls for protectionism, fear of foreigners, affirmation of an identity based on biological purity, community loyalty, respect for authority and nostalgic reference to a mythified past—all things that do not define just a different approach to justice, but an alternative to the very idea of justice*" (Levy 2016).

In short, all of these talking points, along with many others, seem to be organized around the CAGE mythology, without which we cannot think about the creation of wealth and jobs. Those who disseminate it are well aware of the existence of some

---

[3]Reporting to the Prime Minister, the mission of the Economic Analysis Council is "to clarify the government's economic options through the comparison of points of view and analysis". It is, in a way, the French equivalent of the American Council of Economic Advisers.

[4]Note the strange grouping of migrants and refugees together with traders, top athletes and academics, among others, who are able to exercise their talents wherever they want. It appears to us that migrants and refugees have not been made aware of this "capacity" etc.

[5]Some examples are territories that are close to being considered industrial districts (the Vendée farmland, Oyonnax, the Arve Valley, Cognac, etc.) or other territories where private research is particularly important (Belfort-Montbéliard, Pau, etc.).

adverse effects, but each propose corrective measures to reduce them, which should be enough to appease the crowds.

8.3 The Weaknesses of the CAGE Mythology

In our opinion, each of the components of the CAGE mythology is based on certain ill-established stylized facts and presents numerous problems: fuzzy categories, indicators, hypotheses and questionable methods of processing, methodological bias, etc. Let's examine them one by one.

8.3.1 Competitiveness

Concentrating resources in a few global cities to attract the talent of tomorrow is the best way to fight against other territories in the context of deepening globalization and therefore to promote the creation of wealth and jobs.

This *dictate* of competitiveness in a globalized economy comes up against several empirical lessons. Economic activity, first of all, is far from being limited to a set of activities exposed to international competition: if we take again the distinction made by the National Institute of Statistics and Economic Studies between production activities (activities exposed to globalization) and "presential" activities (activities that serve the populations present in a territory, with little or no exposure to globalization), it turns out that the share of the latter in employment in France has increased from 52% in 1975 to 66% in 2014. Secondly, the decline in production employment, especially in industry, is only a relatively small part of the deepening of globalization. The most recent study available published by the Banque de France (Kalantzis and Thubin 2017) shows that the 9-point drop in the industry's share of French GDP over 40 years (10% in 2015 compared with 19% in 1975) is mainly the result of changes in household consumption demand and technical change, with foreign trade accounting for only 13% of the decline.

These first two points are widely shared by supporters of the CAGE mythology, which explains for example that Davezies and Pech (2014) develop the concept of the production/residential system, considering that large cities and hinterlands are complementary, the production activities exposed to globalization tending to be concentrated in global cities, with residential activities being developed outside of these cities. This Manichean representation of economic geography, however, conceals the fact that a lot of economic activities affected by globalization, carried by companies with strong competitive advantages, exist in many territories, including those outside of large cities (Bouba-Olga 2017).[6]

[6]We have shown the limits of the "GDP per capita" indicator to judge the relative performance of French regions (Bouba-Olga and Grossetti 2015). Larceneux (2018) takes up and develops the

Apart from these elements, the very notion of competitiveness or competition between territories raises serious questions. Territories are a partition of physical space, defined based on administrative (regions, departments, etc.) or socio-economic (employment zones, urban areas, etc.) criterion, making it possible to carry out an organized collection of data and to say things about the geography. To forget this leads to caricatural reifications, which lead some people to speak of Toulouse, Paris, Bordeaux etc. as if they were people with specific objectives that are competing with each other. In doing so, we overlook the importance of what connects the territories: the links between a parent company located in one area and one of its institutions located in another, between a customer and its subcontractor, between a scientific laboratory and its industrial partner... and everything that this implies in terms of the movement of resources (people, goods and information). What would BTP Colas construction group be, whose head office is in Boulogne, in the Hauts-de-Seine department, without the thousands of work projects carried out throughout France? How can we think of the Renault Technocentre located in region Ile-de-France, with its high wages, without considering the production plants located in the provinces or abroad? How could the Safran group offer Airbus and Boeing the new generation of LEAP aircraft engines without the engine blades manufactured by Mécafi, located in the Châtelleraudais basin?

There are many more examples: the contemporary economy is characterized by an increasing fragmentation of the productive processes, which are deployed on a large scale, sometimes on a global scale, to manufacture *Made in Monde* products (Berger 2006). A lot of French territories participate in these processes: they divide up the work, specialize in what they know best and let others do what they (whether they are near or far away) know how to do better. The main focus should therefore be to identify these complementary productive and territorial areas and to look at how to prepare them for the transformations that occur and how to support them, rather than to pit territories against each other.

8.3.2 Attractiveness

In the CAGE mythology, the metropolitan advantage results mainly from the ability of densely-populated territories to attract talented, creative, and innovative people... who find within them all of the elements to which they aspire. Theorized by Richard Florida (2002), the sequence has the following propositions:

---

argument, explaining in particular the limits of the regional comparison between GDP per capita and RDB per capita, revived by Laurent Davezies, for example in his study for the metropolis of Lyon (Davezies 2016). Larceneux points out that all wealth created (GDP) is made up of household income (GDI, Gross Disposable Income) as well as business income (EBITDA). The gap between the GDP per capita and the GDI per capita in Paris and the provinces, or between the metropolis of Lyon and its surrounding territories, is largely due to this distribution.

(i) Our societies have a "creative class" (professions related to cutting-edge technology, entertainment, journalism, finance, arts and crafts etc.) who, in today's economy, are the people that develop new ideas, technologies and creative content,

(ii) this "creative class" is sensitive to the amenities, to the spirit of openness and tolerance of the inhabitants, to the social mix, to the cultural diversity … to the set of elements that Florida qualifies as *soft factors*,

(iii) cities that want to develop need to innovate and therefore to attract these creative people on which innovation is based, so they need to offer all of the *soft factors* that they demand.

North American cities were the first to take on Florida's ideas, starting with Toronto, whose university recruited the author of the creative class theory for a sky-high fee. Another exemplary case is the city of Milwaukee, a city marked by its industrial past. The redesign of the city's image and the launch of ambitious redevelopment policies in the 2000s were explicitly designed to attract creative classes, after consultation with Richard Florida. The results, measured on the scale of the agglomeration, are non-existent (there are no more "creatives" amongst the general population than there were 15 years earlier), whilst the targeted investments concentrated in the centre were carried out to the detriment of facilities for the population as a whole in the city's other neighbourhoods (Zimmerman 2008). Despite this, European cities, including many French "metropolises", are moving in the same direction.

Unfortunately, the discourses on attractiveness do not stand up to the test of reality any more than those about global cities. Geographical mobility is weaker than people imagine, it has more to do with metropolitan macro-regional realities and above all, the determinants of mobility are not those advanced by Florida. Regarding the degree of mobility of people, in the case of France, examination of the 2013 census data shows us that 88% of people lived in the same dwelling as they did the previous year; of those that moved, 36% remained in the same municipality, 35% in the same department and 10% in the same region; only 2% have therefore moved to a different region, due more to metropolitan than macro-regional realities: the Atlantic and Mediterranean coastal areas, the centre and the southern parts of the country "attract", whilst the large North-Eastern quarter "repel" (Bouba-Olga 2017). Based on a different data set and geographical area, Rieutort (2017) shows, furthermore, that the population growth of sparsely populated and very sparsely populated areas over the period 1999–2013 is even greater than that of densely populated and very densely populated areas, which is the product of a lower natural balance extensively offset by net migration that is well above average. This low geographical mobility is also found when studying the subset of entrepreneurs, including start-ups: all studies converge towards the same result—that business creators start their businesses where they live (Dahl and Sorenson 2009; Reix 2008; Grossetti et al. 2016). The mobility of productive establishments is also rather weak, if not locally, with regard to moving from one area of activity to another, much less changing departments or regions (Delisle and Lainé 1998).

What are the determinants of people's mobility? To answer this question, Martin-Brelot et al. (2010) interviewed more than 2000 so-called "creative" people in 11 European cities (Toulouse, Amsterdam, Barcelona, Milan etc.).

First result: more than half of the respondents (53.3%) were born in the area where they live or in their immediate vicinity. Creative people are not therefore "hyper-mobile" individuals. There are, however, fairly strong differences between agglomerations, ranging from 31% for Dublin to 76% for Barcelona. Second result: 63.3% of "creative" people live in the agglomeration where they finished their studies, with the percentages ranging from 47% for Toulouse to 91% for Poznan. Third result: when we directly ask "creative" people about the factors that made them choose to live in the city considered, 55% of their answers refer to personal issues ("I was born here", "I have family here", "I have close friends here", "I finished my studies here"), 36% of the answers are related to hard-factors ("I moved here because of my job"), "I moved here because of my spouse's job", "I moved here because of good job opportunities", etc.), with only 9% of the answers corresponding to Florida's *soft factors*. Scott (2010) reached the same conclusions by examining data on the mobility of American engineers over the period 1994–1999: the destinations chosen by engineers are mostly chosen due to local employment opportunities, with amenities playing no major role. Sternberg (2012) dismantles the illusion of local policies intended to establish creative industries.

In fact, members of the "creative class" are just like the rest of the population: they have a history, a family, networks and job opportunities that strongly constrain their spatial choices. Normally, they do not choose a city: they stay in or return to one in which they have already lived, or they accept a better job in an acceptable city.

8.3.3 Globalization

Globalization is a complex phenomenon, which we approach here only with regard to the so-called "global" cities (the "metropolises" of French discourses) in which the deepening of the process of globalization leads to the concentration of most of the creation of wealth and jobs. In France, the very notion of a "metropolis" poses a problem, however, because there is no consensus on its definition and it cannot be associated with a clear statistical category. In the case of France, some analyse data by area of employment (empirical analyses of the geographical economy in particular, as in Martin et al. 2011), whilst others analyse data by region (comparisons of GDP per capita in Davezies and Pech 2014),[7] others (the most numerous) by urban areas but, in the latter case, the urban areas that are considered metropolises are not always the same: Guilluy (2014) uses the first 25, Davezies and Pech (2014) focus on the 5 they consider most dynamic, Askenazy and Martin (2015) provide statistics on 14 of them, France Strategy uses 15 in 2016 and then 12 in 2017 .... This uncertainty is a major problem in any debate because for some people in France "metropolis" means Paris, whilst for others it refers to about a dozen or half a dozen cities, for

[7] The twelve "metropolises" are urban areas with more than 500,000 inhabitants, Paris is in another class of its own, then there are urban areas of 200,000–500,000 inhabitants, 100,000–200,000 inhabitants, etc.

others, metropolisation refers to urban areas and some people have in mind the 22 metropolises established by a recent law.

Apart from this uncertainty, the assertion that "metropolises" enjoy higher growth than other territories is not supported by the facts, when this phenomena is measured rigorously. The affirmation of France Strategy (2017a) that "*the beginning of the 21st century is marked by a movement of concentration of employment in a dozen French cities*" is based on the comparison of the **average** annual growth rate of the 12 largest urban areas with other sets of urban areas ranked by size.[8] Indeed, **on average** the "metropolises" had a higher growth rate of +0.4% per year between 2006 and 2013 compared to −0.2% per year in France as a whole, but this average masks significant disparities: Bordeaux, Montpellier, Nantes, Rennes and Toulouse have been more dynamic than the national average since the early 2000s; the growth of Grenoble, Lille, Lyon and Marseille has been in line with the average; Nice, Rouen and Strasbourg have had below average growth.

It should be noted in passing that it is the same type of questionable simplification that has led many commentators to see in the results of the presidential election a confrontation between the France of big cities that is nomadic, globalized and enterprising where Macron won the vote and a peripheral France that is withdrawn, xenophobic and fragile and that takes refuge in voting for Le Pen, judging the whole of a municipality based on the candidate that came first. A more precise analysis shows that the results in the first round of the two finalists were remarkably stable in all types of territory: Macron won 24% of the vote in large urban areas, 22% in small and medium-sized towns and 21% in rural areas; Le Pen won 21%, 23% and 25% respectively (Gilli 2017).

When the data available is analysed more carefully, avoiding this bias of categorization, to the question "do the metropolises perform better than the other territories?", the systematic answer that we get is "it depends": some "metropolises" are dynamic, whilst others are not; some non-metropolitan territories are dynamic (some, such as Figeac, Vitré, Issoire, Vire, etc. are more dynamic than the most dynamic metropolitan areas), whilst others are not. This is the result obtained when analysing the data on total employment by area of employment (Bouba-Olga and Grossetti 2015). The result is the same for the subset of private employment excluding agriculture by urban area (Bouba-Olga et al. 2016) and it is also the same, for other datasets or other periods, as that obtained by other authors such as Shearmur and Terral (2013), Baude (2015), Poupard (2015), Rieutort (2017). France Stratégie acknowledges this beautifully in its last note by writing at the bottom of page 4 "*Overall, the metropolitan dynamic is obvious*", only to correct itself at the top of the next page by stating "*However, taken individually, these large cities have each had different fortunes*" (*sic*, Stratégie 2017b). In short, not so obvious at all.[9]

[8] Frick and Rodrigues-Pose (2017) also showed, by analysing data on 113 countries over the period 1980–2010, that the link between geographic concentration and economic growth is not obvious, but rather that the quality of specialization, infrastructure and modes of governance are particularly decisive.

[9] Source: http://www.agence-nationale-recherche.fr/investissementsdavenir/documents/ANR-AAP-IDEX-2010.pdf.

The geographical economy proposes other types of processing (Martin et al. 2011; Combes and Gobillon 2014; Combes et al. 2015; etc.), but they also present serious defects and this makes the interpretation of the results obtained questionable. They are based on a hypothesis at the heart of the neoclassical approaches, according to which the salaries paid are linked to the productivity of the individuals who receive them and therefore to judge differences in the performance of the territories it is sufficient to compare the differences in the salaries paid. However, this hypothesis is highly questionable, especially for certain professions (traders, senior executives of major companies, certain artists, top athletes, etc.) whose salaries are disconnected from their productivity which in any case is hard to quantify (Piketty 2013), these professions being very highly concentrated in Ile-de-France. The analysis of geographic differences in salaries is therefore less an analysis of differences in the economic performance of the territories than an analysis of income inequalities. Apart from this problem, the analysis of the geography of salaries that we have used based on a more detailed breakdown of professions than that used by the authors mentioned above shows the weakness of geographical effects: outside of Ile-de-France, when the job is the same the differences in salary are very small. The specific case of Ile-de-France (Paris and Hauts-de-Seine, in fact, for the most part) is partly explained by the concentration of very high wages, partly due to specializations inherited over time and partly due to a cost of living that is about 10% higher than the rest of France (see Bouba-Olga et al. (2018) for developments).

8.3.4 *Excellency*

The notion of excellence, another important component of CAGE, permeates policy discourse, especially with regard to research activity, which is essential for enhancing the capacity for innovation of companies and countries. This obsession with excellence has been considerably strengthened with the publication of the Shanghai ranking, in which French universities occupy a poor position. This ranking, which is rightly widely criticized (Gingras 2009), still inspires policies designed to improve the position of French universities.

Of course, we agree with the idea that we should aim for excellence in this and other fields. What remains to be seen is what is meant by this term. In the minds of many politicians and certain researchers, excellence cannot be separated from the notion of concentration: it is necessary to focus efforts on just a few sites and the most talented people within them in order to avoid, *Horresco Referens*, "spreading ourselves too thinly". The General Commission for Investments (GCI) therefore recommended in 2010 to "*establish 5–10 multidisciplinary centres of excellence in higher education and world-class research in France*".[10] Achieving a sufficient

[10]This does not mean that the number of citations of articles published by researchers located in Ile-de-France has decreased: it has actually increased, but it has increased more slowly than that of researchers located in other regions. This is not to say that they are any more "productive",

density of researchers in an institution, city, or region would enable the quality of the research to be good, as researchers are expected to need many colleagues nearby to exchange ideas and be stimulated in their work.

The problem, again, is that the empirical work available does not validate this hypothesis. Some attempts have been made to link the number of researchers in the same city or region to the average number of articles published per researcher, but such a relationship could not be proven (Bonnacorsi and Daraio 2005). On an aggregated scale, the number of publications in a given city or region is usually almost a linear function of the number of researchers, resulting from developments in higher education and national or local policies. Moreover, the analysis of the evolution of the geography of research on a global scale shows that in almost all countries we are not witnessing a concentration, but rather a deconcentration of scientific production as well as the quality and visibility of the latter, measured by the number of citations (Maisonobe et al. 2017). On the French side, this deconcentration translates into a reduction of the Parisian advantage, with its national share of citations falling from 41.7 to 37% between 2000 and 2010.[11]

The idea that it is appropriate to simultaneously reward the most talented researches within these "agglomerates" is based on an empirical regularity, which is generally called "Lotka's law" (Lotka 1926). If 20% of the researchers are responsible for 80% of publications, why not focus solely on this 20%? This type of advocacy is part of a classic phenomenon of the social world, which is the tendency to accumulate advantages, known as "the Matthew effect", analysed by the sociologist Robert Merton (1968). This type of analysis obscures the fact that the most cited researchers are just the most visible part of a huge collective effort by the entire scientific community.

Unlike the supporters of the CAGE mythology, we therefore consider that excellence is not the monopoly of an elite group of researchers that is better known than their colleagues, but can characterize activities that everyone can strive to achieve. This quality is not measured by the number of citations obtained (which is only and approximately an indicator of visibility), but by the capacity of the results produced to stand the test of time and scientific debate. Therefore, it is essential to support a broad set of laboratories, just as it is important to support a broad set of innovative initiatives, not just a few start-ups incubated in the meeting places of larger cities.

This type of positioning, which is obviously less exciting, often runs up against a misunderstanding: doesn't supporting a wide range of actors risk spreading resources too thinly? Should we not, once again, focus on the "best"? This kind of reaction would be understandable if it was certain that the best of yesterday will also be the best of tomorrow. But one of the distinguishing characteristics of public and private research and innovation in general is radical uncertainty, which makes it impossible to identify the "elite" of tomorrow. To advocate support for a broad set of actors,

---

"performing" or "talented": the number of positions has simply increased more quickly than in Ile-de-France (where it has also increased, but more slowly).

[11]The CAGE mythology in fact permeates the speeches of the leaders of many territories, which on the one hand compare themselves to other territories that they consider competitors and on the other hand tend to consider themselves as "more successful", "more attractive" and "more dynamic" than the smaller territories that surround them.

based essentially on the quality of future projects rather than the reward of past successes, does not correspond to a strategy of "spreading resources too thinly" but rather a "watering the flowers" strategy: we cannot know in advance where the best crop (scientific or economic) of tomorrow will blossom. By watering just one place, we could deprive ourselves of seeing tomorrow's excellence flourish.

8.3.5 The Effects of the CAGE Mythology in Big Cities

The CAGE mythology seems to us just to be as harmful to the inhabitants and the activities of the big cities as to those of the rest of the country.[12] It ends up causing visions in which the reassuring discourse on competitiveness prevents us from sufficiently taking into account socio-spatial inequalities, congestion of transport networks and pollution problems, which can be perceived in this context as the price that has to be paid for performance. CAGE is particularly damaging to innovation if it prevents us from seeing that it is not a mechanical effect of size but rather it is produced by people, many of whom (young researchers or engineers, for example) have incomes that make it difficult for them to access housing, especially when they start having children, that often have to deal with long travel times or cramped housing. This is also true of the companies themselves, which are forced to move away from the centres that are made expensive due to real estate speculation as soon as they need more space for their activities. Some of them, in the fields of communication and culture, for example, can take advantage of the masses of trainees or young, sometimes precarious employees looking for their first experiences, who adapt well to the constraints of life in very large agglomerations and find the company of a generational social enclave within them due to their size. But other companies, such as those in the field of engineering design, which need experienced staff, may find the conditions in "metropolises" such as Paris less favourable than elsewhere.

Whilst the concentration of population can promote social enclaves, social fragmentation and socio-spatial segregation, it also has advantages for couples in which each member is looking for a specific job related to his or her specialities, fans of rare cultural practices and people with unusual alternative lifestyles. The concentration of population makes it easier to find people with similar practices or lifestyles. Experimentation with different lifestyles and cultural practices is not synonymous with economic success and can be respected and valued without resorting to economic justifications. Getting away from a strictly economic justification makes it possible to better perceive the intrinsic issues of these lifestyles and experiments.

[12] In a study for the metropolis of Lyon, the economist Davezies (2016) insists on the "attractiveness" of Lyon, which is giving life to territories that are further and further away, citing the rise in the number of commuters. He forgets to mention that this "attractiveness" is partly the result of relegation: a student couple, to start a family, has to move out of the city to access the property market. People do not always come from far away because they are attracted. It is also, sometimes (often even) because they are repulsed.

8.4 How to Escape from the CAGE?

Overall, the mythology CAGE suffers from several problems. Problems related to the use of fuzzy categories, such as “metropolis”, “creative” or “excellence”; problems with the indicators chosen, which are wrongly assumed to be indicators of success (whether GDP per capita or wages paid); problems related to the methods used (comparisons of resources between heterogeneous sets of territorial entities, transformation of continuous data into binary values as in the case of comments on the results of the elections); problems related to the formulation of unverified hypotheses (the hypothesis of the hyper-mobility of creative people attracted by *soft factors*, more generally an overestimation of sustainable mobility, the hypothesis of a “critical mass” effect in terms of research and that of the increasing geographical concentration of this activity), etc.

In order to analyse the economic dynamics of territories in a more satisfactory way it is necessary, apart from the problems mentioned above, to avoid three biases that are found in the various components of the mythology.

Firstly, the reification bias. As we have already said: territories are not organizations with a specific objective that are fighting against each other, they are large or small portions of a geographical area, defined according to different criteria, which are affected by social and economic processes that go beyond them. The simple fact of creating rankings of the territories based on different indicators and setting the objective of having higher growth in population or employment than that of a territory which is imagined to be a competitor is part of this bias.

Secondly, the bias of presentism, which consists in forgetting about history: rather, we advocate dissociating the analysis of long-term processes, which makes it possible to understand the constitution of local economic systems, from a shorter-term analysis which seeks to explain, with a globally stable geographical structure, the flows of exchange and the relationships that structure the territories within a given period. As far as long-term processes are concerned, it can be shown in this way that the current situation of the Toulouse agglomeration is the result of more than a century of policies, initiatives and contingent events resulting in an economic system that is fairly similar to that of Grenoble, whose history is very different (Grossetti and Zuliani 2013). Another example, the specialization of the Pau region in aeronautics, results from the desire of the French State in the 1940s to distance this industry from the German borders, but also due it being the birthplace of René Lucien (in the canton of Oloron-Sainte Marie), who was then the head of the company Messier, as well as the love for mountains of Joseph Szydlowski, the boss of Turbomeca (Ferru 2018). For shorter-term analyses, it is necessary to change the temporal focus: to analyse the links between actors, to grasp their origin, to understand the realities that affect them, to question the importance of the effect of the border, as well as gaps between socio-economic areas and areas of intervention, etc.

The bias of the closure of the analytical framework, which consists in focusing on a relatively narrow subset of socio-economic processes, which is presented as a closed and formalised system of actors (the territories) that are in competition with each other for access to goods, forgetting the other processes, which are sometimes more decisive, that do not fit into the model. In the CAGE mythology, the focus is entirely on the ability of large cities to attract talent, with the phenomena of spatial segregation being largely overlooked. Askenazy and Martin (2015) have also acknowledged this, explaining that "*the issues of urban segregation (…) are left for later*" *sic*, page 2). However, "metropolises", which are generally considered the most "attractive" geographical areas, can also be "repulsive": the rise in the price of land, in particular, relegates the people with the weakest resources to certain neighbourhoods, whilst it also tends to increase the time taken for people to travel to work.[13]

When we avoid all of the pitfalls identified, it appears possible to describe the current developments in another way and to insist on the issues facing the territories.

The evolution of demand, technical change and the deepening of globalization have led to a profound transformation of the structure of economic activities, with presential activities becoming more and more significant whilst production activities have declined. Logically, from a demographic point of view the most dynamic territories such as the Atlantic and Mediterranean coasts or the South-East of France have taken advantage of this evolution: they attract more population than other regions, which has led to a more sustained development of employment. This "attractiveness" is not without its downsides: it often leads to problems of congestion, pollution and increases in the price of land, which can lead to gentrification processes and/or conflict.

Activity exposed to globalization, however, remains decisive for the creation of wealth and jobs. Contrary to the claims of the CAGE mythology, this activity is not limited to large cities: it is spread over different territories, which over time have been able to accumulate distinctive skills and resources and adapt to the threats and opportunities of globalization, technological change and changing demand.

One of the problems is that the CAGE mythology is very common in the different political currents present in France. The main candidates in the last presidential election either benefited from the advice of some of the authors mentioned or referred to their works or other similar ones. They consider it a more or less adequate description of reality, whether they like, reject it or are trying to transform it. On a political level, our criticism cannot be reduced to a commitment to any one of these currents. Nor is our criticism a defence of an "anti-urban" discourse that would enhance rural spaces: we would be just as critical if the dominant discourse considered medium-sized cities or rural areas vital to the economic future of France. We only dispute a particular reading of the urban reality, a reading of which we have tried to identify the components. Leaving the CAGE can help us to imagine other types of analyses and other types of policies.

---

[13]Silicon Valley, taken as a model for economic development by many, was originally an industrial area located outside of the city of San Francisco. Vicente (2016) recalls that its growth has caused several problems since the mid-2010s: growing gentrification of San Francisco, several thousand homeless people in "The Jungle" of San José, protests by the bus drivers chartered by the Google, Yahoo!, Apple and Facebook, demanding increased salaries, etc.

The challenge, in terms of public action, is, in our opinion, to leave the cult of excellence, attractiveness and "global" cities, which mostly just leads to wasting public money. It would be far better to use the money put in the CAGE to provide territories with generic resources so that the most innovative projects (which by definition cannot be predicted) can develop within them, without assuming that these innovations are reserved for certain types of territories, sectors or people. Compared to other countries, France benefits from a good level of equipment that is well distributed geographically, whether it is for health, training or communications. However, there is some disintegration of this equipment, probably partly because of the diffusion of CAGE. To continue to maintain this equipment and to consider ways to innovate in this area in order to improve it seems essential: not to "do social work" but to support the creation of wealth and jobs.

The other issue, in the sub-domain of economic development policies, is to question the investment needs in terms of specific resources. In this regard, we recommend starting with the productive processes that go through them rather than with the territories, in order to question the transformations resulting from digital development, the energy transition, the growing connection between industries and services, the characteristics of a hyper-industrial society (Veltz 2017), in order to identify what needs to be done, depending on the territorial contexts, which are often very specific.

8.5 Conclusion

In this essay, we have systematically and clearly presented what we perceive as myths which reinforce each other in analyses and discourses about the spatial aspects of economic activity. Of course, it is not a perfectly coherent system identified in the same way by all of the authors. Some are interested in some of the ingredients that we have identified and not in others. Their positions are not identical and present nuances that we have not described in detail due to our focus on revealing the overall structure of the mythology. Similarly, we have focused on the case of France and no doubt there are variations in the discourse depending on the country, but it seems to us that France is not the only country that is locked in the CAGE.

The links that we have tried to highlight between these ingredients are not always systematic and have their own variations. Therefore, let's be clear, confinement in the CAGE is not some kind of conspiracy! It is a diffuse network of beliefs that are more or less shared (and often admitted without too much reflexive examination) which works in our view as a system when we place ourselves at the level of the small group of people (researchers, consultants, technocrats, elected representatives) who are concerned with local economic development, spatial planning, regional policy or the spatial dimension of socio-economic activities.

Regarding each separate myth, many authors (some of whom have been quoted) have, like us, challenged the arguments that are always presented to support them (measurements of local productivity, job creation or mobility) and presented robust and cumulative empirical results that contradict these myths. But this criticism is rarely listened to because of the systemic character of the mythology: yes, this aspect is questionable, but if it is discarded then we would have to abandon other aspects that appear more robust. In our view, the systemic aspect is very important in explaining the resilience of these beliefs. Even if we are not at the level of coherence of a "Kuhnian paradigm shift", we can probably transpose some of his analyses, such as the fact that the accumulation of contradictory data (a "refutation" as defined by Popper) is not enough to make people abandon the belief system. In particular, according to Kuhn's analyses, a system can only be endangered if a competing system emerges.

In the last part of this text, we have tried to consider what could constitute an alternative system for the interpretation of the same phenomena that does not fall into the same pitfalls as the CAGE mythology. But in this regard, interpretation alone is not enough. These beliefs are associated with public policies for which the CAGE mythology provides a coherent framework. To deconstruct and imagine another framework for analysis as we have tried to do is not sufficient if it is not accompanied by alternative courses of action. It is obvious that this is, for everyone working in this field, a long-term undertaking that will require the expertise of many economists, geographers, historians and sociologists. We hope that this text can contribute to the emergence of a multidisciplinary critical community concerned with empirical rigour, that is capable of proposing the new and robust approaches that today's world needs so that it can escape from the CAGE and the circular reasoning that it induces.

Acknowledgements We would like to thank Antoine Belgodère, Bastien Bernela, Emilie Bourdu, Pascal Chauchefoin, Richard Shearmur, Josselin Tallec and Jérôme Vicente for their comments on the first version of this text.

References

Askenazy, P., Martin, P.: «Promouvoir l'égalité des chances à travers le territoire», note du CAE n 20 (2015)

Baldwin, R.: The Great Convergence: Information Technology and the New Globalization. Belknap Press, Cambridge (2017)

Barthes, R.: Mythologies, Paris (1957)

Baude, J.: «Les métropoles créent-elles plus d'emploi?». Agence de Développement et d'Urbanisme de l'Aire Urbaine Nancéenne, octobre (2015)

Berger, S.: Made in monde: les nouvelles frontières de l'économie mondiale. Seuil, Paris (2006)

Bonnacorsi, A., Daraio, C.: Exploring size and agglomeration effects on public research productivity. Scientometrics **63**(1), 87–120 (2005)

Bouba-Olga, O.: Dynamiques territoriales: éloge de la diversité. Editions Atlantique (2017)

Bouba-Olga, O., Grossetti, M.: La métropolisation, horizon indépassable de la croissance économique? Revue de l'OFCE, no **143**, 117–144 (2015)

Bouba-Olga, O., Grossetti, M., Tudoux, B.: Les inégalités spatiales de salaire en France: différences de productivité ou géographie des métiers? (2018). <hal-01679747>

Bouba-Olga, O., Grossetti, M., Tudoux, B.: On voit des métropoles partout, sauf dans les statistiques (2016). <hal-01276897>
Charmes, E.: «Une France contre l'autre?». La vie des idées, (2014). http://www.laviedesidees.fr/Une-France-contre-l-autre.html
Combes, P.-P., Duranton, G., Gobillon, L.: Salaires et salariés en Ile-de-France. Revue Economique **66**(2), 317–350 (2015)
Combes, P.-P., Gobillon, L.: «The Empirics of Agglomeration Economies». PSE Working Papers n°2014-31 (2014). <halshs-01071761>
Dahl, M.S., Sorenson, O.: The embedded entrepreneur. Euro. Manage. Rev. **6**(3), 172–181 (2009)
Davezies, L., Pech, T.: La nouvelle question territoriale. Terra Nova, septembre (2014)
Lainé, Delisle M.: Les transferts d'établissements contribuent au desserrement urbain. Economie et Statistique **311**, 91–106 (1998)
Ferru, M.: Trajectoire de développement d'un territoire non-métropolitain: l'apport des méthodes mixtes pour l'analyse du bassin palois. In: Pecqueur, B., Nadou F. (dir) Mutations économiques et dynamiques territoriales. Transitions, intermédiations, innovations, l'Harmattan, chapitre n°5, à paraître (2018)
Florida, R.: The Rise of the Creative Class and How It's Transforming Work, Leisure, Community and Everyday Life. Basic Books, New York (2002)
Stratégie, F.: «Dynamique de l'emploi et des métiers: quelle fracture territoriale?», no 53, février (2017a)
Stratégie, F.: «Dynamique de l'emploi dans les métropoles et les territoires avoisinants», n°64, novembre (2017b)
Frick, S., Rodriguez-Pose, A.: Big or small cities? On city size and economic growth. CEPR Discussion Papers 12324, C.E.P.R. Discussion Papers (2017)
Geoffard, P.-Y.: «Pas de mondialisation sans redistribution», Libération, 14 Nov 2016
Gilli, F.: «Deux France se feraient face? C'est un peu rapide!». Le Monde, 27 Apr 2017
Gingras, Y.: Le classement de Shanghai n'est pas scientifique. La Recherche **430**, 46 (2009)
Grossetti, M., Barthe, J.F., Chauvac, N.: Innovateurs ordinaires: Une enquête sociologique sur des startups françaises, texte provisoire d'un ouvrage (2016). <halshs-01377992>
Grossetti, M., Zuliani, J.-M.: «La construction d'un système industriel de haute technologie à Toulouse entre logiques locales et logiques nationales». In: Fridenson, P., Griset, P. (dir.) Entreprises de hautes technologies. État et souveraineté depuis 1945, Paris, Comité pour l'histoire économique et financière de la France, pp. 267–281 (2013)
Guilluy, C.: Le Crépuscule de la France d'en haut. Flammarion, Paris (2016)
Kalantzis Y. et Thubin C.: «Les causes de la désindustrialisation en France». Banque de France, Bloc-Notes Eco, 16 Nov 2017
Larceneux, A.: Géographie de la production et des revenus? Remarques critiques sur les analyses de Laurent Davezies, Revue d'Economie Régionale et Urbaine **4**, 737–761 (2018)
Levy, J.: Les riches ont voté Trump, les villes Clinton. Le Monde, 16 Nov 2016
Lotka, Alfred J.: The frequency distribution of scientific productivity. J. Wash. Acad. Sci. **16**(12), 317–324 (1926)
Maisonobe, M., Grossetti, M., Milard, B., Jégou, L., Eckert, D.: The global geography of scientific visibility: a deconcentration process (1999–2011). Scientometrics **113**, 479–493 (2017). https://doi.org/10.1007/s11192-017-2463-2
Martin, P., Mayer, T., Mayneris, F.: Spatial concentration and plant-level productivity in France. J. Urban Econ. **69**(2), 182–195 (2011)
Martin-Brelot, H., Grossetti, M., Eckert, D., Gritsai, O., Kovács, Z.: The spatial mobility of the 'Creative Class': A European perspective. Int. J. Urban Reg. Res. **34**(4), 854–870 (2010)
Merton, R.: The Matthew effect in science. Science **159**, 56–63 (1968)
Piketty, T.: Le Capitalisme au XXIème siècle. Seuil, Paris (2013)
Poupard, G.: Développement local et emploi productif: un monopole des métropoles? Popul. Avenir 2015/5, n° 725, pp. 4–8 (2015)

Reix, F.: L'ancrage territorial des créateurs d'entreprises aquitains: entre encastrement relationnel et attachement symbolique. Géographie, économie, société **10**(1), 29–41 (2008)
Rieutort, L.: La ruralité en France: des potentiels oubliés? Popul. Avenir. 2017/1 (n° 731), pp. 4–7 (2017)
Sassen, S.: The Global City: New York, London, Tokyo. Princeton University Press, Princeton (1991)
Scott, A.J.: Jobs or amenities? Destination choices of migrant engineers in the USA. Pap. Reg. Sci. Wiley Blackwell **89**(1), 43–63 (2010)
Shearmur, R., Terral, L., Polese, M.: La géographie de la croissance d'emploi en France à l'aune de processus nord-américains: vers une théorisation du contexte. Cybergeo: Revue Européenne de Géographie (2013). http://cybergeo.revues.org/25703, https://doi.org/10.4000/cybergeo.25703
Sternberg, R.: Learning from the Past? Why 'Creative Industries' Can hardly be created by local/regional government policies. Die Erde; Zeitschrift der Gesellschaft für Erdkunde zu Berlin **143**(4), 293–317 (2012)
Veltz, P.: La société hyper-industrielle: le nouveau capitalisme productif. La République des Idées, Paris (2017)
Vicente, J.: Economie des clusters. Paris, La Découverte, collection «Repères» (2016)
Zimmerman, J.: From brew town to cool town: Neoliberalism and the creative city development strategy in Milwaukee. Cities **25**, 230–242 (2008)

Chapter 9
On the Survival of Butterflies in the Jungle of Urban Systems

Lena Sanders, Isabelle Thomas and Céline Vacchiani-Marcuzzo

Abstract As suggested by the title, this contribution is inspired by Gilles Duranton's and Andrés Rodriguez-Pose's paper of 2005 entitled "When economists and geographers collide, or the tale of the lions and the butterflies". Comparing economical (the lions') and geographical (the butterflies') scientific approaches, they pointed out their radically different ways of proceeding and the lack of cross-fertilization between these two disciplines. Looking at the geography of cities, our aim is to continue the discussion through pointing out some differences in points of view. First we discuss what seems to be a "necessity" for lions, that of rooting the approach in methodological individualism for understanding the dynamics of a system of cities. Indeed, this seems to be the mainstream position in economy when butterflies explore different ways of flying from one scale to another. We build on philosophers' tools in order to explore more deeply the meaning of methodological individualism and the relevance of different kinds of "objects" when modelling cities' dynamics. We then focus on the question of delineation of spatial objects like cities, building on an empirical approach to discuss how proper scale choice associated to a multiscale approach favors comparison. At last, we expose the feedback of an interdisciplinary experience where the city delineation played a crucial role when developing a LUTI model. As a conclusion, we are convinced that more opening and interactions between butterflies, lions (and tigers!), will help the scientific ecosystem to be more resilient.

L. Sanders · C. Vacchiani-Marcuzzo (✉)
CNRS, UMR Géographie-cités, Paris, France
e-mail: vacchiani@parisgeo.cnrs.fr

L. Sanders
e-mail: lena.sanders@parisgeo.cnrs.fr

I. Thomas
Université Catholique de Louvain, Louvain-la-Neuve, Belgium
e-mail: Isabelle.Thomas@uclouvain.be

C. Vacchiani-Marcuzzo
Université de Reims Champagne-Ardenne, UMR Géographie-cités, Reims, France

D. Pumain (ed.), *Theories and Models of Urbanization*, Lecture Notes in Morphogenesis, https://doi.org/10.1007/978-3-030-36656-8_9

9.1 Introduction

In 2005 Gilles Duranton and Andrés Rodriguez-Pose signed a paper entitled "When economists and geographers collide, or the tale of the lions and the butterflies". Let's briefly recall the demonstration of the two authors, one being a geographical economist, the other an economic geographer. Through analyses of cross citations and behaviors at conferences, they conclude to a relation of "mutual ignorance, rather than outright discord" between the researchers of these two communities. Looking at their scientific approaches, they underline two radically different ways of proceeding: the lions "improving and expanding their tricks", laboring "the same core questions over and over again", and the butterflies "freely flying the fields of knowledge with the aim of tasting the best from every flower they visit". The two authors note that both communities are in fact interested in the same questions (among others, the geography of cities), but that there is very little scientific cross-fertilization.

At the same period (2007), a similar debate took place in the French scientific journal *L'Espace Géographique*: Denise Pumain, Jacques-François Thisse, Isabelle Thomas and Bernard Walliser tried to explore the relationship between New Economic Geography and Geography. Jacques-François Thisse admitted at this occasion that "En ce qui concerne les échelles spatiales, il est exact que la plupart des économistes n'ont pas compris grand-chose à cette question[1]".

One specific point in this debate deserves to be focused on, that of spatial scales that is maybe a major point of discrepancy between lions and butterflies. Indeed, initially (and very broadly said), economists (the lions) were myopic, considering a city as a point on a line (1D), and hence over simplifying urban realities for the sake of rigorous economic theory. At the same time, geographers (the butterflies), more aware of the crucial role of scale, were analyzing multiple urban spatial dimensions (2D and more) gathering nectar from different theoretical and empirical fields. Even if both were dealing with urban matters, they were busy with different aspects.

What is the situation today, a bit more than a decade after these early debates? Steinberg (2015) recently compared publications and citations of German-speaking economic geographers and geographical economists. He observed more cross-fertilizations than expected, especially because young scholars are doing much better. Things are then improving even if joint publications remain rare and geographers are still seldom cited by economists (while the opposite is less true: geographers cite economists). The same holds for non-German researchers. If in the past 20 years, few economists and geographers have successfully worked together and shared some topics, such experiences remain unfortunately to be pinpointed and gaps between both disciplines persist with the advantage of a strong and well recognized (economic) theory for the firsts, and a good knowledge of space, its biases and the inherent quantitative methods for the others. As a consequence, economists have long looked at empirical analyses with contempt. Surely, if these two disciplines could more often listen to each other, research objectives would be for sure more challenging and

[1]"As far as spatial scales are concerned, it is true that most economists have not understood much about this question".

results scientifically more exciting. It is important to insist on that as nowadays, the empirical context has changed. With the emergence of massive geolocalised data, new actors have appeared, coming from a variety of other disciplines than economy and geography (physicists, engineers, computer and data scientists), bringing new methods to analyze the structure of urban space. They are new predators in the jungle. Is it time for a new tale? Or time for the butterflies to be listened to by lions and these new panthers!

This metaphor was adopted at the ERC GeoDiverCity colloquium (2017) to discuss Michael Storper's contribution entitled "Urban systems: the geography of income and population". Michael Storper is, indeed, an example of the a priori impossible hybridity between lions and butterflies. His position at the interface of the economic and geographic worlds is extremely fruitful for studying cities' dynamics and makes him a particularly interesting researcher to discuss with. Referring to the question of why urban systems exist and why cities' sizes are organized in a hierarchical way, Storper discussed in his presentation previous models (Simon 1955; Gabaix 1999; Henderson 1974) and argued for a better account of income in urban models. In the discussion that followed our own presentation, Storper underlined his position as a methodologic individualist, the only tenable position in social sciences according to him and he criticized geographers for not having a theory on individuals' spatial behavior. According to him, human agents have to be considered in an approach going from micro to macro level, otherwise there are just "statistical agents".

Our aim is here twofold. On the one hand, we want to discuss this position of the lions rooted in methodological individualism and in particular the "necessity" to adopt such an approach in order to understand the dynamics of a system of cities. Indeed this seems to be the mainstream position in economy while butterflies' explore different ways of flying from one scale to another. On the other hand, we will focus on the question of delineating spatial objects like cities.

The discussion is based on three complementary points. First we build on philosophers' tools in order to explore more deeply the meaning of methodological individualism when modelling cities' dynamics (Sect. 9.2). Then we build on an empirical approach to discuss how proper scale choice associated to a multiscale approach favors comparison (Sect. 9.3). At last, we expose the feedback of an interdisciplinary experience where the city delineation played a crucial role when developing a LUTI model (Sect. 9.4). We admit that the facts are sometimes a bit exaggerated; it is in order to underline them and to open the debate. Indeed, we are convinced that more opening and interactions between butterflies, lions and tigers, will help the ecosystem to be more resilient.

9.2 How Can Philosophers Help Butterflies? An Ontological Perspective

For urban economics the challenge is to provide micro-foundations to the existence and dynamics of urban systems. In particular, the regular rank-size distribution of cities remains a "mystery" in the sense that no satisfactory model based on individuals' behavior, i.e. based on a methodological individualism approach, has yet been developed (Krugman 1996). There is a large literature in that field and we agree with Storper's point: "it is a distraction to focus on whether a pure Zipf exists or not"; "we should be interested in urban systems, but not overly concerned with a $q = 1$ Zipf" (Storper 2017). But, does a theory of systems of cities necessarily imply developing a theory of individuals' behavior (perceptions, intentions, preferences, actions) in the inter-urban space? That is the question we raise, building on an ontological approach. Our aim is here to discuss this position and defend the idea that it is not necessary, neither the most adequate, to formalize hypotheses at the micro-level of individual agents to explain the hierarchical structure and dynamics of a system of cities. We question whether explanatory mechanisms can be modelled at the city level without considering cities as "statistical agents". To this end we first discuss ontological issues about cities: what is their ontological status, in what sense can cities be considered as autonomous collective entities making sense? We then scrutinize their explanatory potential by briefly comparing two families of simulation models, microsimulation based on methodological individualism principles and a multi-agent system based on autonomous "city-agents".

Let's start with two quotations to illustrate the variety of positions according to methodological individualism (MI):

- "For many economists, a denial of methodological individualism simply signals that a person is confused and does not truly understand economics. In contrast, for many sociologists, supporting methodological individualism signals an anti-sociological reductionist attitude that is bound to overlook many crucial aspects of social reality." (Ylikoski 2014)
- "Political science is divided between those who think that a scientific approach to the study of politics requires methodological individualism and those who consider this idea hopelessly reductionistic."... (List and Spiekermann 2013)

Of course it is a bit of a caricature to start the discussion with these two quotes but they have the merit to highlight the extreme variety of points of view on methodological individualism and to show that the debate goes far beyond a discussion between geographers and economists. Evolving in the mainstream of economy, lions are clearly on the methodological individualism (MI) side when, quite naturally in the butterflies' scientific world, both positions co-exist in the same way as mentioned by List and Spiekermann (2013) for the field of political science. Debates are old and dense, as well in philosophy as in social science but these authors underline that: "A reconciliation becomes possible once we see that there is not just one version of each view, but many, and that being an individualist in some respects is compatible with being a holist in others."

Indeed perspectives have evolved in time and much more balanced positions than the simple opposition holism/individualism have emerged. Udehn (2014) for example underlines that he is much less negative to methodological individualism than he was in earlier work (Udehn 2002) due to the spread of weaker definitions of MI. In sociology, Sawyer (2002) defends a "nonreductive individualism" by distinguishing ontological individualism (1) and methodological individualism (2). In the former case, individuals are the only "things" to exist and "sociological objects and properties are nothing, but combinations of individuals and their properties". Examples of such social objects could be as diverse as a football team, a company's board of directors or a country state. Case (2) states that each phenomenon observed at a collective level *can be explained* by individuals' behavior. A "nonreductive individualism" consists in claiming (1) but not (2) (Sawyer 2002). That means that even if solely individuals exist as social objects from an ontological point of view, social laws are not reducible to individual laws (in terms of intentions, aims, perceptions) and collective entities can be used in an explanatory perspective. In the field of political science List and Spiekermann (2013) formulate that kind of question in following terms: "A key point of contention is the status of collective entities such as states, nations, ethnic groups, cultures, political parties, and other institutions. Are these mere by-products of individual behavior, or are they of independent ontological and/or causal significance?"

In order to move forward in the discussion let us examine Sawyer's propositions (1) and (2) in the case of cities. From an ontological perspective, point (1) of Sawyer goes back to discuss if cities are to be considered as simple aggregates of people and firms or as autonomous entities. Reflecting upon the existence of the state and on its ontological status, Hay (2014) underlines that: "the state is an abstraction – a conceptual or theoretical construct-… Its existence is not obvious nor evident, it has no physical or material presence, it is not visible and it is not directly accessible to any of the senses".

On the one hand, unlike the concept of state, that of city corresponds to a referent with a physical dimension, a materiality which is visible for everyone and which persists in time. It is clearly identifiable in the landscape and in that sense it corresponds, in a way, to a "bona fide" object (objects of common sense) as defined by Smith and Varzi (2000). On the other hand, an aggregate of buildings, pavements, and transport infrastructures is insufficient to define a city and to distinguish it from ontologically different objects as villages or towns. In addition, a city cannot be delineated in the same way as an individual, an island or a lake. In that sense, a city's boundaries correspond clearly, in Smith and Varzi's categorization, to a "fiat" object (constructed object according to an administrative, social or political convention). Thus, cities are constructed entities, in particular their delineation, but they are not pure abstractions, in contrast to social entities as states or political parties, as they have a material referent. One could then defend, and it is our position, that the city is an autonomous entity whose properties are not solely derived from statistical aggregates of individuals and firms. Indeed, a collection of individuals and enterprises are not enough to make a city and an aggregation of individuals' behaviors is not sufficient to understand why a city is expanding or declining, even if one takes into account the relations between these individuals. Nevertheless, from an ontological perspective, it is clear that cities are not "bona fide" objects as are individuals. The discussion points then to the city as a relevant "bona fide" ontological object in the sense that a material referent of common sense exists and a "fiat" object in the sense that its recognition and delineation are based on an abstract mode of construction. From an ontological point of view, the status of the city is then intermediate between "bona fide" objects as individuals, buildings, lakes, and "fiat" objects as households or states.

To carry on the discussion by examining Sawyer's proposition (2) concerning explanation, we can start with Hay's statement regarding the state:

> It is "certainly possible to see the state as a conceptual or theoretical abstraction that lacks a direct real world referent whilst still according it an explanatory power or analytical purchase (as a theoretical construct)" (Hay 2014).

Let's suppose that our project is to explain a city's dynamics within a system of cities. Different candidates to explanation can then be discussed, based on following questions: can such phenomena be explained only relating to individuals' actions or should mechanisms of change be looked for at the level of the cities as autonomous entities? In other terms, should explanation be developed at the level of the individuals living in the city or at a higher scale? Would it be sufficient only to focus on individuals' behavior when getting a job, choosing a place of living or a place to shop? What is the role of the decisions of macro-actors as entrepreneurs, planners, politicians? Does it make sense to formalize hypotheses on the behavior of entities of higher level as cities? The model developed will depend on the researcher's position on these issues.

Microsimulation models for example refer clearly to a methodological individualism philosophy and "incorporate individual behavior and … use theories of individual behavior" (Holm et al. 2002). Population or employment evolution of a neighborhood as well as of a region or a city is obtained by aggregating individuals' positions. The results have shown to be satisfactory for testing the effects of planning scenarios at the scale of an employment area where chain mechanisms shape individuals' behavior when choosing for a job or a housing (a household moving out of its dwelling, creates for example an opportunity for newcomers (Holm et al. 2002; Waddell 2002; Wegener 2011). However, at the scale of system of cities, more complex mechanisms are involved. In that respect, the choice made in EUROSIM (Sanders et al. 2007; Pumain et al. 2009), a simulation model of the European system of cities, was to formalize hypotheses at the level of "city-agents" and their interactions. The underlying hypothesis is that of a relative autonomy of the city rather than the "aggregate consequences of individual behavior". The model is developed with a multi-agent system where the agents are the cities which are characterized by their properties (size, economic specialization) and which interact (through an economic market). The spatial and hierarchical structure of the cities emerges from those interactions at the higher level of observation represented by the system of cities. The key element concerns the formalization of the interactions between the city-agents. Regarding states, Hay (2014) suggests:

> that the state is perhaps best seen as neither real nor fictitious, but 'as if real' – a conceptual abstraction, yet one to we which might still accord a very significant generative and causal power.

This expression "as if real" corresponds to the choice made in the EUROSIM model to formalize interactions. The city-agents are neither of statistical type, neither representative of some urban actors. They are geographic constructs whose interactions operate "as if" each city was represented by a global city-actor. The model was able to explore the consequences of different economic constraints on interactions between European cities.

Such approach makes particularly sense in a long term perspective. Indeed, it is because of the interweaving of a variety of components referring to different scales, in time and space, that the city can be considered as an indivisible entity. The material, cultural, socio-economic dimensions have been built up and transmitted through dozens and even hundreds of generations of people whose main purpose was not to maintain or to make the city grow. Self-organization is then a central process which explains how a city is evolving in a system of cities.

9.3 The Question of "Multiple Realizabilty": From an Empirical Perspective

Following the previous elements, another perspective can be undertaken to add more empirical elements to this cross-disciplinary discussion between economists and other scientists focused on cities and geographers. So it seems important to explore this difference in observation levels and their implications in terms of interpretation. For the geographers interested by system of cities, the contribution of the macro and meso levels scale is high and they are convinced that this approach brings new knowledge both about the evolution of the system and about the possibility to compare different kinds of systems, all around the world. It seems that many analyses about dynamics of urban systems led by economists characterize emergence from the micro level to the macro level while most of the researches in geography on urban systems take place in articulating the meso level and the macro level.

From a geographical point of view, the population size can be considered as a good summary of the effects of micro level mechanisms, both qualitative and quantitative, for studying links and relationships between meso and macro levels. Obviously, all cities evolve according to different growth rates. In all systems around the world, one observes a lot of variability, as well in United States, as in France or in emergent countries, or elsewhere. Individual city trajectories are diversified. But this observation is not in contradiction with the regularities that one can observe at the macro-geographical level of system of cities. Actually we start from the observation that the emergent properties at the meso or macro level can "forget" for the description all the "micro movement" that built them—not keeping the details. Butterflies and lions agree that a methodological mistake would be to believe that it would be necessary to study each small flow or change separately, each individual intention, in order to understand what happens at a higher level of observation.

So, it seems that different approaches of urban systems co-exist and they should be complementary in the debate between economists (or other felines) and geographers to go towards a hybrid way of exploration and thinking. The principle of methodological individualism is difficult to maintain if one is interested in developing an ability to understand the trajectories of the urban entities at a meso level. This debate does not only affect the geography but is more cross-disciplinary. Indeed, List and Spiekermann (2013) argue how this debate exists in political science and how a reconciliation is possible between the two perspectives, the individualist one (i.e. the micro scale) and the holist one.

By analogy, as butterflies, we can question the conditions of a holist position facing the urban systems analysis. "A social system requires explanatory holism if and only if three jointly necessary and sufficient conditions are met: (1) Multiple levels of description: The system admits lower and higher levels of description, associated with different level-specific properties (e.g., individual-level properties vs. aggregate-level properties); (2) Multiple realizability of higher-level properties: The system's higher-level properties are determined by its lower-level properties, but can be realized by numerous different configurations of them and hence cannot feasibly be redescribed in terms of lower-level properties; (3) Microrealization-robust causal relations: The causal relations in which some of the system's higher-level properties stand are robust to changes in their lower-level realization".

The first point has already been discussed above. The two following points, "multiple realizability" and "microrealization-robust causal relations", can be demonstrated by the fact that most of processes which lead the urban systems' dynamics (population settlement process or economic trajectories) are comparable in the world even if the processes at the micro-scale level, i.e. the local level, are different and particular. On a way, this debate can also be related with the one about the opposition, about territorial dynamics, between singularity (or specificity) and universal (or general). The idea from Durand-Dastès (1991)[2] expresses one main challenge to rise when we analyze a geographical space. Moreover, according to Hegel, the comparative approach is a relevant way of thinking the difference between singularity and universalism. "Nature shows us a countless number of individual forms and phenomena. Into this variety we feel a need of introducing unity: we compare, consequently, and try to find the universal of each single case" (Hegel 1830).

So, let us take some empirical examples, in different countries, to illustrate these ideas. The analysis of various urban systems allows us to claim that different causalities, multiple urban morphogenesis, different initial situations, at the micro level, lead

[2]La question des rapports entre le particulier et le général n'est pas propre à la géographie, puisqu'elle se pose en fait dans toutes les disciplines.

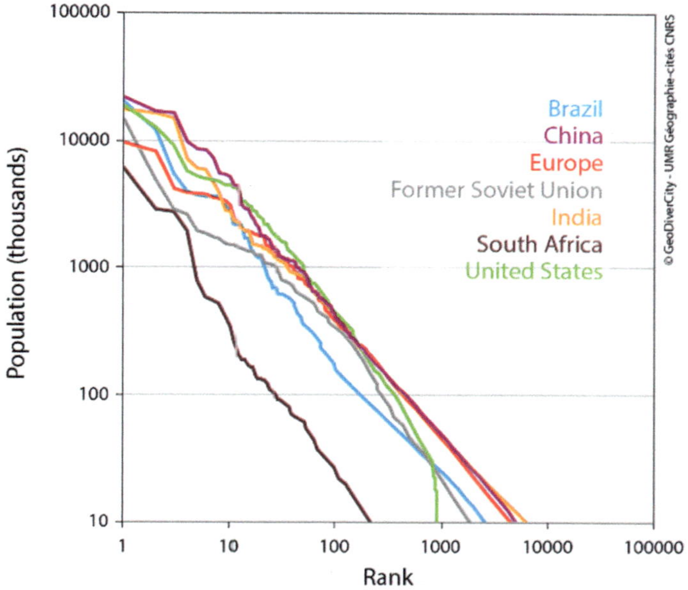

Fig. 9.1 Urban rank-size distributions in different countries (2010) (*Source* Pumain et al., Cybergeo, 2015, 706)

to the same results, to the same process at the macro level. An exploratory and comparative approach between different urban systems demonstrates that some properties are shared among the systems, at the macro level, produced by multiple interferences and by spatio-temporal cross-breeding of processes (Vacchiani-Marcuzzo 2016). For example, the comparison between rank-size regularities among different countries, from the North or the South, reveals a strong similarity at this macro-geographic scale of urban systems (Fig. 9.1).

The differences among the urban systems about their emergence and their formation are real. Some of them are born from a long-standing process of urbanization (as Europe or India) and some are the results of a settlement colony (United States or South Africa). Then, the contexts are extremely different. Moreover, the agents' behaviors on an individual level also vary, in particular due to different weights of informal economic sector from one country to another, especially between emerging and developed countries. Nevertheless, despite of these differences, we observe in all countries relatively close urban dynamics (Figs. 9.1 and 9.2). The similarity of their evolution is remarkable and it illustrates the systemic trend of "self-reproduction" of the territories, whatever their localization.

We can pursue this empirical demonstration through the economic dynamics of urban systems. Indeed, this feature of dependence to the past is also present in evolutionary economic theories (Boschma and Martin 2010; Boschma and Frenken 2006) where the economic transformations observed at various regional levels (country,

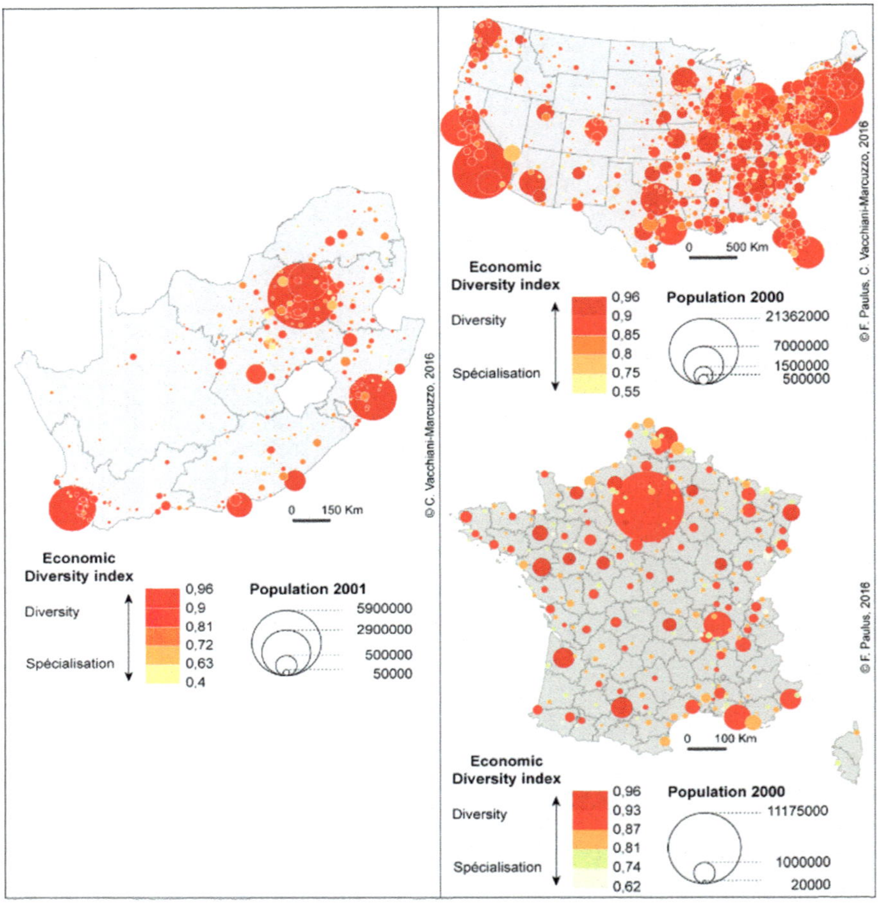

Fig. 9.2 The economic diversity in three urban systems (*Source* Vacchiani-Marcuzzo 2016)

region, city, etc.) come partly from the impacts of small historical events that guide the dynamics of cities. All the urban systems were concerned by the succession of innovation cycles, with temporal gaps. And these cycles left strong footprints in the current structures of urban systems (Figs. 9.2 and 9.3).

For example, the comparison between the structure of activities through three urban systems (South Africa, France and United States) is strongly similar at the beginning of the XXIe century. The empirical and static analysis reveals a common high diversity and complexity of the activities portfolio for the biggest cities among the three systems. This parallel result at the level of urban system is coming from various situations at the more local scale, but they led to the same results. This empirical result is close to the demonstration in political science advanced by List and Spiekermann (2013) about the multiple realizability and the fact that the properties

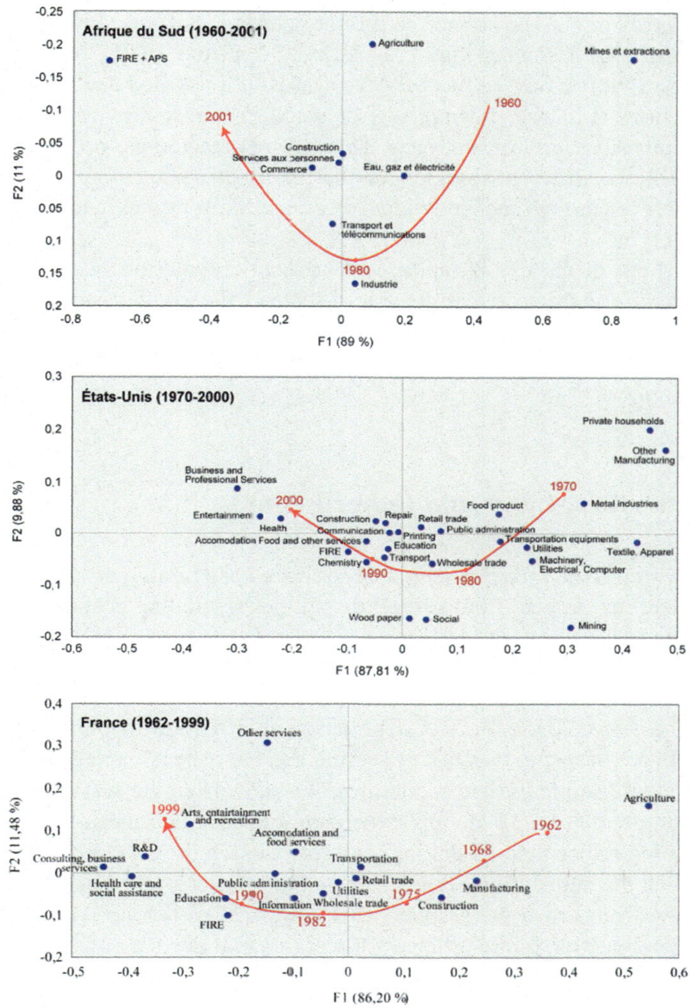

Fig. 9.3 The economic trajectory of cities (South Africa, Unites States, France) (*Source* Vacchiani-Marcuzzo 2016)

observed at a higher level (in our case, the urban system level) can be the results of a multitude of different situations at the lower level (the level of each individual city inside the system).

Another empirical example can be the urban trajectory of different urban systems through the economic cycles of innovation. The comparison since the 1960s of the mean economic trajectory of three different urban systems (Figs. 9.3 and 9.4) reveals a similar trajectory through the transition from an industrial society (*The Old Economy*) to a services industry society (*The New Economy*), main feature observed everywhere.

At the macro-scale, the economic profile reveals a shift or a bifurcation (earlier in the industrialized countries as France or USA in the 1970s, later in emergent countries as South Africa, in the 1980s) towards more skilled and more globalized urban functions (Finance, Insurance, Advanced Producer Services, Research and Development, Arts and Leisure, etc.). This shift, characteristic of the post-fordism economy, follows different rhythms among the urban systems but draws the same main trends, even though each individual city, inside the systems, can follow different trajectories.

Finally, both in the demographic and economic evolution of cities, there is a convergence of the processes at the macro scale of the urban system, whatever the different dynamics at the meso-scale or the micro-scale. This empirical approach allows us to hence the difference in terms of scale in the debate between geographers and other felines.

9.4 Butterflies, Lions and Other Felines

Let us now consider a lived research experience based on a participation to a EU project where researchers from several disciplines (including economists) aimed at applying the UrbanSim *Land Use and Transport Interaction* (LUTI) model to three European cities (Paris, Brussels and Zurich) (Bierlaire et al. 2015). Jonathan Jones further developed this topic in his Ph.D. dissertation and in the consequent scientific papers (Jones et al. 2015a, b, 2017; Thomas et al. 2018); these publications will be the main source of inspiration of this section and have the advantage of opening the debate to other disciplines than economics. We knew from the very beginning of the research project that LUTI models were data intensive and hence often leading to data crunching rather than modelling; as expected, we drowned into data problems among which the delineation of the cities and the choice of the basic spatial units.

We know from urban geographers that comparing cities needs undoubtedly to control the delineation of the urban agglomerations. Indeed, a city is a dot on a map but only for location purposes at a very small scale (i.e. a world map), but at any larger scale a city is always characterized by an extent (surface), a morphology (shape), and also by living masses (inhabitants/households/firms) and these masses can change over time in size and composition. Consequently, each city has a different footprint determined by complex organization of its masses, but also by interacting actors and processes such as its geography, history or governance (e.g. Batty 2005; Derycke et al. 1996; Abdel-Rahman and Anas 2004; Parr 2007; Tannier and Thomas 2013). Despite EU efforts, there is no/very little international agreement about the delineation of functional urban regions and various cut-off values are often basically used (e.g. Gornostaeve and Cheshire 2002; Dujardin et al. 2007; Thomas et al. 2012). Moreover, cities are interconnected, hierarchically organized (Pumain 2006) and hence difficult to clearly isolate. Last but not least, it is rather difficult to delineate a city because its built-up surfaces dilute farther and farther away from the city center within rural areas, thus blurring the morphological and functional limits of the city (Caruso et al.

2007, 2011). This problem becomes even more complex when the polycentric nature of the urban systems is taken into consideration. Hence, the hinterland of a city cannot easily be isolated from another, and all these urban extensions are sources of complex and multifaceted mobility behaviors (e.g. Cervero 2002; Handy et al. 2002).

Though this is stating the obvious in urban geography literature and continuously and by the availability of new data types (crowd-sourced transport geo-spatial information, urban sensors, social network data scrawled on the Internet, land use change at cadastral level, etc.), it is much less so in other disciplines such as for instance economics and transport economics where we see that researchers and consultants have taken a very pragmatic view on urban delineations as well as the choice of basic spatial units: they often forget that their choice may hardly impact the modelling results. This was developed in Thomas et al. (2018), a paper that constitutes the background of this section.

LUTI models are indeed derived from the classical four-step model. The first generation of models led to applications in the US such as MEPLAN (see Echenique et al. 1990) or TRANUS (de la Barra 1989) mainly due to federal regulations requiring land use impacts of new transport infrastructure to be assessed. Their spatial diffusion to Europe is more recent and scarce: LUTI models still appear nowadays to be mostly a pragmatic integration of bits of land use within transport models rather than the opposite, and therefore are weakly connected to urban land use, especially in the aspects related to urban dynamics (Anas 2013). Thomas et al. (2018) concentrated on the lack of attention paid to the definitions of cities when applying these models. Their argumentation develops on a systematic meta-analysis of 21 peer-reviewed publications about applications in European cities. Most papers are not written by butterflies and most authors ignore geographical biases about city size and delineation as well as the grain (size and shape of the basic spatial units and hence the definition of the statistical individual). Thomas et al. (2018) also conducted simulations on a synthetic urban area consisting of 2 city centers (W and E) in a rectangular study

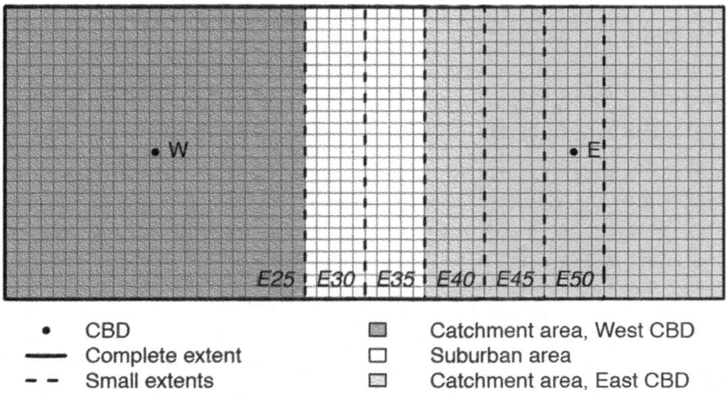

Fig. 9.4 The theoretical case study used for simulation (*Exx* is the urban extents, *xx* referring to the number of columns from the Western extremity; W and E are two different CBDs (*Source* Thomas et al. 2018)

area (Fig. 9.4); the size of the cities was later changed as well as the extend of the W city which is growing further and further right (*Exx* on Fig. 9.4) up to capturing the right hand side city (E). Increasing *Exx* from left to right results in a progressive inclusion of the CBD of E into the studied area.

The reported simulations confirm that the absence of control on city delineation strongly weakens the LUTI result, impacting in particular commuting times. The delineation of the city area plays a major role as well as the size of the spatial units (Jones et al. 2015b, 2017) in transport modelling results and the consequences for not controlling these effects are clearly underestimated in the transport literature.

While practical concerns are perfectly understandable, there is a risk of undermining the effects of choosing the limits of an urban system, especially because models involve complex processes that are more and more detailed geographically and disaggregate in terms of agents, include many non-linear interactions and feedbacks across scales, and actually crunch large and heterogeneous data. Assessing the effect of system boundaries is important to provide certainty and robustness to model findings; this is also true in urban and transport research as soon as one intends to contribute generalizable scientific knowledge about the functioning of city regions. Ideally contributed knowledge should be independent from the dataset, and should also be easily transferable across case studies. Both require transparency in how the geographical limits are chosen and how this choice impacts the spatial distribution and variability of data and whether or not it affects results.

When selecting a city, the larger the study area, the largest the risk of including functional or morphological parts of other cities, especially in densely inhabited regions. Different delineations can therefore change the nature of the studied urban system and automatically determine important parameters of the model including transport outcomes as well as the level of system inertia/response to scenarios. This already was a point in Lee (1973)'s requiem on large-scale models, emphasizing wrongheadedness when implementing gravity processes at all scales. By bypassing the discussion on the size and extent of cities, transport modelers may actually have slowed down the integration in urban theory of their findings related to particular cases and geographies. After almost 50 years of LUTI applications, the theoretical and legitimate promises of a fully interacting land use and transport systems may then not have been fully met. While mature in practical terms, the field still seems to lack of capacity to contribute generalizable scientific knowledge about the functioning of city regions especially because of difficulties in case studies comparability and full transparency in application. On the other extreme, the standard urban economic literature, after Alonso, while informing clearly on generic policies and optimal instruments (cordon, taxes, etc. to tackle sprawl, congestion or negative externalities; e.g. Brueckner 2001; De Borger et al. 2008) stays far from empirical validation beyond stylized facts, and real case calibration and implementation are rare. Hence, urban geographers and planners cannot easily capitalize on transport and land use models. This is somehow a rejoinder to Saujot et al. (2016) who emphasize a gap between theory and the end-users, although our viewpoint is rather that the outcome of LUTI implementations may not be sufficiently general and robust to transfer between cases with a different urban geographical reality.

We could stop here with the example of LUTI models (which are not mainstream economics) and their lack of realistic urban European applications, but (unfortunately) many other models suffer from the same problem. Surprisingly enough, despite early papers about urban modelling, scales and urban borders, this problem is still up to date and even revive nowadays with the invasion of unconventional (massive) data for which several biases are often pinpointed but where spatial aggregation (size and shape of basic spatial units; definition of studied area) is once again ignored or—worse—even re-invented (Kitchin 2013; Miller and Goodchild 2015; Miller 2018). Geographers publish, their research is available on the web and in international geography journals, but it seems that the border between disciplines is hardly crossed. Are discipline jargons the only brake to dialogue? Is it so hard for lions or tigers to listen and understand butterflies?

9.5 Conclusion

Facing urban phenomenon, all social sciences have something to say and are able to provide elements of understanding. Indeed they have complementary methodological savoir-faire. Economists' methods are very useful to formally show how rational individual behaviors produce spatial structures. On their side, data scientists offer stimulating tools to analyze informal massive data, even if, of course, we can't let the data speak by themselves. Geographers produce methods able to take astutely into account scale and space effects. Together they contribute to the building of knowledge about handling space, its complexity and its biases in modelling. Many scientific approaches are relevant and it's essential to combine different theoretical analyses. It is also very fertile to decline empirical process at different scales, according various delineations of the urban object. Indeed, each way of thinking corresponds to a point of view on the studied phenomenon. It is time that felines and butterflies exchange and co-evolve in the fascinating world of interdisciplinarity.

References

Abdel-Rahman, H.M., Anas, A.: Theories of systems of cities. In: Henderson, J.V., Thisse, J.F. (eds.) Handbook of regional and urban economics, vol. 4. Cities and Geography, pp. 2293–2340. Elsevier, North-Holland (2004)

Anas, A.: A response to the guest editorial: economics as the science for urban modelling. Environ. Plan. B **40**, 955–958 (2013)

Batty, M.: Cities and Complexity: Understanding Cities with Cellular Automata, Agent-Based Models, and Fractals. MIT Press, Cambridge, USA (2005). 592 p.

Bierlaire, M., de Palma, A., Hurtubia, R., Waddell, P. (eds.): Integrated Transport and Land Use Modeling for Sustainable Cities, 280 p. EPFL Press, Routledge (2015)

Boschma, R., Frenken, K.: Why is economic geography not an evolutionary science? Towards an evolutionary economic geography. J. Econ. Geogr. **6**(3), 273–302 (2006)

Boschma, R., Martin, R. (eds.): The Handbook of Evolutionary Economic Geography, 576 p. Cheltenham, Edward Elgar Publishing (2010)
Brueckner, J.K.: 2001, Urban sprawl: lessons from urban economics. Brookings-Wharton Pap. Urban Aff. **1**, 65–97 (2001)
Caruso, G., Vuidel, G., Cavailhès, J., Frankhauser, P., Peeters, D., Thomas, I.: Morphological similarities between DBM and a microeconomic model of sprawl. J. Geogr. Syst. **13**(1), 31–48 (2011)
Caruso, G., Peeters, D., Cavailhès, J., Rounsevell, M.: Spatial configurations in a peri-urban city. A cellular automata based micro economic model. Reg. Sci. Urban Econ. **37**, 542–567 (2007)
Cervero, R.: Built environments and mode choice: towards a normative framework. Transp. Res. Part D **7**(4), 265–284 (2002)
Gornostaeve, P.C., Cheshire, G.: Villes et régions urbaines: des comparaisons fiables doivent reposer sur des territoires comparables. Les Cahiers de l'IAURIF **135**, 13–32 (2002)
De Borger, B., Proost, S., Van Dender, K.: Private port pricing and public investment in port and hinterland capacity. J. Transp. Econ. Policy (JTEP) **42**(3), 527–561 (2008)
De la Barra, T.: Integrated Land Use and Transport Modelling, p. 179p. Cambridge University Press, Cambridge (1989)
Derycke, P.H., Huriot, J.M., Pumain, D. (eds.): Penser la ville. Théorie et modèles, 335 p. Anthropos, Paris (1996)
Dujardin, C., Thomas, I., Tulkens, H.: Quelles frontières pour Bruxelles? Une mise à jour. Reflets et perspectives de la vie économique **46**(2–3), 155–176 (2007)
Durand-Dastès, F.: Le particulier et le général en géographie. In: Audigier, F., Baillat, G. (eds.) Actes du 6° colloque de Didactiques de l'histoire de la géographie, des sciences sociales, pp. 207–216 (1991)
Echenique, M., Flowerdew, A., Hunt, J., Mayo, T., Skidmore, I., Simmonds, D.: The Meplan model of Bilbao, Leeds and Dortmund. Transp. Rev. **10**(4), 309–322 (1990)
Gabaix, X.: Zipf's law and the growth of cities. Am. Econ. Rev. **89**(2), 129–132 (1999)
ERC GeoDiverCity colloquium. http://geodivercity.parisgeo.cnrs.fr/blog/ (2017)
Handy, S.L., Boarnet, M.G., Ewing, R., Killingsworth, R.E.: How the built environment affects physical activity: views from urban planning. Am. J. Prev. Med. **23**(2), 64–73 (2002)
Hay, C.S.: Neither real nor fictitious but "as if real": a political ontology of the state. Br. J. Sociol. **65**(3), 459–480 (2014)
Hegel: Encyclopaedia of the philosophical sciences. Logic, Clarendon Press (1975)
Henderson, J.V.: The sizes and types of cities. Am. Econ. Rev. **64**, 640–656 (1974)
Holm, E., Lindgren, U., Kirsten, H., Mäkilä, K.: The SVERIGE Spatial Microsimulation Model. GERUM, vol. 4 (2002)
Jones, J., Peeters, D., Thomas, I.: Scale effect in a LUTI model of Brussels: challenges for policy evaluation. Eur. J. Transp. Infrastruct. Res. (EJTIR) **17**(1), 103–131 (2017)
Jones, J., Peeters, D., Thomas, I.: Is cities delineation a pre requisite for urban modelling? The example of land price determinants in Brussels. Cybergeo **716** (2015a)
Jones, J., Peeters, D., Thomas, I.: Forecasting employment location choices by Discrete Choice Models: a sensitivity analysis to scale and implications for LUTI models. REGION **2**(1), 67–93 (2015b)
Kitchin, R.: Big data and human geography: opportunities, challenges and risks. Dial. Hum. Geogr. **3**, 262–267 (2013)
Krugman, P.: Confronting the mystery of urban hierarchy. J. Jpn. Int. Econ. **10**, 399–418 (1996)
Lee Jr., D.B.: Requiem for large-scale models. J. Am. Inst. Plan. **39**(3), 163–178 (1973)
List, C., Spiekermann, K.: Methodological individualism and holism in political science: a reconciliation. Am. Polit. Sci. Rev. **107**(4), 629–643 (2013)
Miller, H.: Geographic information science II: mesogeography: social physics, GIScience and the quest for geographic knowledge. Prog. Hum. Geogr. **42**(4), 600–609 (2018)
Miller, H., Goodchild, M.: Data-driven geography. GeoJournal **80**(4), 449–461 (2015)
Parr, J.B.: Spatial definition of the city: four perspectives. Urban Stud. **44**(2), 381–392 (2007)

Pumain, D. (ed.): Hierarchy in Natural and Social Sciences, 246 p. Springer, Dordrecht (2006)
Pumain, D., Sanders, L., Bretagnolle, A., Glisse, B., Mathian, H.: The future of urban systems: exploratory models. In: Lane, D., van der Leeuw, S.E., West, G., Pumain, D. (eds.) Complexity Perspectives on Innovation and Social Change, pp. 331–359. Springer, Methodos Series (2009)
Pumain, D., Swerts, E., Cottineau, C., Vacchiani-Marcuzzo, C., Ignazzi, A., Bretagnolle, A., Delisle, F., Cura, R., Lizzi L., Baffi S.: Multilevel comparison of large urban systems. Cybergeo: Eur. J. Geogr. (2015). https://doi.org/10.4000/cybergeo.26730
Sanders, L., Favaro, J.-M., Mathian, H., Pumain, D., Glisse, B.: Intelligence artificielle et agents collectifs : le modèle EUROSIM. Cybergeo: Eur. J. Geogr. Dossiers **392** (2007)
Saujot, M., de Lapparent, M., Arnaud, E., Prados, E.: Making land use—transport models operational tools for planning: from a top-down to an end-user approach. Transp. Policy **49**, 20–29 (2016)
Sawyer, R.K.: Nonreductive individualism. Part I—supervenience and wild disjunction. Philos. Soc. Sci. **32**(4), 537–559 (2002)
Simon, H.A.: On a class of skew distribution functions. Biometrika **42**(3/4), 425–440 (1955)
Smith, B., Varzi, A.C.: Fiat and bona fide boundaries. Philos. Phenomenol. Res. **60**(2), 401–420 (2000)
Steinberg, R.: The publication and citation behaviour of economic geographers and geographical economists compared. In: Review of Regional Research. vol. 35(1), pp. 1–27 (2015)
Storper, M.: Urban systems: the geography of income and population (PPT). In: Colloquium ERC, Theories and Models of Urbanization, Paris, 13 Oct 2017
Tannier, C., Thomas, I.: Defining and characterising urban boundaries: a fractal analysis of theoretical and real Belgian cities. Comput. Environ. Urban Syst. **41**, 234–248 (2013)
Thomas, I., Cotteels, C., Jones, J., Peeters, D.: Revisiting the extension of the Brussels urban agglomeration: new methods, new data… new results? E-Belgeo (2012). http://belgeo.revues.org/6074
Thomas, I., Jones, J., Caruso, G., Gerner, P.: City delineation in LUTI models: review and tests. Transp. Rev. **38**(1), 6–32 (2018)
Udehn, L.: The changing face of methodological individualism. Annu. Rev. Sociol. **28**, 479–507 (2002)
Udehn, L.: Methodological individualism. In: History and Meaning. pp. 464. Routledge (2014)
Vacchiani-Marcuzzo, C.: Is the South African urban system apart? Assessment of an urban hybridity. Habilitation à diriger des recherches, 202 p. Université de Paris 1—Panthéon-Sorbonne (2016)
Waddell, P.: UrbanSim: modeling urban development for land use, transportation and environmental planning. J. Am. Plan. Assoc. **68**(3), 297–314 (2002)
Wegener, M.: From macro to micro—how much micro is too much? Transp. Rev. **31**(2), 161–177 (2011)
Ylikoski, P.: Rethinking micro-macro relations. In: Zahle, J., Collin, F. (eds.) Rethinking the Individualism-Holism Debate. Essays in the Philosophy of Social Science, pp. 117–135. Springer, Berlin (2014)

Part IV
Urban Theories and International Comparisons

Chapter 10
Emerging Cities and Urban Theories: A Chinese Perspective

Fulong Wu

Abstract This chapter begins with a brief summary of recent debate in the 'nature of cities' and then reflects on the major urban theories in the West and their similar representations such as the emergence of new towns and gated communities in China. In addition to seeing the effect of agglomeration, the specific context of development in China is discussed, which plays a critical role in shaping the spatial patterns of urbanisation. We argue that we need to pay attention to the urbanisation processes in order to develop a more nuanced understanding of the outcome of urbanisation in China. China is specific in terms of its political economic setting-ups. But we should not simply attribute its urbanisation to 'state-led' or 'policy-driven' forces. In this regard, China is not unique, which can broaden our perspectives of seeing urban changes.

10.1 Introduction

The word 'city' in Chinese is actually composed of two characters, '*cheng*' (city) and '*shi*' (market). The book '*Spring and Autumn Annals of Wu and Yue*' written in 25–220 AD describes the origin of the city as: 'building the city for the emperor; developing the market for people. This is the origin of the city'. The Chinese definition points out two forces in city building: economic activities and governance. The purpose of this chapter is to understand cities as the outcome of both economic agglomeration and the politics of development. The research question we ask here is: what are the driving forces for the concentration of population and economic activities into the variegated spatial forms known as cities?

Recently, urban theorists have tried to understand the 'nature' of cities through economic agglomeration. Scott and Storper (2015) suggest that the city is an outcome of economic agglomeration and its spatial form is further moulded by the land nexus. They argue that 'agglomeration is the basic glue that holds the city together

F. Wu (✉)
Bartlett School of Planning, University College, London, UK
e-mail: fulong.wu@ucl.ac.uk
URL: https://urban-china.org

D. Pumain (ed.), *Theories and Models of Urbanization*, Lecture Notes in Morphogenesis, https://doi.org/10.1007/978-3-030-36656-8_10

as a complex congeries of human activities, and that underlies – via the endemic common pool resources and social conflicts of urban areas – a highly distinctive form of politics, as we show later' (pp. 6–7). They elaborated that the nature of cities involved 'combining two main processes, namely, the dynamics of agglomeration and polarization, and the unfolding of an associated nexus of locations, land uses and human interactions' (p. 1). They argue that the formation of the city is driven by the force of agglomeration, extending through dispersion. They criticize the way critical urban scholars define the city through various social processes which may be related to urban life but cannot be essentially defined as the city. They suggest that the definition should separate general social process from the process of creating the city.

Their view has triggered wide debates in Urban Studies. In particular, the *International Journal of Urban and Regional Research* devoted a section to debates responding to the paper. Robinson and Roy (2016) criticised the claim for a universal urban theory from the North Atlantic which neglects the diversity and shifting geographies of global urbanization. In another paper, Robinson (2016) advocates a comparative approach to theory building, 'which can help to develop new understandings of the expanding and diverse world of cities and urbanization processes, building theory from different contexts' (p. 187). Other papers in the special issue emphasize the need to understand everyday life and urban experience. Simone (2016) stresses blacks' different experience of cities. He describes the heterogeneity of urban experience, particularly the association of blacks with urban lives. Peake (2016) emphasizes the experience of women in urbanization. Other researchers point out the need to understand the city in relation to policy research. Parnell and Pieterse (2016) argue that rather than thinking of the nature of cities in an abstract form there is a need for basic, descriptive and politically engaged research for Africa. Leitner and Sheppard (2016) call for a 'provincialization of urban theory' and argue that no single theory is sufficient to account for the 'variegated nature of urbanization and cities across the world' (p. 228). More critiques come from a political economic perspective. Walker (2016) examines the urban process as the spatial concentration of economic surplus by ruling classes and the state. He argues that the creation of the built environment in the form of cities is a result of the urban process.

As shown at the beginning of this chapter, China has a long history of city building. Its pace of urbanization, however, has speeded up over the last four decades. Owing to its specific development history and the large scale of city building, the Chinese experience of urban development may bring a new perspective to the understanding of emerging cities. For example, the development of 'edge cities' and post-suburbia has been understood through post-industrial restructuring and flexible accumulation. But the Chinese case indicates the need to understand the role of political leaders in land development, place promotion and a land-finance model of urban development. Similarly, gated communities are widely understood in terms of a lifestyle choice, a rising concern for security, and a preference for private governance. But the Chinese case indicates a more supply side reason. Associated with real estate projects, developers try to create an imaginary western suburbia to brand suburban and rural areas. As will be elaborated in the next section on the model of urban development, gated communities are part of an overall development strategy supported by the entrepreneurial local state, which is incentivised by land-based development from land sales and constrained by its means of mobilising financial resources (because of the restriction on raising money directly from the capital market).

10.2 New Model of Urban Development

China's urbanization has been driven by its development model, which can be characterised as the model of the world factory (Wu 2017). Export-oriented industries have been a key element, integrating the Chinese economy into the global production network, trade and financial flows. Other elements are cheaper land resources and a rural migrant workforce in addition to the well-trained workers in the cities. Through the decentralization of economic decision-making and fiscal reform, local governments are incentivised to promote urban development. They use their control over the land to attract investment. The development of manufacturing industries in turn attracts migrant workers. Rural–urban migration and industrial land development associated with the world factory are leading Chinese urbanization. The major challenges of this model of urban development are both social and environmental. Migrant workers are not treated as urban citizens and thus are excluded from the city. Social exclusion has severe economic implications too, as the consumption capacity for basic needs and services is under-developed. The constraint on domestic consumption forces China to continue to rely on global markets. Second, land development is mainly for capital investment. Competing to release land to attract investment, Chinese cities have seen significant spatial expansion, creating a scattered pattern of urbanization. Large-scale development and spatial fragmentation have led to environmental degradation. Industrial development stimulates the agglomeration of population and services. The entrepreneurial local government actively promotes real estate development because public finance depends upon income from land sales. This business model is commonly referred to as a land-finance model. In order to capture land values, local governments compete with each other in place promotion, because the rise in property values is expected to generate larger land revenues. Increasing land value allows the local state to use the land as collateral to obtain more capital for investment. The actual development is operated through state-owned enterprises or so-called 'local government finance vehicles' (LGFVs). Increasingly, development takes the form of mega-urban projects to maximize land value and access to credit through land-financing.

This model of urban development reveals that agglomeration plays a key role in linking land, capital and labour to create mega-urban projects and new cities. However, the actual operation of this model is determined by not only economic factors but also the specific form of governance. The multi-scalar state, in particular the local entrepreneurial state, orchestrates the conditions for the model. Global capitalism provides the external conditions for this model. The new model of urban development in China is part of a 'global spatial fix'. Over-accumulation in the

West finds its outlet for surplus capital in emerging economies and has expanded its production at the global scale. But increasingly, the new model of development is leading to its own problems of over-accumulation and redundant production capacity in China. With intensifying China–US trade tension since 2018 and a turn towards new trade protectionism, China can no longer rely on the global market to sustain its urbanization and urban development. The need to find a new spatial fix may be a prelude to a change from deregulation to reregulation in urban development. In the next phase of economic development, the city may be the form to implement these regulatory tasks. As such, the thesis of agglomeration only reveals part of the whole dynamics of emerging cities.

10.3 Emerging City-Regions

One important feature of China's urbanization is the development of 'urban clusters' in its highly populated regions such as the Pearl River Delta, the Yangtze River Delta, and the Beijing-Tianjin-Hebei (Jing-Jin-Ji) region. The notion of 'urban clusters' in China is similar to that of the city-region in the Western literature (Harrison 2014). The concept of urban clusters is not a statistical one and refers to loosely linked metropolitan areas. Thus, connectedness between cities is the key for city-region formation. More broadly, cities are connected in a larger regional scale, forming 'mega-city regions' (Hall and Pain 2006). Regional and inter-city transport development facilitates the development of mega-city regions. On city-regions, Scott (2001) argues that agglomeration is driving the development of economic activities not just inside the city but also linking the region where the city is located. Thus, the city-region is a new emerging form of the globalization of economic activities. This view of global city-region development is associated with the view of suburban and post-suburban development in a wider region. More studies nowadays try to understand the politics and governance of city-region formation (Jonas and Ward 2007). Theoretically, the emergence of city-regionalism is understood as 'state transformation', or more specifically the rescaling and territorialisation of the state (Brenner 2004). The theory of state rescaling stresses how the state selects the specific city-region scale to govern its territory. This view of 'state spatial selectivity' applies to China (Wu 2016a). The form of economic regulation has transformed from the national scale before economic reform, to individual cities in the early stage of economic reform, and to the city-region scale after China joined the World Trade Organisation (WTO). Under globalization, new economic activities spread across the boundaries of cities into the region, presenting a new spatial form of the city-region. This is well described by economic agglomeration (Scott 2001; Scott and Storper 2015). But at the same time Chinese cities face fierce inter-city competition, redundant production capacity, excessive infrastructure development and environmental degradation. Thus, city-region formation should also be understood as a project to upscale governance from individual cities to their city-regions (Wu 2016a) in particular through regional cooperation and planning (Li and Wu 2018). In the YRD, this involves both a top-down city-region building process through urban cluster planning and governance and bottom-up collaboration between cities (Li and Wu 2018). Therefore, the emergent city-regions in China are outcomes of both economic agglomeration and institutional and governance changes towards coordinating economic activities.

The development of Kunshan near Shanghai is a good example. As China's globalising city, the municipality of Shanghai covers an administrative region of 6340 km^2, a vast area encompassing the central areas, suburban new towns and rural areas. The metropolitan region itself can be seen as a city-region. But the impact of Shanghai goes beyond its administrative areas and affects cities in Jiangsu and Zhejiang provinces. For example, the former county of Kunshan under the Suzhou municipality is very close to Shanghai. The town of Huaqiao in Kunshan can receive a mobile signal from Shanghai. Kunshan county was underdeveloped. In 1984, it self-funded a development zone which was recognized by the central government as a national level Economic and Technological Development Zone (ETDZ) in 1992. The development of Kunshan utilized technological spill-over from Shanghai and attracted Taiwanese investment in the ITC sector. It is a major site in China for the assembly of electronic products including the iPhone. So, the development of the wider Shanghai region seems to support the thesis of economic agglomeration. However, Kunshan as well as other cities in the YRD has competed with Shanghai. Excessive inter-city competition has triggered two related processes of city-region building: the top-down process initiated by the central government and the bottom-up process created through cooperation between local governments (Li and Wu 2018). Therefore, city-regions are not just economic spaces but also political spaces—in this case they can be seen as a new 'state space' in which the state strives to impose its control while multiple actors participate in this process to gain their own benefits. For example, through the intervention of senior politicians at the provincial and central levels, Shanghai extended its No. 11 metro line into Kunshan, which had been advocated by the city of Kunshan for a long time. Still, city-regional integration remains difficult owing to administrative divisions. In this case, there is a provincial level boundary between Kunshan and Shanghai, which imposes a major obstacle. The central government plays an important role in promoting the integration of cities in the YRD through the YRD regional plan. It specifies the future development corridors and designated spaces for environmental and agricultural land protection. But such a regional plan lacks enforcement power. In reality, local governments utilize the plan for their own benefit.

More recent development in Jing-Jin-Ji reveals the political nature of city-region building. As the capital, Beijing is concentred with government offices, political organisations, the headquarters of Chinese state-owned enterprises, and multinational regional headquarters. The concentration of population in central Beijing is believed to be the main cause of its environmental problems, especially air pollution. The central government has thus striven to decentralize the economic functions of Beijing to a large city-region outside its metropolitan area. The government classified economic functions into 'essential' and non-essential for Beijing's capital role. The decentralization policy is implemented through the relocation of enterprises and the demolition of informal spaces for the migrant population, aiming to reduce the total population of Beijing. The aim is to build the Jing-Jin-Ji city-region to disperse these non-essential functions. Further, Beijing's new town of Tongzhou is designated as an administrative sub-centre which is being built to accommodate the municipal government of Beijing. An even more ambitious plan is to build an entirely new city—Xiong'an in Hebei province.

10.4 Emerging Economic Clusters

While lagging behind in research and development, China has recently promoted 'indigenous innovation' through increasing R&D investment. The spatial concentration of innovation is known as clustering. For example, clustering is a salient feature of biotechnology innovation (Cooke 2004). Cooke (2004) explained the reasons for the development of 'biotech mega centres' in some cities. This is because 'big pharmas' (multinational pharmaceutical companies) cannot manage the complex networks of biotech innovation by themselves. Instead, the city or city-region plays such a role by creating the externalities needed for biotech innovation. The biotech mega centre is a regional innovation system in which biotech companies are situated. In addition to more localised endowments, Zeller (2010) emphasizes the role of production networks and the role of 'big pharmas' in creating complex biotech centres and 'the invasion of large pharmaceutical companies' which create or reinforce the regional knowledge base (p. 2889). A useful example is contracted research organisations (CROs), which perform specific experiments and tasks for big pharmas. Although the effects of policy and key enterprises are important, the understanding of specific development processes is lacking. Initially thinking from the perspective of the regional innovation system (RIS), Zhang (2015) examined the process of building biotech in Shanghai. Although Shanghai is not a major biotech centre in the world, the clustering of biotech firms and pharmaceutical companies is apparent in Shanghai, especially in Zhangjiang High-Tech Park. The phenomenon might be explained by agglomeration and spatial proximity (Porter 2000) or by a regional innovation system which considers a larger territorial environment in which innovations are situated (Cooke et al. 2007; Cooke and Morgan 1998). Other economic geographical explanations look at the interaction between industries, universities and governments as a triple helix model (Etzkowitz and Leydesdorff 2000) or at relatedness and diversity, in particular 'related varieties' (Boschma and Frenken 2006). Now, the understanding of urban development in China can provide useful insights. To rethink the city and innovation, the clustering of biotech in Shanghai and particularly in ZJHP needs to be examined more broadly in the process of urban development dynamics. We need to think of the city as an assemblage to collate various forces, some of which might not be directly related to R&D itself (Zhang and Wu 2019). In other words, innovation needs to be situated in an overall urban development process. Biotech clusters in Shanghai are related to the creation of ZJHP, as are agencies of the central government, the Shanghai municipal government (in its policy of 'Focusing on Zhangjiang'), Zhangjiang Development Corporation and also the model of land financing originated in Pudong New District containing four giant 'functional development corporations'. Zhangjiang is one of these four development corporations in Pudong.

Shanghai's role as China's gateway and economic capital is obviously relevant. The multinationals' strategy to gain Chinese market access and the development of R&D centres and related CROs are important considerations. Seen from emerging cities, the development of economic clusters is part and parcel of the city as 'an arena where all these actors are assembled to play their roles in biotech innovation' (Zhang and Wu 2019, p. 154). Now, Shanghai is being urged by the central government to become a new science and innovation centre for China. Zhangjiang Science City has been created on the basis of ZJHP. Earlier studies emphasize the city as providing 'innovation milieus'—looking at the aspects of, for example, local buzz (Bathelt et al. 2004) or institutional features (as described by RIS). Understanding the city as an arena for innovation actors means other factors that might not be regarded as contributing to an innovation process may actually have critical effects. For example, entrepreneurial governance and land development are two factors related to the development of innovation capacity in Zhangjiang. The Zhangjiang High-tech Park Development Corporation is one of the listed companies in the Shanghai stock market. It plays a role in creating investment funds to support start-ups and developing experimental platforms and incubators for biotech firms. Through its role as the 'primary developer' which receives land resources from the government and uses land banking to capture land values in Zhangjiang, it is also involved as a shareholder in some promising biotech firms. For example, the MicroPort Medical Co. received support from the development corporation fund. After the company was listed in the Hong Kong Stock Exchange, the corporation gained the increased value of its holding in the company. The corporation operates a 'land and share swap agreement' to support high-tech firms, which treats the land from the government as a share in the company. Property sales are also an important source of revenue for the corporation. Zhangjiang has gradually seen a change from a high-tech park to a suburban new town with mixed functions. Through a recent initiative, Shanghai is planning Zhangjiang Science City. According to Charles (2015), the development of science cities in the world has experienced three waves: the first is a cluster of science activities in a campus-based environment. The second stage is the science park with commercialization. The third stage is the combination of science activities and urban activities, using science-led development inside the city. Zhangjiang has been developing since 1992 and has become part of the new urban areas in Pudong. Now, the concept of the science city is envisioned as integrating itself better into the rest of Shanghai. The master plan of Zhangjiang Science City was approved by the central government in 2017, and its area has been expanded from 25 km^2 of high-tech park to 94 km^2 of new areas in Pudong. One key driver is the development of large scientific facilities funded by the central government. The plan of the science city is to convert Zhangjiang from an employment centre to a comprehensive city in itself. Only about 10% of employees currently live in Zhangjiang. To accommodate future science workers, the Zhangjiang Group operates 'talent apartments' and aims to develop more residential areas in the science city. It aims to reach a total population of 700,000 with employment for 880,000. The core Zhangjiang will be one of the municipal sub-centres. It is also hoped that other developers will join in property development. The development of Zhanjiang Science City illustrates the evolution from an economic cluster to an emerging city.

10.5 Emerging Cities

Driven by rapid industrialization, China has seen the development of new cities and new towns, which has created a complex pattern of mixed urban and rural land uses. This pattern of mosaic land uses developed through foreign investment and globalization has been seen across Southeast Asia. McGee (1991) coined a term—*desakota*—from Indonesian to describe the mixture of rural villages and industrial areas. In Southern China, export-oriented rural industrialization has created a novel form of the built environment. A new type of building, known as 'three in one', combines workshop, warehouse and residential uses in the same building. This is in response to the need to accommodate migrant workers in workshops operated by families. The mixture of industrial and residential uses creates environmental and public health problems. Although individual workshops have been developed, public services at the village and town levels are still lacking and insufficient. Because of the lack of suitable accommodation for workers, many bigger factories have to build their own dormitories, creating a new urban landscape of dormitories in former rural areas. Similarly, in peri-urban areas, villages have evolved into enclaves of rental housing to accommodate rural migrants. Farmers extended and even built houses in order to rent them out. Despite different appearances, these self-built environments lack public services funded by the government. Thus, service delivery in these places is not a function of the municipal government but rather is achieved by village collectives from rental income (Xue and Wu 2015). The emergence of manufacturing industries in rural China has led to urbanization and the development of auxiliary residential uses. But an urban institution has not yet been formed. This emergent built environment shows a great degree of informality. The rental market does not guarantee security of tenure and often there is no formal rental contract (Wu 2016b). In part this is because migrant workers are mobile. But more fundamentally, this is due to a housing policy which stresses homeownership and the development of so-called 'commodity housing'. Control over self-built activities in the cities such as Shanghai has led to internal subdivisions into smaller spaces rather than to renovation and upgrading. Informal rental housing in villages faces great uncertainty.

In contrast to informal development, mainstream residential development in China is in planned estates of commodity housing, which resemble 'gated communities' or master-planned estates. These are built to suit the rising aspirations of the new middle class for privacy and a new lifestyle. But the main driver for these emerging landscapes is the Chinese land development model, which relies heavily on real estate development to generate land revenue for local public finance. The development is thus a strategy under entrepreneurial urban governance, often associated with development zones and new towns. The new town is a more advanced form of development zone. Instead of relying only on industrial development, new towns are mixed developments and are driven by the forces of both market and state. They are often associated with the strategic objectives of municipal government, in which real estate development has been introduced as an instrument to generate funding for local public finance. In practice, coordination between industrial and residential development is often problematic, leading to the problem of so-called 'separation between industrial and urban uses'.

These new towns are emerging not just owing to economic agglomeration but also as a result of specific development strategies and institutions. They represent a breakthrough from earlier stages of rural industrialization. Under the strategy of 'one city, nine towns' in Shanghai, Lingang new town has been developed near the deep-water port of Shanghai. The new town was supposed to support the development of heavy equipment manufacturing industries, logistics and activities related to the deep-water port. The entire area of Lingang covers a vast 320 km^2, a large proportion of which was reclaimed from the sea. The design of a German-based firm, GMP, was adopted as a master plan which was allegedly inspired by the Garden City concept. Accessibility is still a major constraint, despite a transit line (No. 11) linking to central Shanghai. Within the new town area, the heavy equipment manufacturing zone and the new town are separated by a long distance. It takes 30–40 min to travel from the new town to the industrial area by car. However, the majority of industrial workers in the development zone do not travel by private automobile. They are rural migrants who have to either rely on shuttle buses or rent informally in nearby areas because they cannot travel far. Heavy manufacturing industries are capital intensive and have not led to the level of population growth expected in the new town. It is difficult for the new town to attract residents from central Shanghai. The population growth of the new town is slower than the target. On the other hand, the heavy industrial area has been developed as planned. It is difficult for these factories to find accommodation for their workers who have to go to nearby rural areas to rent. Some factories have even tried to convert and renovate existing buildings for workers' accommodation. For example, a furniture market, originally developed by rural villages, near an industrial area has been converted into staff living quarters. Clearly, there is a demand for low cost housing near industrial areas. But the way of building the city depends upon specific local conditions and involves complex institutional constraints. In this case, it is related to several contexts: long-lasting urban–rural dualism, the entrepreneurial state which planned an industrial area for expected international industrial relocation (for modern manufacturing), a master plan that imported the idea of the 'garden city', and an institutional arrangement that separates industrial and residential developments. While the industrial area is developed by the municipal corporation—the Lingang Economic Development Group—the new town is built by the Harbour-City Development Corporation owned by the district government. All these governance arrangements affect new town developments.

Chinese new town development shows how cities are built through different drivers. Export-oriented rural industrialization has led to a spontaneous and rudimentary form of cities without much public infrastructure. They are literally a 'regime of dormitory'. But there are also more strategically planned new towns. Agglomeration plays a role as seen in the inflow of migrant workers who are not able to find formal accommodation and have to rent in nearby 'urban villages'. As the name suggests, these forms of 'cities' are partially urban and partially rural. They are informal rental markets connecting with formal industries, but their tenants are not served by formal urban institutions. Finally, strategic new town development relies on state planning intervention and institutional innovation. Their governance adopts a model of 'state entrepreneurialism' (Wu 2018; Wu and Phelps 2011). New towns are governed by the 'development and management committee' and developed by state-owned development corporations.

10.6 Conclusions

China provides a useful perspective to understand the emergence of cities. Chinese cities do not represent a singular model of emerging urbanism. Rather, they contain some constellated elements to understand the multiple forces that lead to the creation of an urban living environment and governance. Emerging cities in China reveal that a new world-scale development process interacts with local institutions and structures, during which new properties and features are created. Therefore, they are emerging in this sense because the development is not fully planned but rather contingent upon agencies and histories. Emergence is thus a complex phenomenon. Chinese cities see the driving force of globalization but local institutions (e.g. household registration—*hukou*—and state land monopoly) are indispensable factors. For example, global R&D chooses Shanghai because of its gateway status, strong economic base, and concentration of science and education resources. The location of biotech firms in Shanghai is due to the development of economic spaces such as Zhangjiang High-tech Park. Its development has been driven by many factors: the central government policy to build an innovation nation, the entrepreneurial local government to build spaces for biotech and IC industries, and the opportunity for market access and contract research for multinational pharmaceutical companies.

The development of new cities has witnessed contrasting approaches applied simultaneously. For example, the state retreated from housing provision and thus left rural migrants entirely reliant on the private rental market. On the other hand, the state controls the land and supports strategic industrial development in places such as Lingang new town of Shanghai and Yizhuang new town of Beijing. Aggressive land acquisition and the absence of housing provision show the seemingly contradictory nature of state entrepreneurialism (Wu 2018).

The China perspective shows the hybrid nature of urbanization led by the market mechanism and the state, which has resulted in different local configurations in different economic sectors and different spaces. Emerging cities require a more holistic approach to understand the dynamics of urban development and the need to accept different possibilities.

Agglomeration effects undeniably exist in the process of urbanization, turning Chinese cities into sites for global production. But the city is being created, remade, shaped and inhabited by multiple actors, whose primary motivation may not be related to agglomeration. In a sense, the city has been rediscovered in the post-reform era because it entertains various possibilities and needs. It is thus important to pay more attention to the process of urbanization rather than resorting to the general notion of agglomeration, as this can help to generate a more nuanced understanding of its outcomes. China is quite specific, as elsewhere, in terms of institutional arrangements related to urbanization and political economic changes. But as shown in this chapter, emerging cities cannot be simply attributed to the state or the market. Although new towns are master-planned and policies play a role in their development, they are not a product of government policies. Their development has been also achieved through market instruments such as development corporations. Their emergence involves both complex market operations and governance. In this regard, Chinese cases can offer a perspective to view these complex relations in contemporary urban changes.

Acknowledgements I would like to acknowledge the funding support from UK Economic and Social Research Council (ESRC) (ES/P003435/1). I also thank Denise Pumain, Natacha Aveline, Clémentine Cottineau, and Elfie Swerts for their constructive comments on an earlier version of this chapter.

References

Bathelt, H., Malmberg, A., Maskell, P.: Clusters and knowledge: local buzz, global pipelines and the process of knowledge creation. Prog. Hum. Geogr. **28**(1), 31–56 (2004)

Boschma, R.A., Frenken, K.: Why is economic geography not an evolutionary science? Towards an evolutionary economic geography. J. Econ. Geogr. **6**(3), 273–302 (2006)

Brenner, N.: Urban governance and the production of new state spaces in western Europe, 1960–2000. Rev. Int. Polit. Econ. **13**(3), 447–488 (2004)

Charles, D.R.: Making 21st century knowledge complexes: technopoles of the world revisited. In: Miao, J.T., Benneworth, P., Phelps, N.A. (eds.) Making 21st Century Knowledge Complexes: Technopoles of the World Revisited, pp. 82–102. Routledge, Abingdon (2015)

Cooke, P.: The molecular biology revolution and the rise of bioscience megacentres in North America and Europe. Environ. Plan. C: Gov. Policy **22**(2), 161–177 (2004)

Cooke, P., Morgan, K.: The Associational Economy. Oxford University Press, Oxford (1998)

Cooke, P., de Laurentis, C., Toedtling, F., Tripp, M. (eds.): Regional Knowledge Economies: Markets, Clusters and Innovation. Edward Elgar, Cheltenham, Glos, UK (2007)

Etzkowitz, H., Leydesdorff, L.: The dynamics of innovation: from National Systems and "Model 2" to a Triple Helix of university-industry-government relations. Res. Policy **29**, 109–123 (2000)

Hall, P., Pain, K.: The Polycentric Metropolis: Learning from Mega-city Regions in Europe. Earthscan, London (2006)

Harrison, J.: Rethinking city-regionalism as the production of new non-state spatial strategies: the case of peel holdings Atlantic gateway strategy. Urban Stud. **51**(11), 2315–2335 (2014)

Jonas, A.E.G., Ward, K.: Introduction to a debate on city-regions: new geographies of governance, democracy and social reproduction. Int. J. Urban Reg. Res. **31**(1), 169–178 (2007)

Leitner, H., Sheppard, E.: Provincializing critical urban theory: extending the ecosystem of possibilities. Int. J. Urban Reg. Res. **40**(1), 228–235 (2016)

Li, Y., Wu, F.: Understanding city-regionalism in China: regional cooperation in the Yangtze River Delta. Reg. Stud. **52**(3), 313–324 (2018)

McGee, T.G.: The emergence of desakota regions in Asia: expanding a hypothesis. In: Ginsburge, N.S., Koppel, B., McGee, T.G. (eds.) The Extended Metropolis: Settlement Transition in Asia, pp. 3–25. University of Hawaii Press, Honolulu, HI (1991)

Parnell, S., Pieterse, E.: Translational global praxis: rethinking methods and modes of African urban research. Int. J. Urban Reg. Res. **40**(1), 236–246 (2016)

Peake, L.: The twenty-first-century quest for feminism and the global urban. Int. J. Urban. Regional. **40**(1), 219–227 (2016)

Porter, M.E.: Location, competition, and economic development: local clusters in a global economy. Econ. Dev. Q. **14**(1), 15–34 (2000)

Robinson, J.: Comparative urbanism: new geographies and cultures of theorizing the urban. Int. J. Urban Reg. Res. **40**(1), 187–199 (2016)

Robinson, J., Roy, A.: Global urbanisms and the nature of urban theory. Int. J. Urban Reg. Res. **40**(1), 181–186 (2016)

Scott, A.J.: Globalization and the rise of city-regions. Eur. Plan. Stud. **9**(7), 813–826 (2001)

Scott, A.J., Storper, M.: The nature of cities: the scope and limits of urban theory. Int. J. Urban Reg. Res. **39**(1), 1–15 (2015)

Simone, A.: It's just the city after all! Int. J. Urban Reg. Res. **40**(1), 210–218 (2016)

Walker, R.A.: Why cities? A response. Int. J. Urban Reg. Res. **40**(1), 164–180 (2016)

Wu, F.: China's emergent city-region governance: a new form of state spatial selectivity through state-orchestrated rescaling. Int. J. Urban Reg. Res. **40**(6), 1134–1151 (2016a)

Wu, F.: Housing in Chinese urban villages: the dwellers, conditions and tenancy informality. Hous. Stud. **31**(7), 852–870 (2016b)

Wu, F.: State entrepreneurialism in urban China. In: Pinson, G., Journel, C.M. (eds.) Debating the Neoliberal City, pp. 153–173. Routledge, Abingdon, Oxon (2017)

Wu, F.: Planning centrality, market instruments: governing Chinese urban transformation under state entrepreneurialism. Urban Stud. **55**(7), 1383–1399 (2018)

Wu, F., Phelps, N.A.: (Post-)suburban development and state entrepreneurialism in Beijing's outer suburbs. Environ. Plan. A **43**(2), 410–430 (2011)

Xue, D., Wu, F.: Failing entrepreneurial governance: from economic crisis to fiscal crisis in the city of Dongguan, China. Cities **43**, 10–17 (2015)

Zeller, C.: The pharma-biotech complex and interconnected regional innovation arenas. Urban Stud. **47**(13), 2867–2894 (2010)

Zhang, F.: Building biotech in Shanghai: a perspective of regional innovation system. Eur. Plan. Stud. **23**(10), 2062–2078 (2015)

Zhang, F., Wu, F.: Rethinking the city and innovation: a political economic view from China's biotech. Cities **85**, 150–155 (2019)

Chapter 11
China's Housing Booms: A Challenge to Bubble Theory

Natacha Aveline-Dubach

Abstract Over the past two decades, Chinese cities have experienced real estate booms displaying clear signs of "bubble" elements, including, *inter allia*, prohibitive residential prices, an accumulation of debt, and severe overbuilding. In 2014, many media commentators claimed that Chinese property markets were about to burst. Yet house prices have started to rise again in major cities, and no significant slowdown has been recorded to date. This chapter addresses the challenges posed by China's residential market dynamics to the bubble theory. Adopting a political economy perspective that breaks with the approaches of real estate economics, it highlights the self-fulfilling logic of the housing booms, resulting from pervasive practices of land value capture by local governments. The paper stresses the inadequacy of the bubble framework to distinguish speculative and "fundamental" explanatory factors of price increases, and provides an alternative reading based on André Orléan's theory of conventions. It is argued that the asymmetric nature of the State's regulation of housing markets—a failure to rein in housing price hikes, yet efficiency in managing downturns—has played a crucial role in shaping the common representation of the market by investors. Beyond the challenge to bubble theory, China's experience of housing booms opens the way for the recognition of alternative paths to finance-led regimes of capital accumulation in the built environment.

11.1 Introduction

Since the 2000, residential prices have undergone a rapid acceleration in China, resulting in a growing divergence between the cost of housing and the dwellers' incomes. In the early 2010s, foreign media expressed concern that a major housing bubble had formed. This sentiment was supported by Chinese top officials' declarations that housing prices had become a source of worry. The buzz in the international media culminated in 2014, when residential prices recorded a small drop in Beijing and Shanghai. All ingredients of an imminent real estate crash seemed to be in

N. Aveline-Dubach (✉)
CNRS UMR Géographie-cités, Aubervilliers, France
e-mail: Natacha.AVELINE@cnrs.fr

D. Pumain (ed.), *Theories and Models of Urbanization*, Lecture Notes in Morphogenesis, https://doi.org/10.1007/978-3-030-36656-8_11

place: extravagantly high residential prices relative to household incomes in major city-regions; a massive oversupply of residential space across the country; the large share of non-monitored ‘shadow banking’ in real estate financing; and wide media coverage even funded in China. The pervasive so-called ‘ghost city’ phenomenon throughout China was regarded as evidence of the dramatic magnitude of the bubble. Then, beyond all expectations, residential prices started to soar again in the biggest cities including Shanghai and Shenzhen, and increases spread to lower-range cities, without being followed by a dramatic price downturn at the time of the writing.

The purpose of this chapter is not to provide yet another test of the bubble hypothesis. Rather, it intends to highlight the challenges posed by China’s residential market dynamics to the bubble theory. It is argued that current housing booms are operating as self-generating processes of urban and economic growth, but that this very dynamic tends at the same time to skew the perception of risks by home buyers, thus creating distortions in the supply/demand balance. In such a context, it becomes impossible to distinguish a clear-cut divergence between actual (speculative) and fundamental values, which theoretically characterizes bubbles. To get a comprehensive understanding of the underlying forces at play in the residential markets, the political economy of China’s urban development must be taken into account, since the current housing booms are deeply rooted in a land-centered economic growth model. Drawing on previous work on China’s unique mode of urbanization (Jiang et al. 2016; Hsing 2010; Chien 2013; Lin 2014; Wu 2015a; Theurillat 2016, to quote a few), this chapter conceptualizes China’s housing booms as outputs of State productivist policies basing action on massive land value capture and exploitation of the rural-urban divide. It contends that the success of this model has anchored a cognitive bias amongst home buyers, or a ‘convention” as defined by Orléan (1999), which encourages them to disregard the warning signs of oversupply in residential markets. To avoid the break-up of this convention, Chinese policy-makers are maneuvering through tight spaces, providing quick responses to market signals while pushing forward new developmental strategies. In this respect, China’s experience provides evidence of alternative pathways to addressing the challenge of speculative behavior in the housing markets.

This chapter is divided into three parts. First, it discusses the underlying assumptions of the bubble theory in finance economics, and their application to property markets. The second part points out the inadequacy of the bubble approach in providing a systemic analysis of the speculative mechanisms in China’s residential sector; it then develops an alternative framework incorporating the housing booms in China’s land-centered model of economic development. The third part underlines the strong commitment of the State in regulating the excesses of this model, while seeking a way forward through a new productivist approach.

The method used has involved the reading of academic articles, publications of the real estate industry, policy documents and newspaper articles. It also draws on fieldwork studies of local real estate markets conducted in a dozen of Chinese cities of various sizes, and follows up on previous publications by the author.

11.2 The Theory of "Speculative Bubbles" and Its Application to Property Markets

The theory of asset price bubbles can be traced back to the 1980s, when a series of financial crises began to undermine seriously the then dominant neoliberal dogma that financial markets are intrinsically self-regulating. Drawing on the rediscovery of Keynes' findings concerning speculative behavior in stock markets, a new stream of literature started to develop upon the assumption that investors' strategies can create, under some conditions, a significant divergence of asset prices[1] from the value that would be an appropriate reflection of their underlying demand and supply positions. Such divergence phenomena were called "bubbles", defined by Kindleberger (1991) as "a sharp rise in the price of an asset or a range of assets in a continuous process, with the initial rise generating expectations of further rises and attracting new buyers—generally speculators interested in profits from trading in the asset rather than its use of earning capacity. The rise is usually followed by a reversal of expectations and a sharp decline in price often resulting in financial crisis".

11.2.1 Speculative Versus Rational Bubbles

Conceptually, a bubble is regarded as a divergence between the actual market price of the assets and their intrinsic, so-called "fundamental" values. However, the characterization of bubbles has proved highly controversial. While many authors contend that bubbles are a purely speculative phenomenon, driven by investors' belief that prices are bound to increase even when this is not justified by fundamental factors, neoliberal economists claim that bubbles can develop even with rational expectations by investors. For example, the Nobel Prize winner Tirole (1985) demonstrated that a "rational bubble" may arise from differential information between traders, and could be kept alive indefinitely without exploding, because rational traders with perfect foresight would ensure a final economic equilibrium. Yet this result is a single solution of a model of equilibrium, based on the assumption that the market has a number of limited agents who are additionally infinitely-lived (the model assumes overlapping generations of traders). Stiglitz (1990) argues that one can always find a unique solution to a rational bubble amongst a large variety of equilibria, and he criticizes the assumption of infinitely-lived agents as being highly unrealistic. In the face of objections from numerous authors, the rational bubble hypothesis has not given rise to a flourishing strand of literature.

Empirical models of "speculative bubbles", on the other hand, have developed strongly, but have also been subject to criticism. Flood and Hodrick (1990) analyzed several models of famous past bubbles and demonstrated that they suffered from the misspecification or from the underestimation of the role of fundamentals. The

[1] An asset is a resource with economic value that an entity owns or controls, with the expectation that it will generate income.

core problem lies in the difficulty of assessing the "fundamental price" of the assets. Financial economists use a wide range of analytical methods to compute this price, but the typical formulae combine two main elements: (i) the discounted present value of the dividend (or rent for real estate) to the owner of the asset during the ownership period (d_{t+1}); and (ii) the expected value that the asset will have at the end of the investment period (q_{t+1}). Flood and Hodrick (1990: 88) provide the following formula:

$$q_t = E_t \, . \, \frac{d_{t+1} + q_{t+1}}{1 + r}$$

where q_t is the fundamental price of the asset, r the discounted rate, and E_t (d_{t+1} + q_{t+1}) the expected value of the future dividend (rents for real estate) and the future price of the asset at time t.

As noted by Stiglitz (1990), this approach faces three problems. The first one lies in the need to forecast future long-term revenue streams while asset prices are generally subject to irregular business cycles. The models respond by generally assuming a regular pattern of revenue streams. The second problem is equally challenging: how can the terminal value of an asset be predicted several years ahead, when financial markets are subject to both endogenous and exogenous shocks? Thus, Stiglitz's question: "how do economists test whether the terminal price can be justified by fundamentals, without having data extending infinitely into the future?" (Stiglitz 1990:15–16). The financial theory assumes that the terminal value is the expected value of future dividend and future price of previous periods, and thus calculates it based on a recursive process of the same equation. Finally, the third problem is to define the discount rates needed to translate future returns into current values. Here again, financial analysts have no choice but to forecast unpredictable conditions based on current interest rates.

11.2.2 Cognitive Bias in the Formation of Asset Prices

A major puzzle that econometric models are also unable to address is the conditions in which a bubble may initiate and terminate (Hui et al. 2012). While the field of behavior finance has provided valuable insight into investor's cognitive bias (see De Bondt 2003 for a review), the French School of Convention was the first to analyze the behavior of financial investors in relation to the successive dynamics of asset prices during a boom-bust cycle.

Orléan (1999) contends that it may be rational for well-informed financial investors to follow the trend of a speculative wave if the market is driven by the perceptions of market players rather than by fundamental values. The reason is that investors make a greater profit by benefiting from the increase in asset prices rather than by betting against the tide of the market to restore equilibrium. Therefore, according to Orléan, the mechanism of price formation relies on a complex set of interactions between fundamental factors and collective cognitive dynamics. The nature of these interactions evolves over time, shaping different sequences in the speculative mechanism. The initial phase of asset inflation is triggered by the expectation of a significant change in the fundamentals of a given economy. Investors endeavor to achieve a common vision of these new market conditions, and gradually converge on a shared perception, a 'convention' (agreement) relying on a fundamentalist diagnosis (for example, the 'Internet convention' caused in the late 1990s by the emergence of new information and communication technologies). During this phase, investors tend to select information that confirms the relevance of the convention, filtering away warning signals. Nevertheless, as conflicting information on the convention accumulates, some investors come to question the convention and to develop strategies that challenge it. The third phase of investor behavior is thus marked by what Orléan calls a 'self-referential crisis', in which all investors lose their faith in the convention and focus on each other's strategies, adopting a herd behavior irrespective of fundamentals, thereby precipitating the fall in assets prices.

By embedding market's endogenous factors within asset price dynamics, the convention framework opens up new perspectives to address speculation mechanisms. It breaks with the key economic postulate that supply and demand are independent from each other. The emergence of a 'convention' leads to interactions between supply and demand, since the escalation in asset prices does not discourage investors' demand from growing. When investors' behavior is affected in such a way that it *transforms the fundamentals*, the convention may even become a self-fulfilling prophecy. In such a situation, investor's expectations are not validated because they are in accordance with fundamentals, but because they provoke behavior that makes them true a posteriori.

11.2.3 Application of the Bubble Theory to Property Markets

So far, the bulk of the bubble literature has explored asset price dynamics in financial markets. This should not come as a surprise, as most of the memorable crashes in history took root in finance. It was not until the 1980s that real estate markets started to experience frequent speculative mechanisms of large magnitude. The main cause of this change was the deregulation of financial markets, and the subsequent internationalization and de-specialization of banking businesses. Deregulation of financial markets originated in the US in the early 1980s, spread to Europe and Japan in the mid-1980s, and then to other East Asian countries. For the first time in history, synchronized boom-bust cycles arose globally in property markets (Mera and Renaud 2016). More was to follow: structural reforms carried out in the finance industry established a wide range of new financial channels dedicated to property investment (private equity, REITs), thereby increasing dramatically the mobility and liquidity of capital invested in real estate (Corpataux et al. 2009; Theurillat and Crevoisier 2013; Aveline-Dubach 2017b). A major outcome of this change has been the rising power of global institutional investors who have become a driving force of urban (re)development projects in key city-regions around the world (Aveline-Dubach, forthcoming). Urban policies have tended to rely increasingly on the imperatives of investor's targeted risk-adjusted returns, thus exposing the urban built environment to the hazards of global finance (Halbert et al. 2014; Savini and Aalbers 2015).

Despite its 'quasi-financial' status as an asset class (Coakley 1994), real estate differs significantly from its purely financial peers. As un underlying asset of financial vehicles, it stands out by being both a *tangible* and *localized* commodity. These characteristics make real estate particularly prone to bubbles. Because of long production lags in property development, real estate is subject to intrinsic cyclical patterns (Barras 1983). Supply is slow to follow demand, and delay in adjustment of prices to fundamentals are potential drivers of market distortions (Ball and Wood 1999). Added to that, population ageing in advanced countries has increased the demand for property as a saving vehicle; with the proliferation of investment channels, episodes of capital over-accumulation in real estate have arisen in many places, entailing threats of construction oversupply. Because real estate investing is more spatially selective than purchase for owner-occupancy, increasing competition for scarce land puts upward pressure on property prices in sought-after, usually central, urban locations. This effect is amplified by the positive externalities generated by public investment in these places to improve cities' competitiveness.

Equally important, if not more so, the large amount of loan borrowing for real estate transactions has a magnifying effect on the amplitude of property boom-bust cycles. Herring and Wachter (2003) explain this effect by the behavior of bank managers. During the upward phase of the booms, banks are encouraged to extend credit as the value of loans collateralized by real estate grows. When the downward phase of the cycle arrives, the drop in prices downgrades the value of banks' collateral and asset holdings. Banks respond by rising interest rates as risk premiums and consider the higher probabilities of shocks. Credit tightening soon transforms into credit rationing for all sectors of the economy, as banks try to rebuild their reserves, which puts further downward pressure on property prices. Needless to say, the effects described above are all the more exacerbated when banking credit is highly leveraged. Housing mortgages have a fairly high loan-to-value ratio, of over 70% worldwide according to a cross-national survey by IMF (quoted in Crowe et al. 2013). It is no surprise then that the vast majority of recent systemic banking crises has been associated with housing boom-bust cycles (two-thirds of the 46 crises analyzed by Crowe et al. 2013). Real estate slumps are also more harmful than stock market crashes: output losses are twice as big according to an estimation by Helbling and Terrones (2003), because they play an important role in collateralizing loans, and they hit a larger number of households.

While real estate assets have noticeable cross-national commonalities, their markets are embedded in highly differing local institutional and social structures (Wood 2004). Therefore, there is a strong idiosyncrasy in the way property prices behave, with asset prices depending, *inter allia*, on local regulations, the physical and social aspects of the urban fabric, credit conditions, cultural norms and practices of property and related players. These characteristics are often not paid much attention in the bubble tests, but they do influence the way analysts design their models.

11.2.4 Empirical Studies on Real Estate Bubbles

Due to the growing volatility in real estate markets over the past decades, a large number of scholars have come to develop econometric tests of bubbles. However, the complexity of these markets has led to a variety of methods for the assessment of fundamental values. Beyond this diversity, three main approaches tend to prevail.

A fist approach consists in applying financial theory to real estate and to estimate housing prices based on forecast future revenue streams. For a housing investment, the price of the asset is calculated through the method mentioned above, which analyses future rental cash flows, and discounts them (with a targeted rate of return) to calculate an estimate of the present value. As discussed earlier, these models rely on highly simplified assumptions.

In contrast to the financial approach, which does not consider the characteristics of local property markets, the two other methods calculate fundamental housing prices based on a regression of actual prices on a set of demand and supply variables. One method focuses on local socio-economic and demographic factors. It combines variables such as households' incomes, demographic change, the employment rate, (regional) GDP per capita, housing starts and mortgage conditions. Another method, based on the hedonic approach, places emphasis on the attributes of property themselves (size, appearance, various features, and conditions), and their location characteristics (accessibility, schools, environmental factors and crime rate, to name a few).

While these different approaches may be combined to produce sophisticated models, the mere fact that they rely on such different ways of looking at intrinsic housing values is sufficient in itself to question their explanatory power. Bubble tests are nevertheless powerful in raising public concern about the potential threats of speculative housing markets for macroeconomic stability.

In parallel to econometric models, qualitative research provides new insights into market perceptions and the behavior of real estate investors. In a frequently cited paper, Case and Shiller (1988) compared the investment behavior of 5,000 home purchasers in four US metropolitan areas with contrasting market conditions, and found that buyers in the booming cities perceived little risk and showed strong investment motives. This was because first-time buyers believed that they should hurry up and invest in a home before the prices became totally unaffordable. These results provide empirical support for Orléan's contention that investors' shared perceptions of the market are a primary driver of asset price formation.

11.3 Housing "Bubbles" with Chinese Characteristics

The rapid escalation of housing prices in China is seen by many commentators as the manifestation of the next real estate bubble of worldwide importance. Indeed, the recent increase in value of China's homes (typically apartments in residential

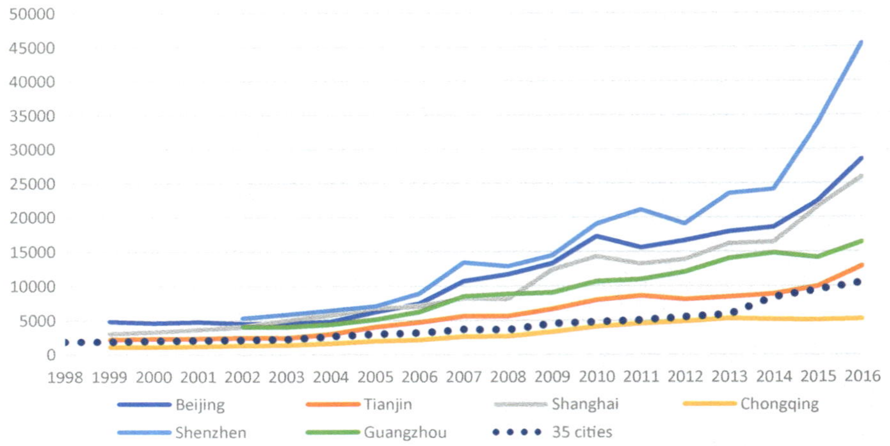

Fig. 11.1 Change in housing prices in major Chinese cities (in CNY per square meter). *Source* National Bureau of Statistics of China

complexes) has been particularly impressive: according to underestimated official data (see below), residential prices in China's largest 35 cities increased by nearly sixfold from 1999 to 2016, and by more than eightfold in Shanghai in Shenzhen (Fig. 11.1). Price hikes have shown a visible acceleration in the past decade, after the adoption of an extremely ambitious stimulus package to mitigate the impact of the global financial crisis in late 2008 (CNY 4 trillion, equivalent to US$ 586 billion). In 2017, the Chinese banking sector estimated the price-income ratios at 32 in Shenzhen, and around 26 in Shanghai and Beijing.[2] Such dramatic price growth has quite naturally drawn the attention of scholars, and Chinese cities have become the new target for testing the bubble hypothesis.

Attempts to develop econometric models nevertheless face serious problems of data availability and accuracy. It must be recalled that China's real estate markets are very young, having existed for only three decades. The first legislation relating to property ownership in the transitional phase to a market economy was implemented in 1988. It enacted the separation of urban land ownership, which remains under State control, and land use rights. These are transacted by local governments at market-determined prices,[3] for fixed periods ranging from 20 to 70 years, the latter for residential uses. It was not before 1999 that official data on housing became available. Existing data series can thus only capture an ascending phase of housing prices, which does not allow prices in China's housing markets to be compared

[2]This figure is taken from China Banking News, July 6, 2017. http://www.chinabankingnews.com/2017/07/06/housing-price-income-ratio-exceeds-10-across-16-chinese-cities/ (checked on September 20, 2018)

[3]The decentralization process operated in the 1980s conferred great freedom on local governments to manage their economies. In 1998, the sale of land-use rights was put under their jurisdiction. The same year, the public residence allocation system was converted into a system of marketization.

across cycles. Furthermore, official data are not quality-adjusted. They are obtained by dividing the total sales revenues of commercial residential buildings by the total floor area sold, and are therefore seriously biased downwards (Wu et al. 2014).

11.3.1 Bubble Prices Versus Fundamentals in Chinese Housing Markets

A number of authors have reported the existence of bubble phenomena in several Chinese cities. Scholars started to observe price deviations from fundamental values as early as in the 1990–2000s in some major coastal cities (for example, Yue and Hongyu 2004; Zhou 2005). However, the majority of tests were conducted with more recent data. Using an independent data set controlling for housing quality,[4] Dreger and Zhang (2013) calculated fundamental values based on macro-economic factors in 35 major cities from 1998 to 2009, and found that actual prices exceeded the fundamental values by 25% on average. They observed a larger magnitude of the bubble in the South-eastern coastal areas and special economic zones. Liu and Sun (2009) used a method based on price-to-income ratios and also concluded that major Chinese cities were experiencing bubbles, with particularly acute deviations of prices in first-tier cities.[5] Some tests focusing on Beijing, Shanghai or Guangzhou provided further evidence of a serious misalignment of prices that can be interpreted as bubbles (Lyu 2010; Chen 2012). Chen and Wen (2017) approached the bubble hypothesis from the perspective of labor market dynamics. According to them, the bubble is the result of rational expectations by investors, with corporate investors (both SOEs and private firms) accounting for a significant share. Their decisions in favor of housing investment have been sustained by strong fundamentals resulting from the reallocation of labor and capital from a less-productive conventional public sector towards more productive private firms—especially following the restructuring of SOEs in 1997. Chen and Wen argue that housing investment is "driven by the expected strong future demand for housing, which is self-fulfilling and rationalized by the fact that the future rate of capital returns, will be sufficiently low in the post-transition stage. Under such an expectation, holding housing today can yield large capital gains tomorrow even if housing has no intrinsic value" (Chen and Wen 2017: 6). Their model thus predicts the deceleration of the currently fast-growing bubble as the surplus labor in rural areas falls.

Other authors have expressed more reserved views about the characterization of bubble. Ahuja et al. (2010) ran a model based on macroeconomic factors (real

[4]Wu et al. developed an independent data set using hedonic prices to control for housing quality. This dataset covers indices based on sales of newly built housing units in 35 major Chinese cities.

[5]China's cities are divided into four categories. The first tier includes the four most developed metropolitan areas: Beijing, Shanghai, Shenzhen, and Guangzhou; the second tier includes most provincial capitals and some very developed prefecture cities. Tier 3 cities include prefecture cities that have medium to high levels of income, while Tier 4 cities are further behind economic development and smaller in size.

lending interest rate, population density, real GDP per capita, and land prices) in 35 cities over the period 2000–2009. Their model rejected the bubble hypothesis for China as a whole, but indicated excessive prices relative to long-term fundamentals in several coastal cities (Beijing, Shanghai, Shenzhen, Ningbo, Fuzhou and Xiamen) and in two inland cities (Wuhan and Kunming). Ren et al. (2012) made a more radical rejection of the bubble hypothesis. Their model did not estimate fundamental values, but yearly residential returns (rental income and capital income) in 35 cities between 1999 and 2009. The forecasted returns were estimated based on a set of macroeconomic factors. The authors did not find evidence of a bubble, but explained the high returns on housing investment by the circulation of capital from rich to poor regions.

Whatever the characterization of China's housing booms, the bubble tests converge on a diagnosis that housing prices are at 'above-fundamental' levels, with large variations across cities depending upon their size and location. China is clearly facing a speculative mechanism that has taken root in its major city-regions, especially the coastal ones, and is moving towards lower-tier cities and inland regions, partly fueled by inter-urban capital flows. Yet these dynamics are far from being purely speculative. As underlined by Chen and Wen (2017), the reallocation of labor from low-productivity to high-productivity sectors has been a major fundamental driver of increases in residential prices. However, such a finding only accounts for part of the formation of housing prices. To fully grasp the underlying forces of the housing booms, it is necessary to re-situate them in the process of China's unique urbanization model. I contend that this model combines three key elements: a productivist approach in State policies, a mechanism of large-scale land value capture, and the exploitation of the deep rural-urban divide.

11.3.2 The Housing Bubble as a Creation of the Developmental State

In the post-Mao era, the main mission of the China Communist Party (CCP) was redefined as "from achieving Communism to achieving rapid economic growth" (Heilmann 2008). China's State thus took inspiration from its neighboring East Asian 'developmental states' (Johnson 1982) to achieve a State-guided economic miracle (Baek 2005). Yet China's policy-making is shaped by very distinctive dynamics, defined by Chien (2010) as an 'asymmetric decentralization' process, whereby the large autonomy granted to local governments to develop their economies is counterbalanced by political centralization under the CCP. Local officials are not elected but assigned by the Party, which means that their career advancement depends on their ability to generate GDP, accommodate FDI, and increase trade (Chien 2007). So far, this has been mainly achieved by providing high-quality urban physical infrastructure and supplying cheap land to export-oriented manufacturing enterprises (in Special

Economic Zones and in peripheral urban areas), thereby releasing cheap labor from the State sector to the more productive private sector.[6]

The real estate industry is also a major pillar of economic growth, and as such has been strongly encouraged to expand, especially in the housing sector (Wu 2012). Home ownership is emphasized by the State as a major determinant of social status (Hu 2013). Housing units are additionally key status goods for marriage purposes in China, due to the gender imbalance resulting from the one-child policy. It is estimated that there are currently 30 million more men than women aged 25 or more, and young males who own a home will have a greater chance of finding a bride (Wei et al. 2012). Buying a residential unit is strongly encouraged by the substantial financial aid from the parents, especially in one-child families (Or 2018). Home ownership also serves as a store of wealth for the old age, in a 'productivist welfare regime' (Holliday 2000) characterized by the underfunded provision of pension and healthcare.

The State has encouraged the purchase of newly built residential units supplied by property developers though several programs, including the Housing Provident Fund Program that involves employers' contributions to employees' mortgage loans at better terms than conventional mortgages. Mortgages delivered by banks to home purchase mainly target newly-built housing units. Over the past ten years, 70% of housing sales in the 35 Chinese major cities have been newly-constructed homes (Zheng et al. 2016).

The productivist policy also involves the strong control of financing channels, so as to allow State-led allocation of capital towards targeted industries via the banking sector. Market finance is restricted. Banking channels are prevalent, with the "big four" state-owned banks controlling more than half of loans and assets.[7] Interest rates on checking and saving accounts are kept below the market rate, to ensure cheap credit to privileged firms—primarily SOEs and, to a lesser extent, major private groups. Stock markets are poorly regulated and dominated by SOEs, the exchange rate is tightly managed, and capital outflows from China are closely controlled. As a result, the availability of financial assets as saving vehicles for Chinese economic agents is very limited, which makes housing a most desirable investment target. Given the wide array of agents that have limited access to banking credit, informal finance has developed through 'shadow banking', a poorly regulated sector including small loan companies and usury-rate lenders, but mostly revolving around the indigenous trust

[6]The number of workers in SOEs stood at 122 million in 1998, and fell thereafter to 76 million in 2006. The remaining SOEs account for 30% of the total assets in the secondary and tertiary sectors. They are bigger, more competitive and more profitable than were former SOEs (Ren 2013).

[7]The "big four" state-owned banks are the Industrial and Commercial Bank of China, the Bank of China, the China Construction Bank, and the Agricultural Bank of China.

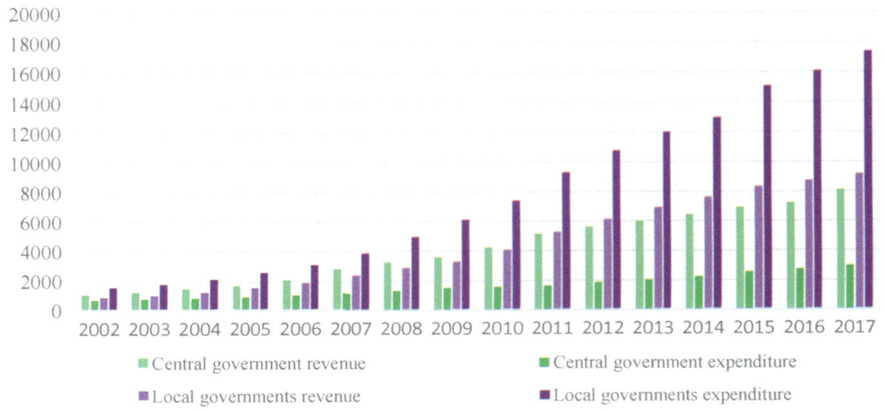

Fig. 11.2 Increase of local governments' funding gap (billion CNY). *Source* National Bureau of Statistics of China

industry[8]. A significant share of informal credit from this composite ensemble is linked to banks (Sherpa 2013).

Local governments have also increasingly relied on shadow banking to deal with the rapidly increasing cost of urbanization. The centralization of tax resources in 1994 left them with a large funding gap that they were prohibited from resolving by borrowing directly in the markets or even from banks (Fig. 11.2). To boost their economies, local governments have invested heavily in the construction of urban infrastructure. They have developed intermediary bodies known as Local Government Financing Platforms (LGFPs), which raise extra-budgetary capital through bank lending or financial markets (equity or bonds) on their behalf. These debt instruments receive explicit guarantees by the collateralization of local governments' land banks. Local debt surged through LGFPs after the adoption of 2008–2009 stimulus package. In 2013, infrastructure construction accounted to approximately 86.8% of total funds invested by local governments.[9]

Despite the development of off-balance sheet funding channels, local governments still have to pay off their debts and find ways to balance their budgets. This challenge has been dealt with by the emergence and progressive generalization of large-scale mechanisms of land value capture, implemented after the 1994 tax reform. Also referred to as "urban-centered accumulation" (Hsing 2010) or "landed urbanization" (Lin 2014; Ye and Wu 2014), the process of value capture takes place in successive

[8]The Trust industry is a unique financial system that has very little in common with Western trusts industries. It is primarily engaged in two main categories of services: private placement investment banking (for both high-net-worth individuals and corporate investors), and conduit business (operating as a conduit to allow banks investing in forbidden asset class). The trust industry has grown rapidly after the GFC. The total value of its assets under management rose from 960 billion 2007 into 23 trillion in mid-2017. Chinese policy-makers are actively working to normalize this industry.

[9]https://www.prometeia.it/atlante/China-government-debt. Checked on September 3, 2018.

stages, which Fulong Wu has described comprehensively: first, land rights for industrial purposes are sold by local states at low prices in urban peripheries to enlarge the manufacturing base of the city and create jobs. This translates into GDP growth, which increases tax revenues and raises land prices as industrial zones urbanize. Then, once urban development reaches a certain level, the land use rights of serviced parcels of land are sold to developers at their highest value for commercial and residential purposes. Revenues from the leases are then used to balance the local budgets and pay off debt for infrastructure construction (Wu 2015b).

To magnify value capture, local states leverage the highly differential treatment of citizen and property rights between rural and urban areas. Migration of rural populations to cities was restricted after 1958 by a system of official registration (*hukou*) that assigns local benefits (from retirement pension, to education, to healthcare) to households, based on agricultural and non-agricultural residency status. After the economic reforms, the system was relaxed to allow rural populations to migrate to cities. But these workers were denied access to urban welfare. They greatly contributed to faster GDP growth in cities by supplying low-cost labor in the manufacturing sector without exerting pressure on local finances. Housing booms have also been fueled by internal migrants' abundant labor force in the construction sector, kept at low wages though informal working contracts under conditions of "tolerated illegality" (Swider 2015). The urban/rural divide was further exploited by massive farm land grabs in the urban fringe areas (Siciliano 2004). Farmers' land use rights could be expropriated and compensated at rates far below the market prices (at the agriculture production value) because land ownership is held by village collectives, and farmers had little power and resources to resist local governments' decisions. Hence, local officials have managed to expand massively the geographic boundaries of their jurisdictions, so as to replenish local land reserves with the aim of developing new industrial spaces and pursuing value capture strategies.

The whole system of land value capture is supported by continuous property investment (either directly or indirectly) from both households and entrepreneurs, and is based on the collateralization of local government land banks. Because the system generates jobs and urban infrastructure, and contributes to increasing incomes, it works partly as a self-fulfilling mechanism that drives up fundamental values (Fig. 11.3). Yet, at the same time, it creates housing vacancies, as investors primarily focus on capital gains, and are encouraged to acquire several properties by the absence of property taxation (see below). One may wonder why home buyers do not perceive the growing risk involved in the accumulation of vacant homes. This puzzling investment confidence can be regarded as the expression of a strong faith by households and firms in the capacity of the current economic model to generate further increases in incomes. More precisely, housing investors share the belief that the exorbitant home prices are justified by the extraordinary achievement of China's economy, under the CCP leadership. This 'convention' draws on investors' inexperience of a property crash in an emerging and constantly rising real estate market. It also builds on investors' awareness that the State has a strong political stake in supporting housing markets. Zheng et al. (2016) pointed out a positive change in investors' outlook in 2009, after the adoption of the Four Trillion Yuan stimulus

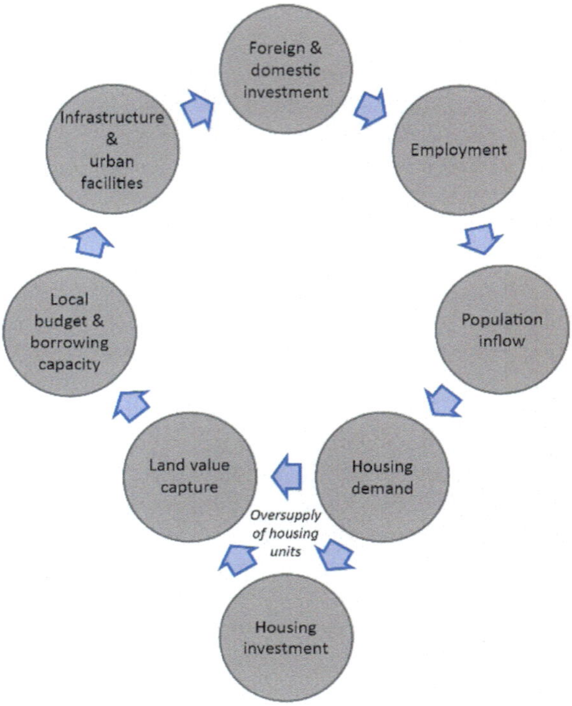

Fig. 11.3 The land-centered economic growth model of Chinese cities. *Source* The author, adapted from Wu (2015b)

package and accommodative measures to cope with the Global Financial Crisis. Furthermore, the convention is strengthened by the centralization of power in the hands of President Xi Jinping, a leader who embodies long-term political stability and the continuity of China's economic performance (Cabestan 2015).

11.4 The State's Management of the Speculative Mechanism

It should be noted that, in face of dangerous speculative elements in the housing market, the Chinese State has managed to avoid a 'hard landing' until now. This is the result of a strategy built up over time, produced by the complex dynamics of central-local interactions, and based on try-and-error, learning-by-doing approaches.

11.4.1 Avoiding Mainstream Financial Channels to Fund the Domestic Real Estate Industry

First and foremost, Chinese policy-makers benefited from the lessons of previous financial crises triggered by real estate boom-bust cycles. Thus, at an early stage of the commodification of the housing markets, they opted for a set of measures seeking to avoid systemic risks in the economy at large, and in real estate sector in particular.

A first step was to impose prudential requirements for mortgage lending, which is mostly operated by the four major commercial state-owned banks. Loan-to-value ratios of home buyers are set at low rates, and although total household debt has considerably increased since 2009, it remains inferior to that of major economies: 49% of the country's GDP in December 2017,[10] compared to 50% on average in the European Union, 78% in the United States, and more than 80% in South Korea and Taiwan.[11] The banking sector is much less involved in the funding for real estate developers, with an official share of 20–30% of total funding. Non-banking capital includes self-funding and revenues from pre-sales, as well as issuance of financial assets (bonds and shares) for listed SOEs and major private developers (Theurillat and O'Neill 2017). However, the largest share of developers' funding comes from informal finance, especially through wealth management products, quasi-real estate investment trusts (Q-REITs), and pawn shops (Sherpa 2013). Through these investment channels, individual and corporate investors find indirect ways to engage capital in real estate by taking shares in property development projects (Theurillat and O'Neill 2017).

The shadow banking sector as a whole recorded marked growth after the stimulus package, surging from less than 10% of the system in 2008 to almost 40% in 2013; in 2016, it amounted to an estimated at CNY 71 trillion, or 118% of GDP (Collier 2017). Given the large size it has reached, and due to its interconnections with banks, the informal sector carries potential systemic risk, but it remains quite fragmented and primarily confined to domestic investors.

Letting the grey finance sector grow helped to avoid developing transnational financial channels to fund the domestic real estate industry. Despite the pressure exerted by global investors to develop mainstream financial instruments in China's property markets, government authorities have continuously sought to "keep foreigners hands off the Chinese land" (Hsing 2006) and have avoided exposing the domestic real estate industry to the vagaries of global finance. Unlike India, which recently established Real Estate Investment Trusts (REITs), China has not operated the securitization of its real estate. A handful of REITs listed in Singapore and Hong Kong have developed China-based property portfolios, but the number of their assets under management is very limited (Aveline-Dubach 2017a). Foreign equity funds have been allowed to penetrate China's real estate sector, but their market share has

[10] This figure does not consider consumption loans that are used to circumvent regulations for housing investment.

[11] https://www.ceicdata.com/en/indicator/china/household-debt--of-nominal-gdp. Checked in Sept. 2018.

become negligible (Aveline-Dubach 2017b),[12] except in some recently commodified subsectors such as infrastructure, logistics and long-term care facilities for elderly. Local regulations demand foreign capital to be anchored in China through joint-ventures with domestic development firms. The benefits of such arrangements are not so much to providing capital to domestic developers—those meeting high foreign standards are major players—but rather from generating an inflow of advanced foreign technical and operational knowledge. By channeling foreign expertise and capital towards various property sub-sectors in this way, Chinese local governments have successfully helped the domestic real estate industry move up the value chain (Aveline-Dubach 2017b). Now that the reliance to foreign partnership is considerably reduced, the risks of external shocks to local property markets have decreased accordingly, confining the land value capture mechanism to a closed, mainly domestic system. It is important to stress that these strong barriers to capital inflows in property markets stand in sharp contrast with overseas investment in trophy property assets by Chinese institutional investors, in accordance with the asymmetrical integration of China's macroeconomic regime into the world economy (Boyer 2013).

11.4.2 Market Monitoring and Public Intervention in Real Estate Markets

As rapidly growing housing markets became unstable, Chinese policy-makers endeavored to provide quick responses to market signals. Since 2003, a succession of measures has been taken, mainly under central government's administrative guidance, alternatively to cool off or stimulate residential markets in line with the changes of local housing prices. Cooling off measures typically include: increases in interest rates, policy guidelines for commercial banks to rein in the pace of mortgage loan issuance, tighter down payment ratios (up to 70% for second homes and cash purchase for third homes in many cities), restrictions on the resale of homes in less than five years. When housing prices recorded a small drop in a given city, the cooling measures are relaxed until the next round. Prices have recently stabilized in major cities, but several local governments or first and second-tier cities have introduced lottery systems for housing sales with government-imposed prices to "curb speculation".[13] This system allows some first-time buyers to purchase affordable housing, but it also encourages backdoor tactics such as extra charges or even corruption.[14]

---

[12]Hong Kong developers are by far the main 'foreign' players in the Mainland. Some groups manage a large number of buildings in major cities and hold extensive landbanks. However, unlike Western operators, they seldom invest in property trough financial vehicles.

[13]China Daily, "8 cities with a lottery system for housing sales", June 4, 2018. http://www.chinadaily.com.cn/a/201806/04/WS5b147119a31001b82571de20.html. Checked on November 15, 2018.

[14]South China Morning Post, "Shanghai authorities get more involved in supervising lottery system used by developers to allocate properties", May 5, 2017. https://www.scmp.com/property/hong-kong-china/article/2093036/shanghai-vows-crack-down-home-sales-irregularities. Checked on November 15, 2018.

A less well-known form of state intervention in the housing markets has consisted of 'extinguishing the fire' when a local property market goes into meltdown, to avoid a spreading of investors' panic throughout the country. The city of Wenzhou, which was hit by a financial crisis in 2011, provides a good example of such strategy. This prosperous coastal city of 3 million inhabitants experienced a boom in real estate. Local investors were mostly SMEs, financed by informal credit, which were pouring capital into an array of risk sectors. By 2011, the average residential price in Wenzhou reached CNY 34,000 per square meter (€4220/m^2 at the current exchange rate), surpassing that of Shanghai and Beijing at the time. When the central bank raised interest rates to contain inflation, credit demand shifted towards the shadow banking sector, pushing up the yearly share of the informal lending market from 15 to 43% (Yufeng et al. 2018). A growing number of Wenzhou's SMEs could not repay their debts, and default rates soared, entailing a financial meltdown, with a 50% drop in housing prices. Given the significance of informal credit delivered through relationship-based and personalized transactions in Wenzhou, the vast majority of capital loss was mutualized by local families. Although the formal banking sector was much less affected, the dramatic situation of local industry prompted the central authorities to step in. SOE banks were asked to grant extended credit to Wenzhou SMEs, and to loosen interest rates. Premier Wen Jiabao and other top leaders even visited Wenzhou to calm the panic.[15] The Wenzhou crisis sent warning signals to Chinese policy-makers, pointing to the dangers of unregulated capital accumulation. Since then, considerable effort has been made to normalize the informal Chinese trust industry and to restructure the liabilities of local governments through debt swaps.

In parallel to direct intervention in the housing markets, the State has acted to satisfy households' demand for profitable saving vehicles by opening up transnational channels for residential investment. In other words, the State has opted to 'export the housing bubble' so as to relieve the pressure on domestic housing markets rather than resorting to developing alternative saving vehicles by liberalizing the financial industry. These investment channels are controlled by the State, so that the volume of capital involved can be fine-tuned in accordance with the movements of domestic housing markets. A handful of key Chinese developers operate in this overseas residential sector, focusing on North America, Western Europe and Asia Pacific. Needless to say, investments through these channels have increased both housing prices and housing vacancy in many recipient markets, and were subsequently not well received by local populations (Moser 2018). In 2017, China's outbound direct real estate investment reached US$ 56.5 billion (including a large share of property purchased by institutional investors), an amount equivalent to the cumulated sum of UK and US cross-border capital in real estate the same year.[16] Although capital outflows in real estate have been recently restricted in China to prevent a weakening of the yuan (Meyer 2018), cross-border investment by Chinese institutional investors

---

[15]Derecet News business, October 18, 2011, https://www.deseretnews.com/article/700189099/Debt-panic-in-Chinas-Wenzhou-may-auger-wider-woes.html. Checked on September 25, 2018.

[16]JLL Global Research, China 12: China's cities go global 2018. This figure includes investment in Hong Kong.

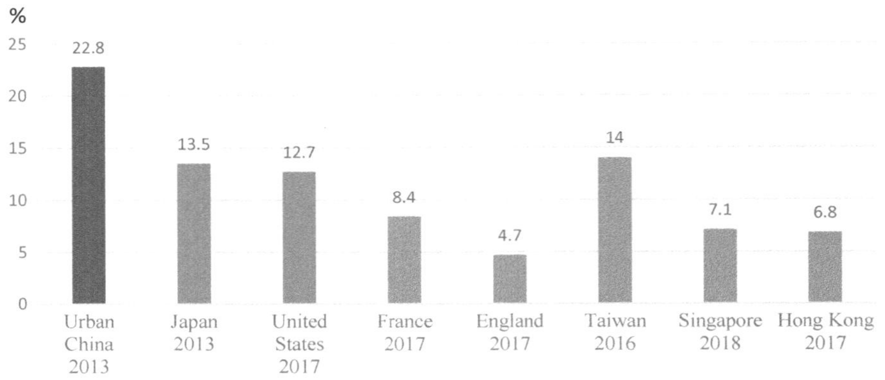

Fig. 11.4 Comparative housing vacancy in several countries (% of housing stock). *Source* CHFS, Japan Statistic Bureau, US Housing Vacancy Survey, INSEE, England's Ministry of Housing, Communities and Local Government statistics, Taiwan Ministry of the Interior's Construction and Planning Administration, Singapore Urban Redevelopment Authority, Hong Kong Data Government Statistics

and households have become a potential threat for the stability of property markets in a number of regions.

The Chinese State is naturally more concerned with the huge accumulation of unoccupied residential units at home. Housing vacancy is unique in China in the sense that the vast majority of unoccupied homes is held by households. According to the latest 'official' estimate provided by the China Household Finance Survey (CHFS), they were 52 million vacant housing units in China in 2013, of which 3.5 million were as developers' inventory.[17] This represents an average 22.4% of the housing stock in China's urban areas, a much higher rate than Japan, the US and Taiwan (Fig. 11.4). Although the CHFS survey was updated in 2015 and 2017, no data have been published at the time of writing. Therefore, the 2013s figure is likely to be significantly underestimated. In particular, the stock of developers' inventories has grown considerably over the past years. Glaeser et al. (2017) used a developers' database (Soufun, for 32 cities) and found that the surface of newly built unsold units had grown nearly threefold (from 4.1 to 11.3 billion sq ft) during the period 2011–2015. Two-thirds of this inventory is concentrated in third- and four-tier cities. This situation reflects a lack of technological skills in smaller cities, along with comparatively lax land control policies (Li 2017). In 2015, President Xi Jinping expressed his will to "reduce the stock (*qukucun*)" in the property market as one of his key objectives. To this end, the government introduced stimulus measures to absorb inventories, such as interest rate cuts and tax reductions. Stock reduction measures also seek to encourage migrant workers to buy unsold housing units through the

[17] CHFS Data Talks, *Trend in the housing market and Housing vacancy rate in urban China*, 2004. This survey was based on interviews of some 28,000 households, carried out in 262 counties and 29 provinces.

relaxation of *hukou* regulations. The government announced a plan to grant urban *hukou* to 100 million people by 2020 in small and medium-sized cities (Losavio 2018). Yet this raises serious challenges, as a large number of rural migrants may not find adequate jobs enabling them to purchase homes in less developed cities.

11.4.3 Setting up an Alternative Productivist Model

Chinese policy-makers are aware that regulating the excesses of the current land-centered economic model is not sufficient, and this has prompted them to promote a new model of development. Since 2014, China's economy has entered a single-digit growth stage known as the 'new normal'. To avoid falling into the 'middle-income trap', a situation in which a country has lost its competitive edge in the export of manufactured products because of rising wages (Gill and Kharas 2007), the Chinese economy has to move up the value chain. The government has launched a very ambitious policy of innovation in technology and services, to keep up with advanced countries and achieve the transition towards high-income. This entails a greater reliance on domestic-led growth through measures targeting a mass consumption society. Rural populations, which represent a high potential for innovation and consumption, are to play an active part in this new development model. High-skilled migrants, in particular, are seen as a potential resource to forge new avenues of urban innovation through hybridization and cultural diversity.

In many cities, former *hukou* regulations are being replaced by points-based systems filtering rural migrants (Zhang 2012). The conditions for being granted permanent urban citizenship depend on city size, and become harsher with size, with first-tier cities and wealthy second-tier cities imposing extremely stringent conditions. Migrants are selected according to criteria such as education, age, type and seniority of employment, type of housing, payment of urban social insurance and length of residence in the city (Losavio 2018). They are put under pressure by the points-based system to enhance their competitive edge over the years. Those migrants who manage to get good scores but do not reach the threshold are granted partial citizenship (incomplete local welfare), whereas lower-skilled migrants are excluded from urban welfare benefits or encouraged to move to smaller cities.

Contrary to the official narrative of social inclusion conveyed by the *hukou* reform, the urban/rural divide is not disappearing, but rather transforming to serve the new productivist model. Lyu et al. (2018) have coined the expression 'innovation-based urbanization' to characterize this new growth model. With the rise in both incomes and skills in the manufacturing sector, the massive exploitation of cheap migrant labor is going to lapse in many cities. China is no longer a low-cost outsourcing destination for global manufacturers but is becoming a hub of global supply chains in which domestic and foreign firms compete to attract Chinese talent. Exploitation of the urban-rural divide through land grabbing at urban fringes is also getting less easy, as a result of government measures to preserve agricultural land and the increased ability of collective landowners to negotiate high compensation fees. More generally, the land value capture mechanism operated by local governments is being challenged by the maturity of property markets, and the subsequent contraction of demand for new homes that will generate shrinking land sale fees in the long run.

11.4.4 Introduce Alternative Financial Instruments to Meet Shortfall in Local Resources

Alternative financing tools have thus been established to move towards a new model of local finance. A new special bond with set quotas was created in 2014 to help local governments raise funds for infrastructure construction in a more transparent way. The same year, public-private partnerships (PPPs)[18] were launched to fund infrastructure projects. They experienced a sharp boom, reaching 14,220 projects with an aggregate value of CNY 17.8 trillion in November 2017.[19] However, approximately 60% of PPP projects turned out to be funded by state-owned entities, with financial arrangements often seeking to circumvent controls on local government borrowing.[20] This went so far as to push the central government to cancel some 2500 PPPs in 2017, involving 18% of estimated PPP capital.

Simultaneously, some cities initiated new arrangements to transfer the burden of transit infrastructure funding to local public metro companies. In exchange, the latter have been empowered to construct and manage high-rise buildings over their station depots, and to draw profit from joint-developments with private developers. While this transit-based type of land value capture has recorded much success in Hong Kong (Aveline-Dubach and Blandeau 2018), it is not likely to alleviate the risk borne by the public sector in China, as a large part of 'private capital' comes from state-owned property developers.

The untapped opportunity of a property tax system remains to be explored. China stands out amongst the world's nations in having no tax on the ownership of private residential properties. In 2018, the Finance ministry announced its determination to roll out a nationwide property tax to be levied by local governments, based on assessed values. However, no clear time schedule has been set. The project is to be implemented very gradually and carefully, following the usual try-and-error approach. Liu (2017) points out that a property tax faces strong resistance by urban households. Over 90% of households own one or more housing units, of which a significant share is vacant, and do not provide earnings. Thus, the taxation is a politically sensitive

---

[18]A PPP is a partnership between a public authority and a private operator aimed at providing a public project or infrastructure, in which there is a transfer of significant risk to the private party.

[19]Reuters, Business news, November 17, 2017. https://www.reuters.com/article/us-china-economy-ppp/china-overhauls-2-69-trillion-public-private-projects-as-debt-fears-rise-idUSKBN1DH0DE. Check on September 20, 2018.

[20]Nikkei Asian Review, May 17, 2018. https://asia.nikkei.com/Opinion/China-must-put-the-private-into-PPP2, checked on September 20, 2018.

project that will probably take many years before being fully implemented, especially in a rapidly ageing society with weak welfare provision.

The development of new financial resources is only part of the needed changes of the current urbanization model. Land use efficiency must be improved to prevent further housing vacancy and solve the 'ghost city' problem. The government has engaged in tackling the issue by actively promoting rental housing tenure. In big and medium-sized cities, property developers are encouraged, by way of easier land supply, to shift their business model away from sales-led development towards construction projects for long-term leasing. One of the major real estate developers, the Vanke group, has even decided to establish the rental residential housing business as a core activity. The potential benefits of promoting rental tenure are many. It would preserve households' wealth, provide means to pay the property tax, improve occupational mobility in big cities, and above all restrain future growth in housing vacancy. However, there is a huge gap between what needs to be achieved and reality. Home ownership is a cornerstone of China's welfare, and this will be exacerbated in an ageing society. It is moreover deeply ingrained in households' representations of wealth and marriage practices. Beyond cultural barriers, ownership tenure remains predominant in legal terms. Rental status has weak legal protection and does not provide similar access to urban resources as does home ownership (such as access to educational facilities, for example). Although local governments are trying to improve the attractiveness of long-term leasing in their jurisdictions, this form of housing tenure is likely to remain limited in scope for a while.

11.5 Discussion and Conclusion

This chapter has highlighted the role of housing booms as crucial engines of urbanization in China's distinctive context of highly decentralized, land-centered, manufacturing-based productivist model of economic growth. Following Wu's account of China's housing cycles (Wu 2015a), the chapter puts the politics of land development, shaped by intrinsic central-local tensions, at the heart of China's macroeconomic policies. The purpose here has been to bring home buyers' speculative behavior into this framework, using the bubble theory. However, this study reveals the limits of the bubble framework to grasp both the 'fundamental' and the speculative aspects of China's housing booms.

Booming residential prices in China's big cities do not merely reflect the rapid demographic and income growth of a transitional economy, but are drivers of self-fulfilling dynamics through the pervasive practice of value capture led by the government. As such, rising prices have contributed to equipping (and, in many instances, over-equipping) Chinese cities with quality urban infrastructure and industrial facilities. This has generated a manufacturing ecosystem that has no equal in other BRICs, and which contributes to enhance China's economy in the global value chain.

On the other hand, empirical research on bubbles has underscored the puzzling simultaneous escalation of residential prices and increasing housing vacancy, which has led to Western media erroneously predicting that China's bubble was about to burst. Yet, bubble analysts have failed to explain households' distorted apprehensions of risk, beyond the lack of alternative saving vehicles and the economic and social value of home ownership. It is argued here, in line with previous work on speculative behavior (Case and Shiller, Orléan) that the *perception* of the market has a strong influence on price dynamics. The striking confidence of Chinese home buyers can be viewed as the result of a 'Xi Jinping Thought Convention' shared by market players, based on the belief that China is destined to a bright future under the guidance of an ambitious, thoughtful leader whose power will remain unchallenged. This convention relies on the strong 'fundamental' assumption that the aggressive innovation policy currently conducted by the government in all domains, and which exceeds levels observed in neighboring East Asian developmental states, will indeed succeed. However, expectations of future income growth are not sufficient to alleviate the currently high risk of housing investment. Other non-fundamental factors strongly contribute to shaping the convention. They include: (i) home buyers' inexperience of major real estate downturns in an emerging and ascendant market; (ii) inaccurate information about property markets: Chinese households have access to information about housing prices through the Internet, but the domestic media tend to downplay local downturns of property markets, and conversely to convey comforting official declarations for investors; and (iii) home buyers' faith in the capacity of the current regime to regulate residential markets. This confidence is based on hard facts. Policy-makers' efficiency in avoiding a downturn in property markets has been effective until now, owing to extensive State influence over the whole urban production chain, including: public land ownership, the predominance of state-owned entities in urban development, State control of the banking sector, State intervention in property markets as well as stringent restrictions on mainstream global channels of property investment. An equally crucial condition of the government's efficiency is the capacity to deliver 'just-in-time' policy responses. This has been made possible by fast and decisive decision-taking in an authoritarian regime in which the whole administrative system is under Party control. It can be assumed that the asymmetric nature of the State's regulation of housing markets—failure to rein in housing price hikes, efficiency to manage downturns—has played a role in shaping investor confidence.

Neoliberal commentators claim that excessive government intervention has an exacerbating effect on the housing cycle. Yet Chinese policy-makers have drawn lessons from the Japanese experience of the 1980s financial bubble and how a slow State's policy response can be very harmful. Accordingly, they are making full use of their multiple levers to keep alive investor confidence. Drawing further lessons from the GFC systemic crisis, Chinese policy-makers have stayed away from the development of financial channels using Wall Street's standards, and instead have allowed the growth of informal finance, with the trust industry as major funder of property and urban infrastructure projects. There is undoubtedly much risk involved in the trust sector's proliferating debt, but the risk is limited by a lower connectivity and a predominantly domestic structure of these funding channels in the real estate sector. China's experience and management of housing booms therefore suggest there may be alternative paths to neoliberal capital accumulation and regulation regimes in the urban environment.

Although this chapter has addressed the linkage between economic growth, urban development and housing production, the critical connections between housing booms and social inequalities would deserve further exploration. Recent years have been marked by a boom in the informal market for "Small Property Right Housing" (SPRH), in which migrant households can buy residential units at much lower than market prices in urban villages. Following the seminal paper on SPRH by He et al. (2019), further research is needed to examine the mutual relationships between regular and informal residential markets, as well as to assess the ability of the SPRH to mitigate the pressure of soaring housing prices on the lowest-income groups.

Obviously, Chinese policy-makers are facing major challenges to maintain balance in housing markets. The two interdependent pillars of the prevailing model of economic and urban growth—the value capture processes operated by local governments and the exploitation of the urban/rural divide—face a challenging environment in big cities, due to prohibitive housing cost on the one hand, and the effects of the new productivist policy on the other hand. Against this background, the government is actively trying to promote alternative financial channels to fund local infrastructure and services. So far, its efforts have not been very successful. The PPP projects have hardly managed to avoid further collusion between local governments and state-owned entities, and the proposal to tax private property is meeting strong skepticism by homeowners. Efforts to promote the rental housing tenure are primarily aimed at restraining the supply of new homes for sale, not at reducing existing housing vacancy. While there is an urgent need to preserve the wealth stored in household's homes, especially in a rapidly aging society with underdeveloped welfare provision, it may take years to gain acceptance for new norms of housing tenure and property taxes by Chinese households. It thus remains to be seen if China's distinctive regime of capital accumulation in its urban environment can be adapted to the country's new growth model.

References

Ahuja, A., Cheung, L., Han, G., Porter, N., Zhang, W.: Are house prices rising too fast in China? IMF Working Paper No. 10/274 (2010)

Aveline-Dubach, N.: La centralité du foncier dans le régime d'accumulation du capital en Chine. Revue de la régulation. Capitalisme, institutions, pouvoirs (2017a). https://doi.org/10.4000/regulation.12262

Aveline-Dubach, N.: Embedment of "liquid" capital into the built environment: the case of REIT investment in Hong Kong. Issues Stud. (2017b). https://doi.org/10.1142/s1013251116400014

Aveline-Dubach, N.: The financialization of real estate in megacities and its variegated trajectories in East Asia. In: Sorensen, A., Labbé, D. (eds.) International Handbook of Megacities and Megacity Regions. Edward Elgar Publishing, Cheltenham (forthcoming)

Aveline-Dubach, N., Blandeau, G.: The political economy of transit value capture: the changing business model of the MTRC in Hong Kong. Urban Stud. (2018). https://doi.org/10.1177/0042098018821519

Baek, S.-W.: Does China follow "the East Asian development model"? J. Contemp. Asia **35**(4), 485–498 (2005)
Ball, M., Wood, A.: Housing investment: long run international trends and volatility. Hous. Stud. **14**(2), 185–209 (1999)
Barras, R.: A simple theoretical model of the office-development cycle. Environ. Plan. A **15**(10), 1381–1394 (1983)
Boyer, R.: The Chinese growth regime and the world economy. In: Boyer, R., Uemura, H., Isogai, A. (eds.) Diversity and Transformations of Asian Capitalisms, pp. 184–208. Routledge, London and New York (2013)
Cabestan, J.-P.: Le système politique chinois: un nouvel équilibre autoritaire. Presses de Sciences po (2015)
Case, K.E., Shiller, R.J.: The behavior of home buyers in boom and post-boom markets. NBER Working Paper No. 2748, National Bureau of Economic Research, Cambridge, MA, USA (1988)
Chen, D.: An empirical analysis of house price bubble: a case study of Beijing housing market. Ph.D. thesis, Lincoln University (2012)
Chen, K., Wen, Y.: The great housing boom of China. Am. Econ. J.: Macroecon. **9**(2), 73–114 (2017)
Chien, S.-S.: Institutional innovations, asymmetric decentralization, and local economic development: a case study of Kunshan, in post-Mao China. Environ. Plan. C: Gov. Policy **25**(2), 269–290 (2007)
Chien, S.-S.: Economic freedom and political control in post-Mao China: a perspective of upward accountability and asymmetric decentralization. Asian J. Polit. Sci. **18**(1), 69–89 (2010)
Chien, S.-S.: Chinese eco-cities: a perspective of land-speculation-oriented local entrepreneurialism. China Inf. **27**(2), 173–196 (2013)
Coakley, J.: The integration of property and financial markets. Environ. Plan. A **26**(5), 697–713 (1994)
Collier, A.: Shadow Banking and the Rise of Capitalism in China. Palgrave Macmillan, Singapore (2017)
Corpataux, J., Crevoisier, O., Theurillat, T.: The expansion of the finance industry and its impact on the economy: a territorial approach based on Swiss pension funds. Econ. Geogr. **85**(3), 313–334 (2009)
Crowe, C., Dell'Ariccia, G., Igan, D., Rabanal, P.: How to deal with real estate booms: lessons from country experiences. J. Financ. Stabil. **9**(3), 300–319 (2013)
De Bondt, W.F.M.: Bubble psychology. In: Asset Price Bubbles: The Implications for Monetary Regulatory and International Policies, pp. 205–216 (2003)
Dreger, C., Zhang, Y.: Is there a bubble in the Chinese housing market? Urban Policy Res. **31**(1), 27–39 (2013)
Flood, R.P., Hodrick, R.J.: On testing for speculative bubbles. J. Econ. Perspect. **4**(2), 85–101 (1990)
Gill, I.S., Kharas, H.: An East Asian Renaissance: Ideas for Economic Growth. World Bank, Washington, DC (2007)
Glaeser, E., Huang, W., Ma, Y., Shleifer, A.: A real estate boom with Chinese characteristics. J. Econ. Perspect. **31**(1), 93–116 (2017)
Halbert, L., Henneberry, J., Mouzakis, F.: The financialization of business property and what it means for cities and regions. Reg. Stud. **48**(3), 547–550 (2014)
He, S., Wang, D., Webster, C., Chau, K.W.: Property rights with price tags? Pricing uncertainties in the production, transaction and consumption of China's small property right housing? Land Use Policy **81**, 424–433 (2019)
Heilmann, S.: From local experiments to national policy: the origins of China's distinctive policy process. China J. **59**, 1–30 (2008)
Helbling, T., Terrones, M.: When bubbles burst. World Econ. Outlook **2**, 61–94 (2003)
Herring, R., Wachter, S.: Bubbles in real estate markets. In: Asset Price Bubbles: The Implications for Monetary, Regulatory, and International Policies, vol. 217(6), pp. 217–230 (2003)

Holliday, I.: Productivist welfare capitalism: social policy in East Asia. Polit. Stud. **48**(4), 706–723 (2000)
Hsing, Y.: Global capital and local land in China's urban real estate development. In: Wu, F. (ed.) Globalization and the Chinese City, pp. 167–189. Routledge, London and New York (2006)
Hsing, Y.: The Great Urban Transformation: Politics of Land and Property in China. Oxford University Press, Oxford (2010)
Hu, F.: Homeownership and subjective wellbeing in urban China: does owning a house make you happier? Soc. Indic. Res. **110**(3), 951–971 (2013)
Hui, E.C.-M., Liang, C., Wang, Z., Song, B.-T., Qi, G.: Real estate bubbles in China: a tale of two cities. Constr. Manag. Econ. **30**(11), 951–961 (2012)
Jiang, Y., Waley, P., Gonzalez, S.: Shifting land-based coalitions in Shanghai's second hub. Cities **52**, 30–38 (2016)
Johnson, C.: MITI and the Japanese Miracle: The Growth of Industrial Policy: 1925–1975. Stanford University Press, Standford (1982)
Kindleberger, C.P.: Bubbles. In: The World of Economics, pp. 20–22. Springer (1991)
Li, M.: Evolution of Chinese ghost cities. Opportunity for a paradigm shift? The case of Changzhou. China Perspect. **2017**(2017/1), 69–78 (2017)
Lin, G.C.S.: China's landed urbanization: neoliberalizing politics, land commodification, and municipal finance in the growth of metropolises. Environ. Plan. A **46**(8), 1814–1835 (2014)
Liu, Z.: Land-based finance and property tax in China. ATI Working Paper, WP-18-04 (2017)
Liu, M., Sun, B.: On the formative mechanism of land price and real estate bubble [J]. J. Financ. Res. 10:004 (2009)
Losavio, C.: Building wealth through a stratified inclusion: the point-based hukou system in Zhuhai. In: Aveline-Dubach, N. (ed.) Pathways of Sustainable Development Across China—The Cases of Hangzhou, Datong and Zhuhai, pp. 109–121. Venise Imago Editor, Venise (2018)
Lyu, J.: The measurement of the bubble of urban housing market in China. Econ. Res. J. **6**, 28–41 (2010)
Lyu, L., Sun, F., Huang, R.: Innovation-based urbanization: evidence from 270 cities at the prefecture level or above in China. Acta Geogr. Sin. **73**(10), 1910–1922 (2018)
Mera, K., Renaud, B.: Asia's financial crisis and the role of real estate. Routledge, London and New York (2016)
Meyer, C.: L'occident face à la renaissance de la Chine. Défis économiques, géopolitiques et culturels. Odile Jacob, Paris (2018)
Moser, S.: Forest city, Malaysia, and Chinese expansionism. Urban Geogr. **39**(6), 935–943 (2018)
Or, T-m: Pathways to homeownership among young professionals in urban China: the role of family resources. Urban Stud. **55**(11), 2391–2407 (2018)
Orléan, A.: Le pouvoir de la finance. Odile Jacob, Paris (1999)
Ren, X.: Urban China. Wiley (2013)
Ren, Y., Xiong, C., Yuan, Y.: House price bubbles in China. China Econ. Rev. **23**(4), 786–800 (2012)
Savini, F., Aalbers, M.B.: The de-contextualisation of land use planning through financialisation: urban redevelopment in Milan. Eur. Urban Reg. Stud. (2015). https://doi.org/10.1177/0969776415585887
Sherpa, D.: Shadow banking in India and China. Econ. Polit. Wkly. **48**(43), 113 (2013)
Siciliano, G.: Rural-urban migration and domestic land grabbing in China. Popul Space Place. 20(4):333–351 (2014)
Stiglitz, J.E.: Symposium on bubbles. J. Econ. Perspect. **4**(2), 13–18 (1990)
Swider, S.: Building China: precarious employment among migrant construction workers. Work Employ Soc. **29**(1), 41–59 (2015)
Theurillat, T.: The role of money in China's urban production: the local property industry in Qujing, a fourth-tier city. Urban Geogr. **38**, 834–860 (2016)
Theurillat, T., Crevoisier, O.: The sustainability of a financialized urban megaproject: the case of Sihlcity in Zurich. Int. J. Urban Reg. Res. **37**(6), 2052–2073 (2013)

Theurillat, T., O'Neill, T.: Financing urban growth in China: a case study of Qujing, a medium-sized city in Yunnan province. China Perspect. **1**, 57 (2017)
Tirole, J.: Asset bubbles and overlapping generations. Econometrica **53**, 1499–1528 (1985)
Wei, S.-J., Zhang, X., Liu, Y.: Status Competition and Housing Prices. National Bureau of Economic Research (2012)
Wood, A.: The scalar transformation of the US commercial property-development industry: a cautionary note on the limits of globalization. Econ. Geogr. **80**(2), 119–140 (2004)
Wu, F.: Housing and the state in China. In: International Encyclopedia of Housing and Home, pp. 323–329. Elsevier (2012)
Wu, F.: Commodification and housing market cycles in Chinese cities. Int. J. Hous. Policy **15**(1), 6–26 (2015a)
Wu, F.: Planning for Growth: Urban and Regional Planning in China. Routledge (2015b)
Wu, J., Deng, Y., Liu, H.: House price index construction in the nascent housing market: the case of China. J. Real Estate Finance Econ. **48**(3), 522–545 (2014)
Ye, L., Wu, A.M.: Urbanization, land development, and land financing: evidence from Chinese cities. J. Urban Aff. **36**(s1), 354–368 (2014)
Yue, S., Hongyu, L.: Housing prices and economic fundamentals: a cross city analysis of china for 1995–2002 [J]. Econ. Res. J. 6:78–86 (2004)
Yufeng, C., Zhipeng, Y., Guan, H.: The financial crisis in Wenzhou: an unanticipated consequence of China's "four trillion yuan economic stimulus package". China: Int. J. **16**(1), 152–173 (2018)
Zhang, L.: Economic migration and urban citizenship in China: the role of points systems. Popul. Dev. Rev. **38**(3), 503–533 (2012)
Zheng, S., Sun, W., Kahn, M.E.: Investor confidence as a determinant of China's urban housing market dynamics. Real Estate Econ. **44**(4), 814–845 (2016)
Zhou, J.-K.: Real estate price fluctuations and speculative behavior—an empirical analysis for fourteen China cities. Mod. Econ. Sci. **4**, 19–24 (2005)

Chapter 12
What Is Emerging? Understanding Urbanisation Dynamics in BRICS Countries Through a Geographical Approach, the Case of Russia and South Africa

Solène Baffi and Clémentine Cottineau

Abstract In this chapter, we discuss the emerging features of urbanisation in two BRICS countries (Russia and South Africa) by bringing together the different meanings of the concept of emergence in development economics and in complexity science. The objective of this cross-investigation is to identify the features of urban theories which only apply to the first industrialised countries (but do not apply to emerging powers such as the BRICS countries) and to identify the emerging patterns of growth and structure in urban systems deemed as extreme: e.g. the former Soviet Union and South Africa.

12.1 Introduction

Two nations transitioned most drastically in the 1990s. First, in 1991, Russia emerged from more than 70 years of a socialist regime (the Soviet Union). Soon after, in 1994, South Africa put an end to about 45 years of apartheid. Both countries, in doing so, embraced neoliberal economic principles alongside democracy. In the background, China was moving towards a more market-based economy, whereas India and Brazil became production and trade giants, forming a novel collection of nations put hastily together under the acronym BRICS by Goldman Sachs in 2001. This aggregation was initially to serve financial analysis purposes only, but it stroked a chord with

---

S. Baffi
Department of Geography and Environmental Studies, Stellenbosch University, Stellenbosch, South Africa
e-mail: solene.baffi@gmail.com

S. Baffi · C. Cottineau (✉)
Université Paris 1 Panthéon-Sorbonne, UMR 8504 Géographie-Cités, Paris, France
e-mail: clementine.cottineau@ens.fr

C. Cottineau
CNRS, UMR 8097 Centre Maurice Halbwachs, Paris, France

Centre for Advanced Spatial Analysis, University College London, London, UK

D. Pumain (ed.), *Theories and Models of Urbanization*, Lecture Notes in Morphogenesis, https://doi.org/10.1007/978-3-030-36656-8_12

Fig. 12.1 The BRICS leaders in 2014. From left to right: V. Putin, N. Modi, D. Rousseff, J. Xi and J. Zuma. Photo by Roberto Stuckert Filho, Agência Brasil, Wikipedia Commons

those who had noted the increasing economic and political power that these nations gained in the 2000s. These countries themselves later embraced the idea and started meeting regularly (Fig. 12.1).

I. Turok confirms the definition of BRICS countries as "*fast-growing, middle-income, emerging powers to rival Europe and North America*" (Turok 2014, p. 123), but he also suggests they have urbanisation features in common. In particular, he sees in them a new historical relationship between development and urbanisation: i.e. the growth of cities without economic growth. This situation reverses the historical example of western countries' concomitant growth of urbanisation and economic development. It suggests that either the cost of rural-to-urban migrations (housing and service provision, congestion, unemployment, etc.) outpaces the usual economic benefits associated with the agglomeration of people, firms and capital (Duranton and Puga 2004; Rosenthal and Strange 2004) in BRICS countries where populations are way more numerous and cities are growing very fast, either that national economic growth has historically been the driver that fueled the growth of cities rather than the opposite, as Polèse (2005) argues. More generally, as these countries have become the focus of new analytic light, it is legitimate to wonder if the scientific models commonly used to study, to understand and to predict urbanisation dynamics which are predominantly based on early western early urbanisation (Robinson 2006) are similarly rendered wrong or less useful by the recent experience of BRICS countries. If so, what is emerging instead?

In this chapter, we give a go at these questions, starting with the problematic definition of the concepts we are working with. Indeed, both *emergence* and *models* have polysemous meanings, notwithstanding *urbanisation*. The first section will expose two options for each concept that we will refer to when exploring their relationships. In the following sections, we present two sets of model examples: a set of models which are contested and/or falsified when applied to the urban conditions of BRICS countries, and a set of models which are corroborated and still useful. They are completed by a fourth section on "models to re-invent", in which we argue that the new urban experience in countries of the former socialist block and the "Global South" should not be only a terrain to test existing concepts and theories about cities but should become examples from which to start to generalise new regularities and more encompassing urban theories.

Throughout the chapter, we refer to BRICS countries but with a special focus on its most unusual members: South Africa and Russia. It has become rather consensual to admit that urban theories have to be re-worked if and when they do not help us understand emergent urbanisation in China and in India (as well as in Brazil to a lesser extent). However, the cases of Russia and South Africa in the group of BRICS seem less influential in this regard, as they can easily considered 'too specific', Russia because of its socialist past and South Africa because of apartheid. Their very inclusion in the BRICS is even sometimes compromised (when South Africa is outed by the acronym BRIC, or when Russia is "forgotten" along the way, as in Turok's (2014) single-country description of urbanisation and development of BRICS countries from the Global South).

12.2 Working with Polysemous Concepts: Emergence and Models in Urbanisation Studies

Because of the use of the same words for different meanings, our research question could span over quite different fields. Some elements of definition will help disentangle these meanings and clarify the analysis.

12.2.1 Emergence

Although the use of "emergence" is very common in scientific writing in general (on average one occurrence every three pages in the content of our bibliography) there are two very different formalised uses of the concept which are of relevance to our subject. The first one refers to developmental economics, mostly through the expression "emergent economies". The second one refers to the multidisciplinary field of complexity science, where it qualifies objects ("emerging pattern", "emerging property") or is qualified itself ("strong emergence", "weak emergence").

The first use of "emergence" stems from the economics field, as the term was coined in the financial sector in the 1980s. It then described developing countries showing a rapid economic growth combined with an important local market and which offered secured investment opportunities. Nowadays, the term does not strictly refer to an economic understanding. Indeed, other common features bring together the "emerging countries", such as the completion of the demographic transition, or the stability of their domestic political life (Gabas and Losch 2008). However, although it is commonly used, the term "emerging" remains very seldom defined. A few geographers tried to identify the commonalities among emerging countries in order to understand to what extent the notion possesses an analytical meaning. Besides the intense economic growth, these countries show a transformation in their economic structure leading to more added value in the production sphere (often linked to industrialisation) and the apparition of a middle-class. This last point entails as well deep social transformations through the adoption of globalised standards and the production of hybrid cultural products (Erkip 2003; Fleury and Houssay-Holszchuch 2012). Also, emerging countries present high urbanisation rates and specific spatial configurations. In particular, the economic growth and foreign investments often lead to increased inequalities and polarization processes within urban spaces, resulting in urban fragmentation (Bret 2011). Eventually, the geopolitical influence is another characteristic that can be attributed to emerging countries, as they happen to form a counter-power to the G7 members (Gabas and Losch 2008) and therefore to represent the voices of "not-yet-fully-emerged" countries of the South among international organisations. In the case of Russia and South Africa, their inclusion in this group also results from their geopolitical reach, both at the regional and international levels (Darbon 2008; Jaffrelot 2008). However some scholars question the relevance of the notion of emergence as yet another way of categorising non-developed countries. By definition, "emergence" indicates a momentary stage, the transition from one stage to another (Morange 2011). While recent research underlines the necessity to grasp cities and territories of the South not from the lens of models developed in the North, but according to their own trajectory (Shatkin 2007; Parnell and Robinson 2012), scholars such as Lorrain (2011) argue against using the term "emerging" in a normative way.

Regarding the second use of "emergence" in complexity science, it represents a rather central concept. Emergence is associated with the appearance of new properties and/or patterns with increasing levels of organisation and complexity for systems in which elements interact with their environment, in time and with each other (Lestienne 2015). More critical to the definition, the emergent properties of higher-level objects can be *explained by* but not *assigned to* the lower-level elements. For example, in the agent-based model proposed by Schelling (1971), city-wide segregation patterns occur through the residential moves of the agents as a response to their neighbourhood composition, even when none of them is individually strongly intolerant to residing next to agents from a different group. In the case of systems of cities, Pumain (2006) considers the hierarchy of city sizes, their regular spacing and functional differentiation to be emergent properties of the system of individual cities. In the research agenda of complexity science, being able to reproduce the emergence of some macro-properties through the simulation of generative mechanisms implemented as rules of actions of micro-elements of a system is considered the first step (Marchionni and Ylikoski 2013) if not the gold standard (Epstein 1999; Machamer et al. 2000) for causal explanation. Finally, Chalmers (2006) introduces a distinction between 'strong' and 'weak' emergence: in the strong case, the emergent phenomenon is not strictly *deducible* from the low-level entities in interaction, whereas in the weak case (which corresponds better to the emerging phenomena observed in social sciences), "*truths concerning that phenomenon are* unexpected *given the principles governing the low-level domain*" (ibid., p. 1). In the social science cases we are studying in this chapter, the emergence encountered is usually of the weak kind.

12.2.2 Model

Models are everywhere in science, and a plethora of definitions already exist to describe their properties and functions in the different fields. In this chapter, we want to emphasize two broad aspects of scientific models: their representational and descriptive value on the one hand; their normative function on the other hand.

The model as a *representation* can be defined as an attempt to describe, in a simplified and/or abstract manner, an element of reality (the target). The representation has a purpose and can be guided by theory, which makes it possible for different models of the same elements of reality to co-exist in different forms. The model can be a mental, conceptual, numerical, graphical or physical representation. Its simplification means that only the main structure, core dynamics or organisation principle of the target reality is included in the model. In urban studies, examples of models range from the mental maps of Lynch (1960) to the 'rank-size rule' function attributed to Zipf (1941) to agent-based simulations. In this chapter, we will consider statistical models, network models and simulation models. They will be called contested or falsified when the regularity or theoretical relationship they represent is not observed empirically in one of our case study. They will be called useful when the observation corroborates the model.

The model as a *blueprint* is closer to its meaning in the art and fashion world. It corresponds to a normative idea of what reality should be, either in an idealized way, either in the future if one follows the blueprint. In urban studies, this type of models is mostly present in architecture and urbanism ('what it should look like') or in the policy world of planning ('what we expect to see once the policy is implemented'). The potential inadequacy of such futuristic or normative plans with local conditions led scholars such as V. Watson to talk about "fantasy plans" and "urban utopias" (2015). By doing so, she underlines how the global visions and the aesthetics of the city promoted by consultants and developers in Africa and India are often oblivious of the reality on the ground. Also, the choice of specific models, due to the underlying norms and narratives, involves a political dimension or, at least, power relationships. In this chapter, we will consider these types of models as contested if the vision of what the reality should be differs between different urban actors or if the assumptions the model lies on are not met by reality. They will be called useful when they allow to bring in new ideas and solutions to problems in BRICS cities.

12.2.3 Urbanisation

Following Pumain (2006), we identify two nested levels of urban organisation: that of cities as individual entities bounded by a physical or social landscape defined as 'urban', and that of the system they form by interacting together, which is bounded by national or macroregional boundaries. The system level and the cities level—the former emerging from the latter in a complexity science lexicon—are characterised by different attributes, scales and temporalities. They also imply two meanings for the concept of urbanisation.

At the macro-scale, urbanisation is the quantitative and qualitative shift from a settlement system where most citizens live in rural settings—where agricultural activities are dominant, the average density is low and communities are socially homogeneous—to a settlement system where most citizens live in urban settings—whereby heterogenous citizens perform specialised and complementary tasks in denser built-up environments. Urbanisation in this sense is usually measured by the percentage of population living in cities, and the pace of urbanisation describes the rate of positive change in that direction. The model of the urban transition (Zelinsky 1971) formalises a typical urbanisation process as a logistic growth of the percentage of urban dwellers with time. Although the urban transition model does not allow a decrease of this percentage (only its asymptotic convergence towards a limiting maximum), decreases in the percentage urban can be observed and have been coined counter-urbanisation (Berry 1976).

At the scale of a single city, urbanisation can be thought of as a diffusion process of urban characteristics to surrounding former rural locations through the extension of built-up area, the creation of new satellite towns, and/or the attraction of a new commuting workforce. Urbanisation in this sense has physical consequences (the construction of buildings, roads, amenities etc.) as well as social consequences (higher density and diversity of contacts, increased possibilities of economic specialisation, wider marketplaces etc.). Again, there is a myriad of models of urbanisation (concentric sprawl, satellite towns etc.) but fewer of de-urbanisation (or shrinkage), although the latter does happen (Oswalt 2005), generally not symmetrically (Martinez-Fernandez et al. 2012). This phenomenon affects many Japanese, German (Oswalt 2005) and Russian (Cottineau 2016) cities, but also cities in rapidly urbanising countries such as China [with Daqing or Pingxiang (He et al. 2017) for example], due to their obsolete economic specialisation (mining) or network isolation.

In the remainder of the chapter, we exploit the polysemous character of these three concepts to question the legitimacy and usefulness of commonly used models to study, to understand and to predict urbanisation in BRICS countries, focusing on Russia and South Africa. Later on, we suggest the opportunity to develop new (de-centered) models directly from the BRICS.

12.3 Example of Contested Models of Urbanisation in Emerging Contexts

As stated in introduction, some models fail at representing the process of urbanisation in emerging economies, whereas others are contested because they represent blueprints from a western background that do not apply in BRICS urban contexts. In this section, we review two of these, one at the scale of individual cities (the 'compact city' model), and the other one at the level of systems of cities (the 'Central places' model). Each time, we use the demonstration to extract some BRICS-specific regularities which explain the model failure and reveal the potential for developing new models.

12.3.1 The 'Compact City' Model

The 'compact city' model is interesting as a case study because it functions both as a representational model to describe a high level of densities distributed in a concentrated way within cities, in contrast to sprawled cities, as well as a normative model of urban design. In the latter case, its use aims at reorganising polycentric and dispersed cities around pedestrian and public transport-friendly blocks of multiple land use (Cervero 2002), following the examples of historical European cities and the principles of the New Urbanism (McCann and Ward 2012), in a quest for higher sustainability (Burgess 2002). These urban design principles were formalised in the late 1970s/early 1980s in the United States and summarised by the term Transit Oriented Development (TOD), which became, since then, a central tool for urban planning in many cities in the world. We will not question the generic weaknesses of this model,[1] only the cases where it is contested for reasons specific to BRICS urban conditions.

As a way of describing the distribution of density in cities, the urban compactness is looked at mostly in terms of measurement and comparison. In terms of measurement, it relates to the proximity between different urban functions (retail, residence, etc.), to the absence of gaps between urban structures, or to the distance between the observed shape of the city and the ideal-type of the circle by dividing the maximum distance between two individuals by the radius of the circle corresponding to the same area. In any case, it relies on several assumptions which are not necessarily met in BRICS countries in general, and in our case studies in particular. The first assumption is that morphological and human densities are well defined and that the residential population is quite stable between two random days of the week or between two random months of the year. This does not hold in Russia for example, because of the

[1]For example, although "energy consumption is larger in a rich, motorized, sprawled, diffused and polycentric city. However, the compact city model seems to cause other problems; for instance, the degree of sprawl is negatively correlated with the number of fatalities related to transport." (Le Néchet 2012, §80).

dacha phenomenon. The "*mass of small wooden shed-like houses standing cheek-by-jowl, each on its own plot of land, that encircle the country's towns [...] might be mistaken for the spontaneous suburbs of the Global South*" (Nefedova and Pallot 2013, p. 91), but they are actually neither the new homes of rural migrants, neither the results of suburbanisation: they are the week-end and seasonal secondary home of long-time city dwellers. Besides having provided a much needed food production safety net in the 1990s, they hold a real advantage at providing a strong link between town and country for the citizens of large Russian cities. However, in terms of density definition and measurement, they constitute a strong bias: morphologically, they are permanent, but socially, their settlement is temporary. Therefore, should the Russian cities be regarded as very compact on the basis of population density (Bertaud 2004) or as sprawled cities with 'post-surburbia' (Golubchikov et al. 2010)?

The second assumption that compactness measures rely on is that the actual resident population is easily measurable. This is not the case in countries like China, where rural migrants tend to live in cities without being officially registered and counted in official statistics.

Finally, the analytical scope of the compact city model collides with the characteristics of peripheral areas in the cities of the global South. Indeed, these areas catalyse major changes and transformation due to their position of interface between urban and rural spaces, their specific insertion in the globalization processes, and their diversity (Chaléard et al. 2014). In South Africa for instance, peripheral areas can refer to gated communities, gigantic shopping malls, squatter camps, as well as industrial areas or agricultural areas. The hybridity of the peripheries in the cities of the South is thus particularly visible in South African cities, where the process of suburbanisation is also strong. Suburban areas have a long history in this country, as they started blooming in the aftermath of the Second World War as a simultaneous consequence of the diffusion of cars among the White population, and the enforcement of the Group Area Act, which divided each city along racial lines. Since the opening of the country to globalization and international market forces, development projects multiplied in these peripheries, for instance in the form of mega-malls or even casinos. Looking at these specific urban spaces, Mabin (2006) notes that the lifestyle and standards developed in suburban places are often undermined, the idea of urbanity being strongly associated with density—and the model of the compact city. Yet, according to him, the complexity of the South African post-apartheid identity and urbanity lies exactly in the diversity existing in the peripheral areas, which, furthermore, show the strongest densification rates. Also, according to ideas articulated by Pieterse (2006), the challenge in post-apartheid consists mainly in (re)creating links and connections between communities. Within cities whose peripheries concentrate most of the growth and very diverse communities, it is crucial not only to favour the densification of the city centre and the connections between the core and the peripheries, but also to enhance the circulation between the peripheral areas themselves.

As a policy, the 'compact city' model mixes arguments of sustainability (reducing the impact of car emissions or reducing land waste) with elements of historical contingency (for example, how European cities developed at a time of slow transportation means). This urban design model has circulated globally with success to become part of most urban governments' planning toolkit. However, its local implementation faces challenges. First, "*the question of the relevance of compact city policies for developing countries also has to take account of the significant differences in the levels of urbanisation and rates of urban growth between developed and developing countries and between different world regions*" (Burgess 2002, p. 11). Thus, despite global recipes being promoted by architects and urban design consultants (McCann and Ward 2012), the model-blueprint should to be adapted to the different paces of urbanisation (or urban shrinkage in the case of Russia for example). Second, it has to account for local regulations and building standards, which can be more complex outside of the western world. The example of the master plan of Perm in Russia is interesting in that respect (Zupan 2015). Perm was the first city to adopt the 'compact city' model has its planning strategy in 2008. The city hired a Dutch firm to develop a long-term master plan which would implement the 'European compact city' model, although this plan would only be indicative, having no legal enforcement basis. After long debates between the firm, the local planners and the residents, the plan was rejected, partly as it does not comply with national planning rules or the land ownership distribution. Indeed, "*the master plan for Perm' draws a glossy picture of a future 'clean' and 'rich' city, a development that is not mainly dependent on which principles of urban planning are used but instead requires a significant economic upturn, public investment and an appropriate appreciation of the existing urban heritage and history*" (Zupan 2015, p. 48). This failed attempt shows that heritage (the Soviet land market and neighbourhood organisation into *microraion* in this case) cannot be simply replaced by new models just as easily, as the capitalist city rather "*nestles in the socialist city*" (Wagenaar 2004, p. 9) or most likely socialist and capitalist elements co-evolve in the post-socialist city (Golubchikov et al. 2014).

The Transit Oriented Development (TOD) can be considered as another version of the 'compact city' model, as it entails the densification of the urban space through a joint action on transport and housing. A few characteristics are identified in the literature to describe what TOD lies on: an efficient, integrated and reliable public transport system; a high quality public realm which prioritises non-motorised transport and ensures high accessibility of public transport infrastructures; a mix of residential, retail, commercial and community uses; medium to high-density development within comfortable walking distance of the public transport stations; and reduced amount of available parking bays for cars (Bickford and Behrens 2015). The urban planning guidelines that constitute TOD explain why Bus Rapid Transit (BRT) often spawns TOD, in both global cities of the North and the South (Cervero 2013). Nevertheless, the spread of these urban models questions their adaptation in local contexts, such as South African cities. Given the deeply entrenched spatial consequences of apartheid, densification is a priority in order to enhance both densification and accessibility in the post-apartheid city, where a majority of urban dwellers live far away from opportunities and resources. However, even though the principles of BRT and TOD offer appealing solutions to foster densification, the understanding regarding their implementation remains limited. Indeed, Bickford and Behrens (2015) identify the divergent interests and capacities of the multiple stakeholders involved as an obstacle to implement TOD. In particular, the use of land represents one of the main challenges: for some, the land located near the transport infrastructures should be dedicated to mixed-use and mixed-income housing, while for others the interest of land lies in its financial value for private investors and developers. Indeed, at this stage, private actors remain reluctant to invest in public transport or development lying on public transport, given its poor condition. In the context of post-apartheid cities, the use of land crystallises the contradictions inherent to the ambition to simultaneously implement redistributive and inclusive measures and to attract private investments and actors. On top of this main consideration, Bickford and Behrens (2015) also recall the lack of understanding of most stakeholders, and the necessity to enhance greater collaboration between practitioners to implement such a complex urban strategy.

Beyond the difficulties to adapt exogenous urban models in the cities of the BRICS, the circulation and adoption of models is meaningful and indicates the strategic and geopolitical agenda of the cities of the BRICS. For instance, the version of the BRT adopted in South Africa is clearly marketed as being inspired from the models developed in South America, in particular Bogota and Curitiba. The research of Wood (2015) unravels this narrative and enlightens the reasons underlying this choice. Indeed, using South American cities as a reference case could be questioned, since other cities in Africa and in India also implemented BRT systems, and present urban contexts closer to South African cities. The choice instead relates to the will to demonstrate the adoption of best practices by South African cities, as South American cities are considered to be innovative laboratories and cutting-edge with respect to transport planning. Beyond this ambition to assert the position of South African cities within the international hierarchy, Peyroux (2016) underlines how the agenda of Johannesburg evolved over the past years. Through the adoption and circulation of urban models, the goal is not so much to implement an international strategy aimed at positioning the city as a global city, but to show through partnerships an opening towards the cities of the BRICS countries and to assert Johannesburg's position as a catalyst place to develop South-South relationships.

12.3.2 Optimising Models of Urban Settlements

At the level of systems of cities, the two most infamous models are optimising models: Zipf's law of cities' size distribution was explained by its author as following the 'principle of least effort', whereas Central Places' size and location follow the optimisation of accessibility and transport costs. This last model dates back to the ideas of Reynaud (1841), the formalization of Christaller (1933) and its extension by Lösch (1944). It aims to represent and explain the spatial distribution, size and functions of cities within a given region or country based on the distribution of population and a hierarchy of functions based on their frequency of need (or threshold for usage). For example, bakeries are needed every day, so there will be present in all towns and accessible within short time-distances to everyone. Embassies on the other hand will be present only in the larger towns and serve a wider area and population. The optimisation is spatial (areas served by towns and cities are hexagonal) and hierarchical (hexagons for low level services are nested within the bigger hexagons of infrequent service areas). Despite its many critics (the model is a static equilibrium, its assumption of isotropy never holds, etc.), its prediction of a regular spacing of cities, and an increasing average distance between cities of similar increasing size is considered a rather good depiction of many western systems of cities (Germany, USA, France). In the case of BRICS cities, we find four reasons why it does not (and should not) provide a good description of the urban settlement.

The Central Place theory (but it applies to Zipf's law as well) has been developed to describe systems at equilibrium, or at least at a point of urbanisation where the number and location of cities is rather stable, like in contemporary Europe. In the context of fast urbanising countries with high numbers of new towns emerging every year, it is hard to use the model adequately. The number of new towns, which is anecdotic in recent years for Europe, reaches paroxystic values in contemporary India, China (where a third of cities present in 2010 were not cities in 2000), and high percentages in South Africa (Table 12.1). This fast pace of urbanisation in emerging countries is not consistent with a rapidly evolving nested structure of servicing areas and corresponding transport networks. The model is thus of little help to understand where urbanisation will proceed in the future.

Secondly, the Central Place model does not account for exogenous causes for city location (such as extractive industries around mineral ores), which play a major role in many BRICS economies. The rise and fall of the numerous mining cities of China, South African and Russia are thus unaccounted for in the model.

Thirdly, such optimising models assume free mobility flows between places, although BRICS countries have exhibited a wide spectrum of mobility restrictions over the years: on rural dwellers in China, on Blacks, Coloured and Asian citizens during the South African apartheid, on Soviet prisoners and strategic scientists,

Table 12.1 New towns and cities by city systems: BRICS versus Europe

% of new cities (n new cities)	1960s	1970s	1980s	1990s	2000s
Brazil	8% (114)	7% (112)	6% (107)	4% (68)	4% (69)
Russia (FSU)	17% (233)	14% (217)	12% (227)	3% (61)	2% (40)
India	38% (1307)		11% (426)	19% (885)	9% (486)
China	15% (149)		2% (22)	84% (5236)	32% (2948)
South Africa	29% (25)	24% (26)	27% (39)	39% (89)	14% (35)
Europe	*9% (315)*	*6% (235)*	*3% (114)*	*2% (86)*	*0% (2)*

Source From Cura et al. (2017), Annexes 1–5. N.B. Cities in each system are defined as evolutive urban agglomerations over 10,000 inhabitants. FSU stands for Former Soviet Union.
European figures (in italics) were added for comparison, but Europe does not belong to the BRICS countries

etc. Nowadays, transport inequalities within and between cities still reproduce certain patterns of selective mobility. This is particularly striking in the case of South Africa, due to the aggravated social and spatial effects of the apartheid policy, and the permanence of the mechanisms used to implement this policy. Especially, the economic system implemented during the apartheid lied on a paradoxical principle that is the employment of a cheap and numerous African manpower to maintain the economic prosperity of the White population, and the removal of this manpower away from the White city, i.e. the economic centres of cities. Within this perspective, more and more restrictive measures were implemented to limit the mobility of the African population within the city, and to the city. Within the city, the removal of the non-White population translated into the planning of the townships, located far away from the city centre. At the national level, the individuals who were denied access to the city (which was mainly depending on the possibility to justify for employment in town) were relocated in 'homelands'. The homelands were autonomous territories located in the margins of the country. The absence of economic activities in these territories, where subsistence farming dominated, explained the dependence of the inhabitants on the main cities and employment centres (such as the mines) to obtain seasonal jobs. Thus, the South African economy depended on the migration of an important part of the manpower. Although this migration system existed even before the apartheid, the difference comes from the evolution in the intensity and the institutionalisation of the system. Also, the removal of the African population towards the homelands impacted directly on the transport system: the geographic mobility became more extended, to the point that scholars such as Lemon (1982) used the term "frontier commuting" to evoke the regular moves of individuals between homelands and cities. As a consequence, the transport system, and more particularly the railways, orchestrated the regular commutes of workers from places located dozens or hundreds of kilometers away from the urban centres. By doing so, it blurred the traditional levels of urban observation, meaning the system's level and the city's level. Nowadays, in spite of the demise of the apartheid regime, social and spatial structures often linger. In the case of the mobility system, the persistence also results from the long-lasting imprint of the rail infrastructures, which are poorly adaptable.

Finally, the Central Places of Lösch and Christaller (or the cities along the rank-size distribution) are at best independent of each other, at worst duplicates. These models do not account for the interdependencies that link cities with one another. This is a problem in general for the models, but the legacies of some BRICS countries make it even more problematic. In the case of Russia for example, the legacy of strategic inter-dependence of Soviet urban centres in the production chain (Snyder 1993) had produced a system of cities which are not duplicate of self-sufficient places, but a system of inter-dependent centres, where one city's monopoly production was dependent on another city's intermediate production (with all the bottlenecks involved). In South Africa, the geography of apartheid presents another deformation in the distribution of activities and functions between cities. Besides the removal of the population groups according to the racial hierarchy of apartheid, the functional structure of cities was disrupted. With the implementation of an artificial separation within cities, two—or more—commercial centres were to compete for the same demand. But in this context, the centrality is biased, as the functions located in the "white" city benefit from a hierarchical advantage and a better accessibility. Thus, the discrepancy between the "white" city and the township (or the homeland) worsened and exacerbated the dependence of the non-White populations on the white city which concentrated the offer for various types of services and activities. A. Lemon's comment (1982) illustrates this observation: "*To the inconvenience of long-distance commuting is added the need to travel to "white" towns for most goods and services*". Browett and Fair (1974) support this observation by showing how restrictions resulting from the apartheid policy (control of mobilities towards cities, restricted access to employment or the regulation to open business activities etc.) contradict the implicit rationality of the individual that is underlying in Christaller's theory or the principles of the market law. Finally, besides the dramatic consequences of the geography of apartheid, and its implications in terms of concentration of economic activities and excruciating mobilities, the spatial imbalance created conflicts with the logic of hierarchical distribution of economic functions as described by Christaller for instance.

As we have seen, despite some success in the global circulation of models, some models are not very useful to understand the emergent urbanisation patterns of BRICS countries, or only as benchmarks to evaluate the distance from which our case studies stand. This does not hold true for all kinds of models, and some approaches prove useful to study urbanisation in emergent contexts.

12.4 Example of Useful Models of Urbanisation in Emerging Contexts

In this section, we review examples of models and research strategies which have been designed for generic contexts and help understand the specific urban developments of South Africa and the Former Soviet Union.

12.4.1 Transport Network and Territories Interactions in South Africa

In the South African case study, we aimed at analysing the interaction between transport networks (especially railways) and cities, both at the urban system level and intra-urban level. However, notably because of the spatial implications of the apartheid, the South African case happened to show significant variations compared to other countries when traditional models are used to grasp this interaction. It led to the necessity to combine different models in order to assess the urbanisation process in South African in its complexity.

Generally, it is acknowledged that the urbanisation process comes with the development of transport networks, according to a mechanism of mutual adaptation (Cattan et al. 1999; Bretagnolle 2009). In South Africa, the concomitance of urbanisation and the implementation of the railways, especially during the 19th century is particularly striking and questions the interaction existing between the rail network and the urban system. Moreover, this network shows several specificities at the international level which makes its observation worth. Indeed, it represents by far the largest network in Africa, as it accounts for 40% of the total of railways on the continent. It is also one of the only railway network still used nowadays in Africa, after a massive wave of privatisation and closure of railway lines in the 1980s and 1990s. To assess the evolution of the railway network and its interaction with the urban system, a distinct database was built, as part of different research projects.[2] The database Harmonie-cités SA records the opening and closing year of each leg of the railway network between 1860 and 2016, and therefore enables to trace back its evolution. A second database, Gares interurbaines SA, identifies the urban railway stations for each date of the national census between 1911 and 1991. They have been described in-depth in Baffi 2016.

A first analysis regards the evolution of the shape of the network, in order to compare it to other countries (Table 12.2).

Regarding the pace of growth, it is visible that the network grows rapidly at the beginning of the 20th century, and again around the 1980s. The connectedness index grows continuously within the first decades, and stagnates from 1951 onwards, in spite of the growth of the network. Similar analyses led in other countries by Kansky

Table 12.2 Characteristics of the South African railway network over the 20th century

Year	1911	1921	1936	1951	1960	1970	1980	1991
Length of the network (km)	11,376	14,051	16,655	19,057	19,179	19,248	20,575	20,050
Variation rate of the network's length (%)		2.13	1.14	0.9	0.06	0.03	0.67	−0.26
Connectedness index β[a]	1.01	1.01	1.02	1.02	1.02	1.02	1.03	1.03
Maximum connectedness index γ[b]	0.34	0.34	0.34	0.34	0.34	0.34	0.34	0.34

[a]In order to compute the connectedness index, the following formula is used: $\beta = L/V$ (L being the amount of lines that constitute the network, and V the amount of vertices

[b]In order to compute the index of maximal connectedness, the following formula is used: $\gamma = L/[3(V-2)]$

Source Database Harmonie-cités SA (Baffi 2016)

[2]The database Harmonie-cités SA was built as part of the ANR Harmonie-cités (project leader: A. Bretagnolle) and the database Gares interurbaines SA was built as part of the ERC Geodivercity (project leader: D. Pumain).

(1963) enable to compare the South African network to other rail networks. Contrarily to most African countries, the connectedness index in South Africa is already above 1 in 1911, the early achievement of a basic level of meshing. Nevertheless, the index almost does not change throughout the 20th century, while the maximal connectedness index, γ, remains exactly the same.

These trends are specific to the South African case study, and show the consolidation of a tree-pattern network, built with penetrating lines and few connections. The tree-pattern shape is commonplace in former colonial countries, as Taaffe, Morrill and Gould recalled (in Haggett 1973), and more specifically in Africa (Debrie 2007). Indeed, most railway networks were built to connect the colonial harbours to the hinterlands so as to export raw materials. According to this model, the expansion of the network entails a second phase, characterised by the development and the multiplication of connections between cities; however it did not really happen in South Africa. This specificity relates to the extreme political and social laws implemented in the country from the 1950s, and the specific function allocated to the railway in the implementation of the apartheid geography. After the 1950s, most of the railway lines were built between the homelands and the mining areas, in order to maintain the mobility of a cheap manpower—i.e. the segregated populations. The growth of the network into a tree-pattern shape is therefore representative of the spatial principle on which the apartheid regime relied: to control the mobility of populations considered as undesirable, but also essential for the economic system (Baffi 2016).

At the mesoscale of the city, the use of the railways as a planning tool and as a pivot in the functioning of the apartheid city also led to a specific model of urbanisation. The ambition to identify these specificities led several scholars to distort existing models elaborated in other contexts. Davies (1981) theorised a model of the apartheid city directly inspired from the segregation city model of the School of Chicago. At this scale, the function of the railway tracks was clearly articulated by the apartheid planners. Along physical elements such as mountains and rivers, or functional spaces such as industrial areas and graveyards, railway tracks were key elements to divide the urban space and organise the distribution of the population groups. Moreover, after the 1950s, most of the railway lines built in cities opened to connect the newly planned townships with the economic centres. From then on, the train became the transport mode of the segregated populations. Indeed, also at this scale, public transport enabled the daily commutes of the manpower living in the townships towards the employment areas; and in general to control the movement of the different population groups (Baffi 2014). Hence, the tree-pattern shape progressively developed at the urban scale and became a key feature of the functioning of the apartheid city. It resulted in a structure called spatial mismatch. Initially, this notion was introduced by J. Kain in 1968 to assess the gap existing in U.S. cities between the localisation of the lower-income populations in the inner centre, and the localisation of low-skilled employment areas in suburban areas. Bénit (1998) and Naudé (2008) drew on this descriptive model by suggesting that the apartheid city presents an inverted spatial mismatch, the economic activities being located at the centre, whereas the labour pools are relegated to a distance, in spaces deprived from economic activities. Hence, whereas the apartheid city cannot be grasped through traditional models, their distortion brings analytical elements to understand its structure and functioning.

This is even more obvious in the case of the post-apartheid city, where new dynamics overlaid from the 1990s onwards with the concomitant opening to the market law and the implementation of redistributive measures. Firstly, with the lifting of the apartheid zoning, the White population moved towards suburban areas, as did economic activities. However, these new development areas are often located even further away from the townships (Turok 2001). Secondly, some of the redistributive measures introduced after 1994 tend to reinforce the inherited spatial mismatch in the context of a globalised economy. Indeed, simultaneously to the re-opening of South Africa on the international stage, the government implemented a massive programme to provide social housing. However, in a market-driven economy, the affordable land for such a programme was located on the periphery of cities, which often happened to be next to the existing townships. As a direct outcome, travel time and commuting costs are exceptionally high for South Africans. According to a survey made by the OECD in 2011, South Africa shows the longest average commuting time of all OECD members, with 56 min to commute to work. Already during apartheid, surveys showed that commuting costs represented on average more than 10% of households' incomes (Walters 2013). According to Turok (2016), workers can spend between 20 and 35% of their income in commuting, which is way above the international standards. Also, while South African urban dwellers have a high mobility (regarding the amount of trips) compared to other urban dwellers worldwide, it does not entail an improved accessibility to basic resources. From similar observations, Lucas (2013) even erects accessibility and reliable public transport as a basic human right in South Africa. This last point represents a striking example of the specific definition of mobility in South African cities, and the difficulties to mobilise international indicators—but also planning guidelines and best practices—to appreciate and improve South African urban dwellers living conditions.

In conclusion, it appears that various models were necessary to analyse the interaction between transport network and urban systems in South Africa. It results on the one hand from the hybridity of the South African territory (Vacchiani-Marcuzzo 2016), which show features common to industrialised countries but also pioneer territories as well as inequalities representative of the global South. On the other hand, the distance to the existing models enable to identify the specific function allocated to transport to implement the segregative geography of apartheid. Also, from the different examples mentioned, analytical, theoretical and descriptive models were used either to assess the specificities of the South African case, or as a combination to appreciate a complex territorial configuration.

12.4.2 Combinatorial Agent-Based Models in the Former Soviet Union

For the case of the former Soviet Union, we found in a previous study (Cottineau et al. 2015; Pumain and Reuillon 2017) that a type of modelling strategy which worked in understanding urbanisation patterns over the long time was the *combinatorial approach* or *multi-modelling*, where models are modular in nature and composed of different mechanisms implemented from alternative and competing theories. As a first step in this multi-modelling approach, we implemented a core agent-based model of interurban interactions based on urban scaling (of demand and supply with the urban population) and the gravity model (to estimate the potential of a pair of cities to exchange), and five alternative (or complementary) mechanisms implementing different theories of urban size differentiation:

1. a mechanism accounting for interaction benefits (*Bonus*) which provides cities whose interactions are numerous and diverse with some form of spillovers which translates into an economic increase controlled by a parameter b.
2. a mechanism accounting for situation advantages (*Fixed Costs*) which excludes from interactions the pairs of cities whose potential for exchange is too low, being too small and/or too far apart from one another. Indeed, a fixed cost f being applied for interactions to take place, it is sometimes not profitable with regard to the amount anticipated for the exchange.
3. a mechanism accounting for site effects (*Resource*) which provides cities with some extraction revenue (or debt) multiplicator r if they happen to be located on natural deposits such as coal or oil.
4. a mechanism accounting for territorial effects (*Redistribution*) which mutualises wealth within regions and Republics through taxation at a rate t of wealth and redistributes it based on the population shares of regional cities.
5. a mechanism of regional lagging effects (*Urban Transition*) which accounts for the initial position of regions in the urban transition and simulates the opportunity for rural migration in each city by extrapolating the urban transition on a calibrated logistic curve by region. The result is a "rural multiplier" m which is added to the population at the previous step of the model.

The 64 different combinations of model (based on the selection and assemblage of the five mechanisms described above) are calibrated systematically against empirical data (Cottineau et al. 2015) to provide insights of the "most probable" set of mechanisms to explain the observed urban evolution at yearly steps. The models, calibration results and analysis tools are provided online with the VARIUS application.[3] In the following, we analyse the hierarchy of explaining factors for the two time-periods of before and after the collapse of the Soviet Union.

Between 1959 and 1989 (the last census data before the transition of 1991), the single mechanism which provides the closest simulation to empirical data is the mechanism of regionally differentiated *Urban transition*. This indicates that rural

[3]https://github.com/ClementineCttn/VARIUS.

Table 12.3 Hierarchy of explaining mechanisms in (post-) Soviet model of urbanisation

Best model with…	Soviet period: 1959–1989	Post-Soviet period: 1989–2010
1st add. mechanism	Urban transition (0.0142)	Resource (0.0052)
2nd add. mechanism	Bonus (0.0131)	Fixed costs (0.0047)
3rd add. mechanism	Fixed costs (0.0126)	Redistribution (0.0041)
4th add. mechanism	Redistribution (0.0123)	*Urban transition (0.0041)*
5th add. mechanism	*Resource (0.0131)*	*Bonus (0.0041)*

Source https://github.com/ClementineCttn/VARIUS. In parentheses: normalised distance to empirical data
0 means that all cities' population was simulated perfectly at each census. In italics, we highlight the best performing mechanism

migrations were probably the most significant driver of differences in urban growth for Soviet cities (Table 12.3). Next, the two-mechanism model closest to empirical data contains, on top of the mechanism of regionally differentiated urban transitions, the mechanism of bonified exchanges (*Bonus*), with a lower overall distance to data. Thus, places which were at the centre of large and diversified exchanges have probably benefited most from growth. The *Fixed Costs* mechanism appears as the third mechanism to be added to reach a lower distance to data on a 3-mechanism model, indicating that situational characteristics had a minor but probably significant effect on Soviet urban systems. Surprisingly, *Redistribution* appears only as a fourth-order mechanism and reduces the distance to data only marginally. The image we have an equalising Soviet system was thus not probably translated into its urban dynamics, at least at the regional level. Finally, the urban differentiation related to *Resource* extraction is not a good explaining model, because as fifth mechanism in the model, it increases rather than decreases the distance to data, making the calibration worst when in the model.

This hierarchy of explaining mechanisms is even more interesting when compared to the next historical period. Overall, the calibrations are better as the simulation lies closer to empirical data per year and per city, but only three mechanisms reduce this distance significantly. The first one is the *Resource* extraction. From last, it becomes the most important mechanism to simulate urban dynamics after the collapse of the Soviet Union. This reflects the well-known fact that post-Soviet economies are highly dependent on raw materials exportations (Garanina 2009). The second mechanism is the *Fixed Costs* one, which excludes small and isolated cities from the network of urban interactions. This simulated process reflects the challenges of many shrinking cities of the far North and East which have struggled since the end of subsidies on a location and climate that was not suited for profitable (or enjoyable) living (Kontorovich 2000; Cottineau 2017), some of these cities having been developed through the work of prisoners and/or komsomol enthusiasts. Finally, *Redistribution* seems to explain to a lesser extent post-Soviet urban dynamics, but not the *Urban transition* or *Bonus* mechanisms.

All in all, we found so far that simple recipes applied to BRICS contexts do not help understanding urban dynamics of these countries, but that other modelling strategies might alleviate some of these issues. Most importantly, we think that some models need to be (re-)invented directly from the experience of BRICS urbanisation. We treat these models in the following section of the chapter.

12.5 Models to (Re-)Invent

Instead of trying to verify and falsify models built on the experience of western urbanisation, there is an opportunity, with emergent BRICS countries, to highlight routes for the development of new models of urbanisation, which could be useful to understand these cases and others, around demographic and social sustainability for example.

12.5.1 A Model for Urban Shrinkage

One aspect on which urban Russia is ahead of many developed countries is that of demographic shrinkage. Depopulation—in the form of a 'second demographic transition' (Van De Kaa 1987)—and counter-urbanisation (Berry 1976) have been predicted for Europe and the USA, many times and long ago, but only in Russia did the level of demographic shrinkage reach a level where the majority of cities were losing population. Indeed, although the number of urban agglomerations rose from 926 in 1989 to 962 in 2002 and 981 in 2010 and the total urban population stagnated short of 100 million, the number of urban agglomerations loosing population grew after the transition to reach 500, i.e. about 70% of the total number of agglomerations (Slepukhina 2014; Cottineau 2016), mostly because of a natural deficit (deaths outnumbering births) not compensated by a sufficiently positive migration balance. The challenges in terms of housing and service management in a shrinking city are becoming better known (Oswalt 2005; Averkieva 2014), but the cumulated effects over a majority of cities are still unpredicted. Similarly, the relation between the hierarchy of cities and the total urban population has only been approached through the lens of growth, around the question: are (growing) urban systems evolving towards more differentiation and city size unevenness? With the case of Russia, we have shown that a model which reproduces well the growth trajectory of cities could also simulate their shrinkage, in both the directions of a hierarchisation or a de-hierarchisation (Chérel et al. 2017). Therefore, such models need to be explored in order to understand how urban shrinkage unraveled in post-transition Russia, so as to provide tools to analyse and project demographic shrinkage in other countries, such as Japan, Germany or China.

12.5.2 A Model for Trust in Urban Planning for Sustainable Development

A domain in which Russia and South Africa face a common challenge with respect to models of urban policies is that of trust. Indeed, in the past, the local and central governments of both countries have used urban planning and urbanisation schemes to promote an ideology harmful to large sections of the population as well as the environment. Moreover, in the period following the transition, "*many Russian scholars and mayors argue that, under the current legal environment, waiting for federal grants and subsidies or getting involved in federal investment projects is much more profitable than undertaking something by yourself at the local level* (Yasin 2012; Zubarevich 2014)" (Cottineau and Frost 2018, p. 278). As a consequence, there is a strong resistance from the population to the various attempts at planning cities as well as high expectations when planning decisions are implemented. For instance, "*the refusal of the master plan in Perm' should be understood as an opportunity for the city to develop concepts based on local assets and requirements*" (Zupan 2015, p. 49). If BRICS countries could come up with new models of political involvement for urban citizens, they will be able to provide useful solutions, including for the North where elections are becoming less and less of a way to express views about society, planning and cities and more of a duty ignored by many. And indeed, when policy is designed with the collective interest in mind, "*it seems that carefully planned urban growth can strengthen prosperity and well-being. Finally, the BRICS experience indicates the importance of steering urban development onto a more compact and sustainable path because of the damage caused to the natural environment by negligence*" (Turok 2014, p. 132).

12.5.3 New Models to Blur the Norms? Urbanisation from the South

Several scholars recently highlighted the importance of the circulation of models developed in the global South (Verdeil 2005) and the circulation of models between cities of the South (Peyroux 2016). In the latter case, the circulation of the Bus Rapid Transit is a case in point, as the research of Wood (2015) shows. Besides the choice to develop the South African service according to the model elaborated in South American cities, the role of the paratransit within this model warrants our attention. Indeed, in Cape Town, the implementation of the BRT service, called MyCiti, started in 2011 as a way to introduce a safe and reliable transport mode as well as a tool to reform the paratransit industry (Schalekamp and Behrens 2013). However, in 2017 a revised version of the Integrated Transport Plan or ITP (the 12-year framework document presenting the guidelines of the metropolitan government regarding the transport planning) indicates a new strategy regarding the integration of the minibus. The change of approach was theorised by Ferro et al. (2013) as the difference between a "de facto" hybrid transport model, such as the one developed in Curitiba, and the "de jure" model, such as the one developed in Accra. In the scenario adopted in Cape Town and Johannesburg in 2011, members of the minibus industry were given the choice either to integrate the BRT operations, or to accept a compensation in exchange for them to leave the industry. Besides the ambition to implement a safe, reliable and functional transport service, the design of the service also aimed at a complete removal of the paratransit on the BRT routes. However, seven years later, the minibus are back on the roads along the initial route, as MyCiti service cannot cope with the increasing demand.[4] Moreover, the design of MyCiti, inspired from Bogota's Transmilenio, turns out to be unadapted to the mobility patterns in Cape Town. The density being considerably lower in Cape Town, the feeder routes are running at a huge loss, making the whole system economically unviable.[5] This explain why the City of Cape Town decided to move on from the "de facto" hybrid model towards the "de jure" hybrid model. It results from the assessment of the expensive costs associated to the operation of the MyCiti feeders, but it is also made possible due to the lessons learnt and the consolidation of the model used to formalised the minibus operators and the minibus associations into formalised corporations. More generally, this shift reveals how BRT can be a contested model when it implies a poor understanding of the transport industry—especially when it includes paratransit—and between the actors themselves (Schalekamp 2017). Thus, the new version of the ITP indicates that minibus could alternatively provide feeder service without being absorbed into the MyCiti service but as a complementary service.

This revision introduces a considerable shift in the underlying planning vision. Although public policies have tried for decades to take the paratransit off the road, this new vision acknowledges the crucial role of the minibus in the functioning of a global city. This vision relates to the consideration articulated by Jaglin (2014) regarding the diversity of service delivery that characterises many cities in the global South. She argues that rather than an avatar of under-development, the heterogeneous ways of delivering service answer the complexity of the urbanisation process in the South. In this light, the service delivered by the minibus industry might be more appropriate to low-densities, so as the on-demand service answers better the need for flexibility. This approach contrasts with the model of the universal service historically elaborated in industrial countries and underlying most of the urban models implemented in the global South. Besides, as Graham and Marvin (2001) observed, the segmentation and diversification of service delivery happens progressively in industrialised countries, where the unique offer inherent to the universal service model is less and less adapted to the social needs, economic trends and political transformation. In this perspective, one could then imagine that urban models currently developed and implemented in the cities of the South could be transferred to industrialised countries. For instance, the reflections conducted internationally on shared mobility, in the search of more

[4]Interview made by S. Baffi with the Manager for the industry transition, Transport and Urban Development Authority, City of Cape Town (8/03/2018).

[5]https://ewn.co.za/2016/05/25/MyCiti-bus-services-costs-city-R400-million.

sustainable transportation solutions, might lead to consider paratransit as a viable model.

This assumption seems to be even more accurate with the rapid diffusion of Information and Communication Technologies (ICTs) in the South. It gives rise to a reappropriation of models elaborated in North, such as on-demand transport applications based on the model developed by Uber (the example of the Go-Jek app for motorcycles in Jakarta presented by Lee (2018) is one of them), as well as numerous innovations in various sector (health, education, trade, governance) (Ninot and Peyroux 2018). Many of these innovations are developed as solutions for local challenges; however, some of them answer global issues, such as the mobile banking service M-Pesa. This service enables users to transfer money and pay with their mobile phones, even without access to a bank account. Initially created in 2007 in Kenya, this service is nowadays widespread over Africa, Asia (Afghanistan and India) and even in Eastern Europe. This example might be a harbinger that tomorrow's models might be currently forged in the South, as well as an indicator of the labile characteristics of urban norms.

12.6 Conclusion

Even if commonly used, the notion of "emerging countries" remains fuzzy and lacks a theoretical definition. Yet, when it refers to the BRICS countries, it designates demographic giants and most of the biggest countries in the world. Even though we do not dispute the opportunistic or descriptive use of the acronym when it was initially coined, our assumption is also that the traditional models used to assess urbanisation dynamics might not be completely relevant in such countries. Through the analyses conducted in Russia and South Africa, we identified the limits of several analytical and theoretical models to assess the main features of the urbanisation process, in particular to delineate its intensity, spatial extent and mechanisms.

However, some models turned out to be useful by pointing out the exceptional characteristics of these emerging countries, the distance to the normative framework then becoming meaningful. As theoretical exceptions or statistical residuals, the urbanisation dynamics identified in the BRICS countries even show a heuristic value, by distorting the classical understanding and the reach of models. Besides, through this study, we tried to enlighten to what extent the combination of models and levels of observation happened to be fruitful to grasp the complexity of the two countries. Indeed, in Russia and South Africa, urbanisation took place according to heterogenous processes (pioneer fronts, colonisation, authoritarian political regimes and controlled movement of populations). Also, in both cases, the existence of intertwined temporalities in the urban fabric led to use the term "post" to characterise the cities, may they be post-apartheid, or post-socialist. In the temporal sequence of the post-socialist transition, Sýkora and Bouzarovski (2012) identify urban transformations as the longest process of change, following institutional (short-term) and social (medium-term) transformations. The prefix 'post' recalls then the necessity to deal with moments that are dead in time, but not in space (Houssay-Holzschuch 2010) by multiplying the scientific angles and scales of analysis. Thus, emerging countries require different or new sets of methodologies, in order not just to overlay models but to articulate them.

Eventually, the necessity to forge new rationalities and methodologies from the BRICS echoes the assumption formulated by Myers (2014) while arguing for "unexpected comparisons". Whether Russia and South Africa might be considered in many ways as specific cases, they also offer a plain analytical level to consider other territories. Therefore, we argue that it is precisely due to the complexity of the "emerging" territories, and their resistance to the normative theoretical frameworks, that they represent laboratories to understand the future trends of urbanisation, as well as creative spaces to elaborate global solutions.

References

Averkieva, K.: 'Shrinking cities', Opyt izoutcheniya Rossii. Ira.urban Summer School about urban reconfiguration of the post-soviet space, 20 Mai 2014 (2014), Rostov-sur-le-Don (Russie)

Baffi, S.: Chemins de civilisation? Le rail dans les politiques territoriales en Afrique du Sud. L'Espace Géogr. **43**(4), 338–355 (2014)

Baffi, S.: Le chemin de fer et la ville dans les processus de territorialisation en Afrique du Sud: de la séparation à l'intégration territoriale? Doctoral dissertation, Université Paris 1-Panthéon Sorbonne (2016)

Bénit, C.: Gouvernement urbain et production de la ségrégation: quelles leçons de la ville d'apartheid"? Une comparaison Johannesburg - Los Angeles. Rev. européennes Migr. Int. **14**(1), 159–192 (1998)

Berry, B.J.L. (eds.): Urbanization and Counter-Urbanization, vol. 11. Sage Publications, Urban affairs annual review. Beverly Hills, London (1976)

Bertaud, A.: The spatial organization of cities: deliberate outcome or unforeseen consequence? UC Berkeley: Institute of Urban and Regional Development (2004). Retrieved from https://escholarship.org/uc/item/5vb4w9wb

Bickford, G., Behrens, R.: What does transit oriented development mean in a South African context? A multiple stakeholder perspective from Johannesburg. In: 34th Southern African Transport Conference (2015)

Bret, B.: Un questionnement géographique sur les pays émergents. Bull. l'Assoc. Géogr. Fr. **88**(3), 227–240 (2011)

Bretagnolle, A.: Villes et réseaux de transport: des interactions dans la longue durée (France, Europe, États-Unis). Habilitation à diriger des recherches, tome 3. Université Panthéon-Sorbonne-Paris 1, 203 p. (2009)

Browett, J.G., Fair, T.J.D.: South Africa, 1870–1970: a view of the spatial system. S. Afr. Geogr. J. **56**(2), 111–120 (1974)

Burgess, R.: The compact city debate: a global perspective. In: Compact Cities, pp. 21–36. Routledge (2002)

Cattan, N., Pumain, D., Saint-Julien, T., Rozenblat, C.: Le système des villes européennes, p. 193. Economica, Paris (1999)

Cervero, R.: Built environments and mode choice: toward a normative framework. Transp. Res. Part D **7**, 265–284 (2002)

Cervero, R.: Linking urban transport and land-use in developing countries. J. Transp. Land Use **6**(1), 7–24 (2013)
Chaléard, J.-L., The Anh, D., Huamantinco, A., Koffi-Didia, A.M., Mesclier, É., Monin, É., Moustier, P., Ninot, O.: Spécificités des systèmes de productions agricoles et d'élevage à la périphérie des métropoles du Sud. Réflexions à partir de cinq cas. In: Chaléard, J.-L. (ed.) Métropoles aux Suds, le défi des périphéries? pp. 225–242. Karthala, Paris (2014)
Chalmers, D.J.: Strong and weak emergence. In: The Reemergence of Emergence, pp. 244–256 (2006)
Chérel, G., Cottineau, C., Reuillon, R.: Using models to explore possible futures (contingency and complexity). Urban Dynamics and Simulation Models, pp. 81–95. Springer, Cham (2017)
Christaller, W.: Central Places in Southern Germany. Translation into English by Carlisle W. Baskin in 1966 (1933)
Cottineau, C.: A multilevel portrait of shrinking urban Russia. Espace populations sociétés. Space populations societies, (2015/3–2016/1) (2016)
Cottineau, C.: Peut-on estimer la singularité des villes (post-)soviétiques ?. Revue d'Économie Régionale & Urbaine Janvir (1):5 (2017)
Cottineau, C., Frost, I.: The Russian urban system: evolution engaged with transition. In: Rozenblat, C., Pumain, D., Velazquez, E. (eds.) International and Transnational Perspectives on Urban Systems, pp. 263–284. Springer, Singapore (2018)
Cottineau, C., Reuillon, R., Chapron, P., Rey-Coyrehourcq, S., Pumain, D.: A modular modelling framework for hypotheses testing in the simulation of urbanisation. Systems **3**(4), 348–377 (2015)
Cura, R., Cottineau, C., Swerts, E., Antonio Ignazzi, C., Bretagnolle, A., Vacchiani-Marcuzzo, C., Pumain, D.: The old and the new: qualifying city systems in the world with classical models and new data. Geographical Analysis **49**(4), 363–386 (2017)
Darbon, D.: Afrique du Sud: l'utilité comme ressource d'appoint d'une stratégie de puissance émergente. In: Jaffrelot, C. (ed.) L'enjeu mondial. Les pays émergents, 384 p. Les Presses de Sciences-Po., Paris (2008)
Davies, R.J.: The spatial formation of the South African City. GeoJournal **2**(S2), 59–72 (1981)
Debrie, J.: Géohistoire d'un tracé technique: frontières et réseaux en Afrique de l'Ouest continentale francophone. Flux **70**(4), 49–58 (2007)
Duranton, G., Puga, D.: Micro-foundations of urban agglomeration economies. In: Handbook of Regional and Urban Economics, vol. 4, pp. 2063–2117. Elsevier (2004)
Epstein, J.M.: Agent-based computational models and generative socialscience. Complexity **4**(5), 41–60 (1999)
Erkip, F.: The shopping mall as an emergent public space in Turkey. Environ. Plann. A **35**(6), 1073–1093 (2003)
Ferro, P.S., Behrens, R., Wilkinson, P.: Hybrid urban transport systems in developing countries: portents and prospects. Res. Transp. Econ. **39**, 121–132 (2013)
Fleury, A., Houssay-Holzschuch, M.: Pour une géographie sociale des pays émergents. Echogéo (21) (2012)
Gabas, J.-J., Losch, B.: La fabrique de l'émergence en trompe-l'oeil. In: Jaffrelot, C. (ed.) L'enjeu mondial, les pays émergents, 384 p. Les Presses de Sciences-Po, Paris (2008)
Garanina, O.: What beyond oil and gas? Russian trade specialisation in manufactures. Post Communist Econ. **21**(1), 32p (2009)
Golubchikov, O., Makhrova, A., Phelps, N.: Post-socialist post-suburbia: growth machine and the emergence of "edge city" in the metropolitan context of Moscow. Geogr. Environ. Sustain. **3**(1), 44–55 (2010)
Golubchikov, O., Badyina, A., Makhrova, A.: The hybrid spatialities of transition: capitalism, legacy and uneven urban economic restructuring. Urban Stud. **51**(4), 617–633 (2014)
Graham, S., Marvin, S.: Splintering Urbanism: Networked Infrastructures, Technological Mobilities and the Urban Conditions, p. 479. Routledge, London (2001)
Haggett, P.: L'analyse spatiale en géographie humaine, p. 390. Armand Colin, Paris (1973)

He, S.Y., Lee, J., Zhou, T., Wu, D.: Shrinking cities and resource-based economy: the economic restructuring in China's mining cities. Cities **60**, 75–83 (2017)
Houssay-Holzschuch, M.: Crossing boundaries. Vivre ensemble dans l'Afrique du Sud post-apartheid. Habilitation à Diriger des Recherches, vol. 3, 270 p. (2010)
Jaffrelot, C.: L'enjeu mondial, les pays émergents, p. 384. Les Presses de Sciences-Po, Paris (2008)
Jaglin, S.: Regulating service delivery in southern cities: rethinking urban heterogeneity. In: Parnell, S., Oldfield, S. (eds.) The Routledge Handbook on Cities of the Global South, pp. 434–447. Routledge, London (2014)
Kansky, K.J.: Structure of Transportation Networks: Relationships Between Network Geometry and Regional Characteristics, p. 155. University of Chicago, Chicago (1963)
Kontorovich, V.: Can Russia resettle the Far East? Post Communist Econ. **12**(3), 365–384 (2000)
Le Néchet, F.: Urban spatial structure, daily mobility and energy consumption: a study of 34 european cities. Cybergeo Eur. J. Geogr. (2012)
Lee, D.: How Ojek became Go-Jek: disruptive technologies and the infrastructure of urban citizenship in Indonesia. Int. J. Urban Reg. Res. (2018). (Spotlights on Disruptive technologies)
Lemon, A.: Migrant labour and frontier commuters: reorganising South Africa's black labour supply. In: Smith, D.M. (ed.) Living under apartheid, 256 p. George Allen & Unwin, London (1982)
Lestienne, R.: Dialogues sur l'émergence. Le Pommier (2015)
Lorrain, D.: Métropoles XXL en pays émergents, p. 404. Presses de Sciences-Po, Paris (2011)
Lösch: Die Raümliche Ordnung der Wirtschaft. Gustav Fischer, Jena (1944)
Lucas, K.: Making the connections between transport disadvantage and the social exclusion of low-income populations in the Tshwane Region of South Africa. J. Transp. Geogr. **19**(6), 1320–1334 (2013)
Lynch, K.: The image of the city, vol. 11. MIT press (1960)
Mabin, A.: La suburbanisation en Afrique du Sud: entre héritage et innovation. In: Lombard, J., Mesclier, É., Velut, S. (eds.) La mondialisation côté Sud. Acteurs et territoires, 496 p. IRD Éditions & ENS, Paris (2006)
Machamer, P., Darden, L., Craver, C.F.: Thinking about mechanisms. Philos. Sci. **67**(1), 1–25 (2000)
Marchionni, C., Ylikoski, P.: Generative explanation and individualism in agent-based simulation. Philos. Soc. Sci. **43**(3), 323–340 (2013)
Martinez-Fernandez, C., Audirac, I., Fol, S., Cunningham-Sabot, E.: Shrinking cities: urban challenges of globalization. Int. J. Urban Reg. Res. **36**(2), 213–225 (2012)
McCann, E., Ward, K.: Policy assemblages, mobilities and mutations: toward a multidisciplinary conversation. Polit. Stud. Rev. **10**(3), 325–332 (2012)
Morange, M.: "Émergence locale" et régénération urbaine au centre-ville du Cap. Bull. l'Assoc. Géogr. Fr. **88**, 348–360 (2011)
Myers, G.: From expected to unexpected comparisons: changing the flows of ideas about cities in a postcolonial world. Singap. J. Trop. Geogr. **35**, 104–118 (2014)
Naudé, W.: Is there a spatial mismatch in South Africa's metropolitan la bour market? Cities **25**, 268–276 (2008)
Nefedova, T., Pallot, J.: The Multiplicity of Second Home Development in the Russian Federation: A Case of 'Seasonal Suburbanization'?, pp. 91–121. Ashgate Publications, USA (2013)
Ninot, O., Peyroux, E.: Révolution numérique et développement: une trajectoire singulière. Question Int. **90**, 44–52 (2018)
Oswalt, P.: Shrinking cities, volume 1: International research. Ostfildern-Ruit: Hatje Cantz (2005)
Parnell, S., Robinson, J.: (Re)theorizing cities from the Global South: looking beyond neoliberalism. Cities **33**(4), 593–617 (2012)
Peyroux, E.: Circulation des politiques urbaines et internationalisation des villes: la stratégie des relations internationales de Johannesburg. Echogéo 36 (2016)
Pieterse, E.: Building with ruins and dreams: some thoughts on realising integrated urban development in South Africa through crisis. Urban Stud. **43**, 285–304 (2006)
Polèse, M.: Cities and national economic growth: a reappraisal. Urban Stud. **42**(8), 1429–1451 (2005)

Pumain, D.: Hierarchy in Natural and Social Sciences. Springer, Dordrecht (2006)
Pumain, D., Reuillon, R.: Urban Dynamics and Simulation Models. Springer, Berlin (2017)
Reynaud, J.: 'Villes'. Encyclopédie nouvelle, Charles Furne, Paris (1841)
Robinson, J.: Ordinary cities between modernity and development, London, Routledge 204p (2006)
Rosenthal, S.S., Strange, W.C.: Evidence on the nature and sources of agglomeration economies. In: Handbook of Regional and Urban Economics, vol. 4, pp. 2119–2171. Elsevier (2004)
Schalekamp, H.: Lessons from building paratransit operators' capacity to be partners in Cape Town's public transport reform process. Transp. Res. Part A **104**, 58–66 (2017)
Schalekamp, H., Behrens, R.: Engaging the paratransit sector in Cape Town on public transport reform: progress, process and risks. Res. Transp. Econ. **39**, 185–190 (2013)
Schelling, T.C.: Dynamic models of segregation. J. Math. Soc. **1**(2), 143–186 (1971)
Shatkin, G.: Global cities of the South: emerging perspectives on growth and inequality. Cities **24**(1), 1–15 (2007)
Slepukhina, I.: Russian cities at the crossroads: getting lost in transition or moving towards regeneration. Doctoral dissertation, Politecnico di Milano (2014)
Snyder, T.: Soviet monopoly. In: Williamson, J. (ed.) Economic Consequences of Soviet Disintegration, Washington D.C., pp. 176–243 (1993)
Sýkora, L., Bouzarovski, S.: Multiple transformations: conceptualising the post-communist urban transition. Urban Stud. **49**(1), 43–60 (2012)
Turok, I.: Persistant polarization Post-apartheid? Progress towards urban integration in Cape Town. Urban Stud. **38**(13), 2349–2377 (2001)
Turok, I.: The urbanization–development nexus in the BRICS. In: Parnell, S., Oldfield, S. (eds.) The Routledge Handbook on Cities of the Global South, pp. 122–138. Routledge, London (2014)
Turok, I.: South Africa's new urban agenda: transformation or compensation? Local Econ. **31**(1–2), 9–27 (2016)
Vacchiani-Marcuzzo, C.: L'Afrique du Sud est-elle un cas à part? L'analyse d'une hybridité urbaine. Habilitation à Diriger des Recherches, Université Paris 1 Panthéon-Sorbonne, 190 p. (2016)
Van De Kaa, D.J.: Europe's second demographic transition. Popul. Bull. **42**, 1–57 (1987)
Verdeil, E.: Expertises nomades au Sud. Eclairages sur la circulation des modèles urbains. Géocarrefour **80**(3), 165–169 (2005)
Yasin, E.: Scenarios for Russia in the long-term period. A new impulse in two decades. In XIII International Academic Conference on Economic and Social Development, Moscow, April 3–5 (2012)
Wagenaar, C. (ed.): Ideals in Concrete: Exploring Central and Eastern Europe. NAi Publishers (2004)
Walters, J.: Overview of public transport policy developments in South Africa. Res. Transp. Econ. **39**(1), 34–45 (2013)
Watson, V.: The allure of 'smart city' rhetoric, India and Africa. Dialogues Hum. Geogr. **5**(1), 36–39 (2015)
Wood, A.: The politics of policy circulation: unpacking the relationship between South African and South American cities in the adoption of Bus Rapid Transit. Antipode **47**(4), 1062–1079 (2015)
Zelinsky, W.: The hypothesis of the mobility transition. Geogr. Rev. **61**, 219–249 (1971)
Zipf, G.K.: National unity and disunity. Principia Press, Bloomington (1941)
Zubarevich, N.V.: Cities as the centers for the modernization of the economy and human capital. Socio. Res. **51**(4), 3–27 (2014)
Zupan, D.: Local debates on 'global' planning concepts: the 'compact European city' model in postsocialist Russia—the case of Perm'. Europa Reg. **22**(1–2), 39–52 (2015)

Chapter 13
More Urban Constructions for Whom? Drivers of Urban Built-Up Expansion Across the World from 1990 to 2015

Eric Denis

Abstract Since the 1990s, the conversion of land for urban use has never been more intense. Occurring under all latitudes, transcending the economic disparities as well as the political regimes, urban built-up expansion appears very much articulated to the global financial turn, which occurred after the progressive destatization of the money creation, which accelerated and globalized after the debt crisis of the 1980s. Using the Global Human Settlement Layer dataset on the degree of urbanization, we can assess that, worldwide, the built-up surface expanded by 243,000 km^2 between 1990 and 2015 and the average urban density has declined slightly from 2775 inhabitants per km^2 to 2756. After describing the national level dataset used and our methodology, we expose the regional nuances in the global built-up spreading and their relations to population growth. Then in a third section, we correlate those national trends with a set of significant economic indicators. We articulate it also to the individualization and progress of housing condition. Population growth remains the dominant factor determining urban built-up spread, but countries differ. Built-up growth did not respond linearly to the final demand for housing and therefore to demographic pressure. Building is a material good whose value is also interdependent with economic growth. The universal relationship between size and density specific to scaling laws is verified once again in the case of cities. However, economic density or concentration of wealth tends to take precedence over population mass to explain urban built-up expansion.

13.1 Introduction/Facts and Hypothesis

The conversion of land for urban use has never been more intense than it has been since the 1990s. Occurring under all the latitude, transcending the economic disparities as well the political regimes, urban built-up expansion appears very much articulated to the global financial turn, which occurred after the progressive destatization of the money creation, which accelerates and globalized after the debt crisis of

E. Denis (✉)
CNRS, UMR Géographie-Cités, Paris, France
e-mail: eric.denis@parisgeo.cnrs.fr

D. Pumain (ed.), *Theories and Models of Urbanization*, Lecture Notes in Morphogenesis, https://doi.org/10.1007/978-3-030-36656-8_13

the 1980. Deregulation lead to a tremendous expansion of credit but also to a chronic financial instability. Land and real estate becomes, in that context, a banking asset supporting a speculative rise of money creation widely disconnected of the productive machine needs. Hence, land driven development becomes a major motor engine of growth in many countries, in particular for the transitional economies. With gold, land constitutes one of the last warranties for the money creation. Land at the edge of cities represents also a frontier of non-fungible value to conquer and to introduce to the urban market and convert to private and fungible wealth. Conversion of land to urban uses is not the simple support of economic growth, it constitutes a major component of it, including the response to the need of productive spaces (for industries, offices, trades and logistic platforms) as well for housing and for, an expanding part, as a financial asset.

The de-densification and sprawl, is also the complex product of two structural changes: the parental de-cohabitation and the increase of standard of living. Children no longer stay in their parents' home at the time of marriage and houses and apartments tend globally to become larger. In other words, the structural and accelerated decline of density has to do with the dynamic of individualization, nuclearisation of families, as well the aspiration to individual bedrooms for kids. The habitation households claim to settle in tend to enlarge and the desire for individual housing becomes more and more a shared dream. The economy founds new opportunities in these personalizing trends when people are less bound in local solidarity group. They become more dependent of services to assist and eventually stimulate and create their growing individual aspirations.

Sprawl is a multidimensional phenomenon that could not be precisely evaluated using the set of data aggregated at the level of country we use.[1] Only the relation between density and growth of the built-up gives us here an indication of the urban trend. In our chapter, sprawl is viewed as the combination of the intensity of the relation between population density decline and increase of construction footprint.

Two global dependent indicators highlight our assumption: first-of-all, worldwide, the built-up surface expanded by 243,000 km^2 between 1990 and 2015, almost the full area of the United Kingdom; secondly, the average urban density (cities, suburbs and urban clusters) has declined slightly from 2775 inhabitants by km^2 to 2756. The Global Human Settlement Layer (GHSL) Degree of Urbanization 2018 geodatabase analysis confirms what Seto et al. (2011) or Angel et al. (2015) highlighted using a vast sample of cities across the world.

[1]Thanks to Thomas Kemper and Michele Melchiorri for pointing to me, reading this paper draft that it is difficult to assess precisely in what extend the urban built-up expansion contributes to the densification of existing urban perimeters and what is related to sprawl (expansion of the urban perimeters with low density). A key publication on sprawl (Tikoudis et al. 2018) points to the complex apprehension of sprawl that we cannot assess precisely with the summary country tables. A paper to come will precise further the extent of the sprawl using the detailed data per city and the gridded data. It requests to mobilize data to quantify the fragmentation vs. compactness of urban perimeters, the number of population centres (high density and polycentricity) and the share of urban population in low-density area.

The built up annual growth is globally faster than the demographic one, by a factor of 1.17 between 1990 and 2000 and 1.13 between 2000 and 2015. The built up annual growth was established at 1.69% during the first phase and 1.39% during the second.

In the same time, between 1995 and 2014, global wealth grew by 66%, when the GDP at constant price of 2010 almost doubles, from 44.9 to 77.6 trillion. This global enrichment is materializing in the extension of the urban footprint. Furthermore, this capitalization in land and building assets is a major driver of the global wealth expansion. The total population grew only by 28% during the same period.

Our analysis confirms the strong correlation between urban built-up footprint distribution (log) by country and population (log); established at 0.88 in 2015.[2] They are definitively intricate grandeurs as are logs of GDP and built-up ($R = 0.55$). Richest countries have a more extended urban footprint. After Henderson (2003), it confirms the canonical and strong relation between logs of GDP per capita and share of urban population ($R = 0.85$) but population demand matters.

Based on the Degree of Urbanisation Model (DEGURBA) developed in the frame of the "Global Human Settlement Layer" (GHSL) project of the European Commission's Joint Research Centre, in 2015, 84% of the world population leave in areas covered of continuous built-up at 50%, or cities plus suburbs and urban clusters. In other words, in densely inhabited areas and intermediate density areas, using a 1 square kilometre grid to articulate built-up footprint and population, as we will detail it in the following methodological section.

We should here relativize the pace of the built-up sprawl, in the sense that the urban agglomeration remains the most efficient way to use the terrestrial surfaces and to accommodate more than 7 billion inhabitants and most of their productive activities and daily commuting. All the negative externalities associated to agglomeration, notably congestion and pollution can be deeply mitigated using the present technologies and implementing appropriate rule of law. In fact, the total built-up spreads only on 0.6% of the terrestrial surfaces. The agglomeration process is not negative in itself. Even if we consider only the ecumene out of the land mass belonging to countries, it did not reach 1%. It occupies 1.7% of the land mass and 2.7% of the ecumene, if we consider the 1 km^2 grid where population and built-up information are merged in the "Degree of urbanisation" (DEGURBA) model and dataset (Dijkstra et al. 2018). Nevertheless, de-densification of urban environment characterizes the current trend. It leads to an extensive soil conversion.

In order to expose in detail and correlate precisely the global trend of built-up expansion we are highlighting here, in a first section we describe the national level dataset used and our methodology. Then, in a second section, we expose the regional nuances in the global built-up spreading and their relations to population growth. In a third section, we correlate those national trends with a set of significant economic indicators, such as the share of fix capital and labour, FDI, cement production, share

[2]We should consider here that the two variables, built-up and population are not fully independent. The estimated population within agglomerate is evaluated in relation with the density of built-up in 1 by 1 km^2 grid.

of construction sector in GDP. We articulate it also to the individualization and progress of housing condition. In conclusion, we expose some element related to the multiplication of localities and its consequence on urban hierarchies showing that, beside the expansion of urban units and mega urban region formation, we have to consider the less visible process of small towns' emergence.

13.2 A New Information Source on Global Urbanization

Measuring urbanization at the world scale was until recently almost impossible. We were comparing very much different objects defined as urban units. There was no harmonised dataset overpassing the various national definitions of the urban sector. The measurement of urbanization taking into account individual settlements and morphological agglomerations was until recently very rough and their spatial coverage was limited. Up to now, the most used datasets such as the UN-Habitat city population series are incomplete and biased by extremely heterogeneous definitions managed by independent administration of some 260 countries. François Moriconi-Ebrard (1993) demonstrated clearly the limitation of such series for a comparative study of urbanisation, generating major distortions. The international efforts of harmonization are yet very limited. Nevertheless, after Quito Habitat III Conference in 2016 and the New Urban Agenda commitment, World Bank, OCDE and EU agreed to promote standard to generate harmonized geographical datasets. One of the major output is the development of an open source complete and multisource dataset supported by Eurostat. Its fundamental concept is based on a three folds' definition recognising: (i) city, (ii) urban clusters and suburbs, and (iii) rural areas, villages or dispersed population. It distinguishes densely populated area, from intermediate and thinly populated areas[3] (Dijkstra and Poelman 2014).

The dataset we use The Degree of Urbanisation from 1975 to 2015 last release 2018 (DEGURBA 2018) appropriates this threefold definition. It constitutes a fundamental output of The Global Human Settlement Layer project supported by the European Commission through the GEO Human Planet Initiative who generates it. It provides by countries and by morphological agglomerations multiple metrics described in Florczyk et al. (2018). It delineates precisely 10,322 urban centres (agglomerate units) between 50,000 inhabitants and 46 million in 2015 (Guangzhou in China). It is a unique source correlating population and build-up footprint. The Global Human Settlement built-up areas (GHS-BU) provides the concrete evidence of the presence of human concentration and their activities as demonstrated by Melchiorri et al. (2018). The fundamental is to gather the tangible traces or observations of material transformation of the human environment—in other words, to detect and

[3]The only alternate yet promoted by EU and the OCDE after Quito Habitat is the Functional Urban Area (FUA) method. Previously known as larger urban zone (LUZ), it cumulates two major limitations that renders it inappropriate to conduct worldwide comparison: it details only the largest or metropolitan urban areas and it implies to gather harmonised data on daily commuting which are not available for many countries.

qualify the existence of built-up or building.[4] A pure fact (the materiality of construction) grounds this approach of the urban. It is not in first instance supported by an institutional fact (administrative limit of city depending of human/social agreement).

The starting point is to consider, in order to achieve a unique instrument to measure the dynamic of settlement across the world, that only one approach can be used, based on the location of tangible footprint of human presence. It gathers two complementary gridded geographical layers: in one hand, the morphological spread of settlements using satellite imagery at 40 m resolution.[5] The built-up grid gives a gradient of construction from zero to 98% based on the detection of constructions at 40 m. In a second hand, a population grid at 250 m resolution is generated using local population at municipal and district levels provided by censuses. Overlaying the two layers generalized in a 1 km^2 grid, it becomes possible to provide a harmonised and dynamic vision of the distribution of population articulated to the settlements they live in. The dataset is available for different dates: 1975, 1990, 2000, and 2015. This approach, based on mapping the territory through a grid, avoids distortions caused by using LAU2 or communes/municipalities varying tremendously in size and/or shape. Currently the GHSL-DEGURBA approach is certainly overestimating the urban population as it uses too aggregated data for population (mainly at district level) that are then too highly incorporated in urban.[6]

The 1975 GHSL assessment underestimates the existing built-up areas because of two main factors: (i) worse sensor characteristics, and (ii) the presence of large data gaps (no data available) in the 1975 collection. Mud and disperse type constructions are difficult to detect using 1975 satellite imageries. The underestimation is around 20% globally, but it becomes extremely precise and comparable from 1990, with standardized observation and dataset in 2000 and 2016. Technical papers appreciate how the GHS-BU as well its one km^2 gridded overlaying generalization are consistent. GHS appears always equivalent or more precise than any other sources. It has

[4]The concept of “buildings” formalized by the GHSL are enclosed constructions above ground which are intended or used for the shelter of humans, animals, things or for the production of economic goods and that refer to any structure constructed or erected on its site (Pesaresi et al. 2013). For the GHSL concept, refugee camps, informal settlements, slums and other temporary settlements and shelters are included in the notion of built-up area.

[5]An open source automated processing workflow analyses an exhaustive global Landsat mosaic to provide a continuous built-up mosaic at 40 m resolution. It uses advanced automated machine learning and quality control comparing the results with existing open source layers such as MODIS, DMS/OLS night time lights, (Smith 2017) as well OpenStreet Map. Several researches assessed also the quality through comparison of various sources (Leyk et al. 2018; Lu et al. 2014). They validate that is the most advanced model of detection of built-up—errors are very limited. It is an open source in constant development, incorporating more and more advanced imageries, Sentinel in particular, and automated processes.

[6]Geopolis global database (http://e-geopolis.org/) is by far the most precise sources for a count of urban population by unit. Its unique methodology is taking into account the smallest census units, but the acquisition of built up footprints is dependent of human interpretation and scarcely updated. They cannot be used to conduct a precise work on built-up changes. In Geopolis, all the morphological agglomerations with more than 10,000 inhabitants are considered as urban. A morphological agglomeration is a built-up patch with less than 200 m between constructions (following UN recommendations). It uses open sources Google earth imagery.

been compared to the Gridded Population of the World, version 4.10 (GPWv4.10)[2], the Global Urban Footprint (GUF) initiative (Esch et al. 2017), DMPS/OLS night light imagery (Li and Zhou 2017), as well the 30 m resolution global land cover (GlobeLand30) generated by Jun et al. (2014). Because of its multi-temporal availability, it is achieving more than the Human Built-up and Settlement Extent at 30 m resolution too, but limited to a 2010 unique layer (HBASE; NASA-SEDAC).

13.3 Urban Expansion and Population

If urban morphological expansion is a global phenomenon, regional, national and local difference exists. They can be apprehended, first of all, in assessing the materialization of new constructions using the 30 m resolution data, and secondly in appreciating the de-densification using the one km^2 gridded dataset.

In 2015, the urban built-up footprint occupies some 570,000 km^2 worldwide (Table 13.1). With 175,000 km^2 of constructions added since 1990, it expanded by some 44%. This spread equals the urban built-up of the whole Northern America in 2015! The African continent experiment the most important expansion: there, the urban built-up almost doubled. In contrary, Europe experiment the slower built-up spread. Here appears that, at the macroscale, the physical expansion of the urban world is, in first instance, correlated to the demographic pressure (44% vs. 42). Nevertheless, the relation is not uniform; it differs from region to region. For instance, the European built-up expansion is, in fact, important vis-à-vis its population stagnation (18.5% vs. 1.6). Here it should be make clear that the built-up layer capture only the constructions. It ignore street networks and open spaces. It did not represent the entire urban perimeters. By coupling built-up and gridded population, as explained above, DEGURBA's approach provides a measure of functional urban perimeters. The urban built-up footprint is precisely the measure we are interested to study as

Table 13.1 Distribution of the urban built-up in square kilometres in 1990 and 2015, and changes

	Urban built-up			Variation 1990 to 2015		Urban population change
	1990	2000	2015	in km^2	in %	
World	394,842	465,745	569,942	175,099	44.3	42.2
Africa	32,826	43,230	62,138	29,312	89.3	101.7
Asia	138,813	171,802	226,471	87,658	63.1	39.5
Europe	100,075	108,485	118,627	18,552	18.5	1.6
Latin America and Car.	34,113	40,269	45,739	11,626	34.1	47.2
Northern America	81,555	93,809	107,888	26,333	32.3	35.8
Oceania	7470	8165	9097	1628	21.8	52.8

Source GHSL

a unique indicator of the dynamic of construction and investments driven towards buildings.

Among 166 countries we consider here, twenty only contribute 74% of the built-up expansion between 1990 and 2015 (Table 13.2 and Fig. 13.1). The variability between them is tremendous: China built-up growth is two times higher than its urban population progression and USA 4.2 times higher, but India, even it is the third country in absolute terms for urban morphological expansion, it is much below its urban demographic growth; 8.2% versus 19%. China alone concentrates about one quarter of the world urban built up expansion. USA constitutes also a marker of the variability with its 25,000 km^2 of added built-up associated to an extremely limited demographic urban growth: 3.3% compared to 14.2% for the built-up imprint. Urban Europe and Japan experiment also conversions of land use at a much higher rate than their demographic stagnation would request.

Table 13.2 The 20th largest contributing countries to world's urban built-up expansion from 1990 to 2015

Rank	Country	Built-up		Urban population		Ratio B.U. Gro./Pop. growth
		Growth	Weight	Growth	Weight	
1	China	23.61	16.30	11.53	19.59	2.05
2	United States	14.23	17.69	3.35	3.80	4.25
3	India	8.24	5.17	22.28	19.33	0.37
4	Indonesia	4.60	3.44	3.73	3.75	1.23
5	Nigeria	3.44	1.89	4.23	2.56	0.81
6	Brazil	2.05	2.90	2.83	2.68	0.73
7	South Africa	1.88	1.61	0.94	0.68	2.01
8	Mexico	1.83	1.71	2.03	1.65	0.90
9	Japan	1.62	4.34	0.21	1.88	7.83
10	Italy	1.36	2.01	0.13	0.74	10.09
11	Germany	1.33	3.03	0.08	0.95	17.07
12	France	1.31	2.07	0.21	0.66	6.18
13	Russia	1.23	3.16	−0.37	1.85	
14	Turkey	1.16	1.01	1.23	1.04	0.94
15	RDC	1.15	0.60	1.83	0.92	0.63
16	Ghana	1.07	0.59	0.60	0.35	1.78
17	Vietnam	0.95	0.64	1.27	1.39	0.75
18	Malaysia	0.91	0.81	0.59	0.40	1.54
19	Pakistan	0.90	0.65	4.27	2.79	0.21
20	Thailand	0.90	0.78	0.56	0.77	1.60
	Sum	73.77	70.39	61.54	67.78	1.20
	World	100.00	100.00	100.00	100.00	

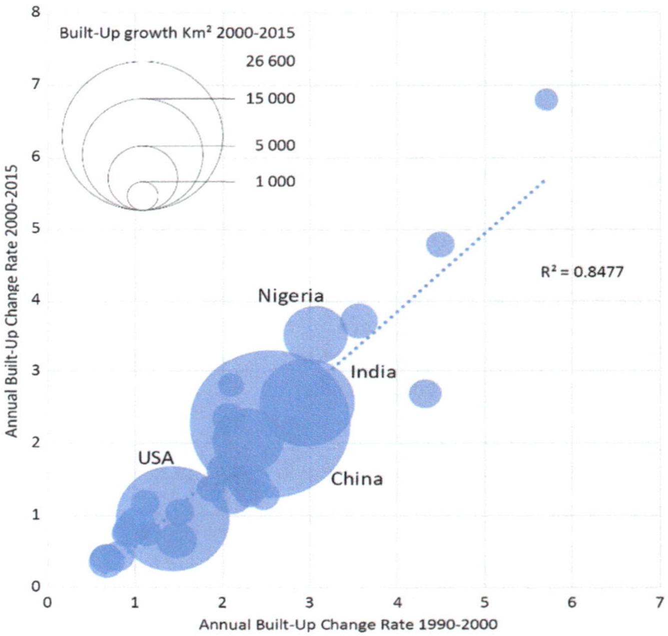

Fig. 13.1 Built-up annual growth rate during 1990–2000 and 2000–2015 for the 30 largest contributors to change (>80% of the total)

Regarding global regional differentiation, Asia experiments the fastest increase of its share of urban built-up followed by the African continent (Table 13.3). Nevertheless, their share of built-up is not yet proportional to their high population weight. Density their remains necessarily higher, but it tends to decline. The share of all the other regions declines. Remarkably, the trajectory of North America and Europe are diverging. Europe urban built-up continues to expand more rapidly than its urban population growth while, in Northern America, this difference tends to reduce.

The urban sector captured with the Global Human Settlement methodology distinguishes two complementary territorial configurations in reference to the 1 km^2 gridded approach: (i) densely populated areas or cities and their immediate peripheries and (ii) intermediate density areas composed of towns and diffuse suburban milieus (Pesaresi et al. 2016). The first category called Urban Centre regroups more than 13,000 morphological agglomerations defined as contiguous spatial units of the 1-km^2 grid cells (four-connectivity gap filling)[7] with more than 50% of build-up and/or a minimum threshold of 1500 inhabitants per cell. Urban Centre should have a

[7]The goal for the high-density clusters is to identify urban centres without any gaps. Therefore, enclaves needed to be filled. If the central square is not part of a high-density cluster, it will be added to a high-density cluster if five or more of the eight surrounding cells belong to a single high-density cluster. This rule is applied iteratively until no more cells can be added.

Table 13.3 Regional repartition of urban built-up and population from 1990 to 2015

	Share urban built-up (%)			Share urban population (%)			Built-up/Pop. ratio		
	1990	2000	2015	1990	2000	2015	1990	2000	2015
World	100.0	100.0	100.0	100.0	100.0	100.0	1.00	1.00	1.00
Africa	8.31	9.28	10.90	10.9	12.4	15.5	0.76	0.75	0.70
Asia	35.16	36.89	39.74	64.0	64.1	62.8	0.55	0.58	0.63
Europe	25.35	23.29	20.81	12.2	10.5	8.7	2.07	2.22	2.38
Latin America and Car.	8.64	8.65	8.03	8.0	8.2	8.3	1.08	1.06	0.97
Northern America	20.66	20.14	18.93	4.4	4.4	4.2	4.66	4.63	4.47
Oceania	1.89	1.75	1.60	0.4	0.4	0.5	4.33	4.05	3.40

(Built-up/Pop ratio = built-up share divided by population share) minimum of 50,000 inhabitants. The second category named Urban Cluster includes some 300,000 peri-urban territories and towns composed of contiguous one-km^2 grid cells having more than 300 inhabitants per square kilometre and a minimum of 5000 inhabitants by agglomerate unit (8 connectivity). Through this dichotomy of urban perimeters, we are able to better qualify what is belonging to the sprawl. It will occur in the lesser dense urban areas bellow 1500 inhabitants per kilometres. Nevertheless, using here only the country summary tables, we are not able to qualify precisely the amplitude of the sprawl, as some of the urban clusters are not contiguous peripheries of urban centres. Some of them constitute forms of diffuse or disperse urbanization rather than sprawl per say.

The density (inhabitants per square kilometre) is globally stagnating but this average is made of contrary trends (Table 13.4a, b). It declines continuously in Asia as well in Europe, in denser cores as well as in their peripheries. Other drivers than population accommodation are at stake in this trend. The deconcentrating trend concerns first-of-all the cities, their cores and immediate peripheries but it expands at a slower pace on the diffuse urban sector. Conversely, densification of high-density areas characterizes Africa, Oceania and Latin America. At a slower pace, the North

Table 13.4 **a** Inhabitants per square kilometre (1-km^2 grid—DEGURBA), **b** annual variation of density (inhabitants per square kilometre)

a

	High density area			Low density area		
	1990	2000	2015	1990	2000	2015
World	5422	5338	5345	1676	1633	1551
Africa	5516	5547	6063	1988	1943	1852
Asia	7095	6792	6431	1974	1868	1695
Europe	3702	3615	3575	1169	1110	1056
Latin America	5404	5447	5823	1643	1637	1638
Northern America	1749	1763	1825	760	774	797
Oceania	1681	1843	2211	1130	1188	1329

b

	High density area		Low density area	
	1990–2000	2000–2015	1990–2000	2000–2015
World	3	1	−4	1
Africa	3	34	−4	−6
Asia	−30	−24	−11	−12
Europe	−9	−3	−6	−4
Latin America and Car.	4	25	−1	0
Northern America	1	4	1	2
Oceania	16	25	6	9

Table 13.5 Square meters added per year and per inhabitant

	Urban area		High density area		Low density area	
	1990–2000	2000–2015	1990–2000	2000–2015	1990–2000	2000–2015
World	1.39	1.12	2.57	1.77	1.21	1.05
Africa	1.64	1.31	2.33	1.46	0.95	1.12
Asia	1.01	0.94	1.08	0.94	0.90	0.92
Europe	1.58	1.25	0.90	0.81	2.38	1.80
Latin America and Car.	1.48	0.71	1.66	0.82	1.15	0.48
Northern America	5.53	3.58	6.80	4.57	3.54	1.76
Oceania	3.15	2.14	4.79	2.64	0.80	1.33

Source Built-up GHSL; reference Population 2000 and 2015

America's paradigmatic sprawl seems to decelerate too and, eventually, agglomerates re-densify there.

With the number of meters added per inhabitant and per year, we appreciate more precisely how the expansion process slowed down vis-à-vis the population whatever the urban sector, dense or less dense (Table 13.5). The trend is much contrasted from region to region: the densification in Africa, Oceania and Latin America diverges with de-densification in Asia and Europe, especially for the dense areas. Northern America tends to stagnate. In this last region, the sprawl has not stopped but only decelerated, and yet Northern America experiments the highest increase in square meters added per inhabitant 3.6. Nevertheless, the subprime crisis reflects in the way the annual built-up footprint growth is slowing down, compared to the respective population growth (Table 13.6).

Convergence of built-up growth in densest and lesser-inhabited urban sectors characterizes the global trend apprehended at the macro-region level (Table 13.6). A gradual decrease of the physical expansion drives this conjunction of trends.

Everywhere, annual built-up rate of growth declines except in Africa low-density areas. It has certainly to do with a deficiency in land governance and physical planning leading to an extensive dynamic of sprawl by juxtaposition of individual houses.

Table 13.6 Annual growth rate of population and built-up for the two urban sectors (%)

	High density area				Low density area			
	Population		Built-up		Population		Built-up	
	1990–2000	2000–2015	1990–2000	2000–2015	1990–2000	2000–2015	1990–2000	2000–2015
World	2.30	1.81	1.86	1.47	1.23	0.88	1.40	1.18
Africa	3.80	3.50	3.32	2.44	2.32	2.12	2.00	2.45
Asia	2.44	1.79	2.22	1.87	1.29	0.75	2.05	1.85
Europe	0.44	0.10	0.56	0.49	−0.10	−0.08	1.01	0.68
Latin America & Car.	2.70	1.99	1.90	1.01	1.43	0.83	1.27	0.54
Northern America	2.46	2.00	1.80	1.26	0.48	0.47	0.85	0.42
Oceania	3.29	2.07	1.24	0.83	0.70	1.41	0.26	0.52

13.4 Beyond the Relationship Between Population Growth and the Built Environment: A Diverse World

If the decrease in population growth goes together with the decelerating of built-up densification[8] and expansion, the relation is far from being uniform. In fact, for the entire urban footprint of the 156 countries with significant figures, the correlation coefficient (Pearson) between the annual population change and the annual built-up growth for the 2000–2015 is solid, established at 0.507. It is sensibly higher than the previous period of observation (1990–2000) when it was at 0.470. Population dynamics still matter in the way cities are expanding nevertheless the correlation between density and built-up growth tends to decline (0.367–0.224).

In the dense urban sector also, globally, the dependence of trajectories tends to increase during the last period of observation (R = 0.683 against R = 0.528 for the previous decade). It is high and stable in less dense areas (R = 0.560 and 0.552).

The relation varies from region to region and tends also to consolidate or stays stable. For instance, in Africa for the last period the correlation is 0.421 (0.310 previously). In Latin and Central America, it declines very little from 0.410 to 0.384. Asia, excluding the Middle East countries, exposes a strong and increasing relation passing from 0.734 to 0.797. In Europe, it goes up very significantly to reach 0.832, from 0.568 previously. In this last, slow population change, even decline sometime, goes hand-to-hand with a limited built-up expansion.

The heterogeneity of the relation (only ¼ of the variance of one growth variable can be statistically explained by the other) can be appreciated when mapping residuals of the linear relation between built-up and population growth for the last fifteen years observed (2000–2015). The map opposes a major part of Africa and Asia to Europe and America including Latin America (Fig. 13.3).

This is not a simple economic and wealth dichotomy nor a demographic regime that drive the contrast we observe on the residuals mapping at the world scale (Figs. 13.2 and 13.3). The relationship is not linear ($R^2 = 0.25$), nor even polynomial ($R^2 = 0.27$) but appears as almost random (Fig. 13.2) reflecting a complex, various, contextual and multifactorial explanation of built-up expansion. The maximum contrast opposes, in one hand, countries with high population growth articulated to an extremely great built-up increase to, in the other hand, countries where urban land conversion and construction is very limited and sometime extreme population growth but more often a sluggish population growth. Globally it reflects a dichotomy between countries characterized by a rapid urban transition combining physical expansion and demographic growth and nations with a limited urban development.

An accelerated urban sprawl affects East Africa from Ethiopia (4.7% per year compared to 2.7 for urban population) to Malawi including Burundi and Rwanda (5.9% vs. 2.7). South Sudan appears here too as a post war speculative (re)construction

[8]By densification, we understand here more continuous urbanization measured at 38 m. The approach did not provide direct information about vertical densification and the transition from detached houses landscape to apartments. Nevertheless, increases of density are indicators of more people in some areas more often linked to building having more floors and apartments.

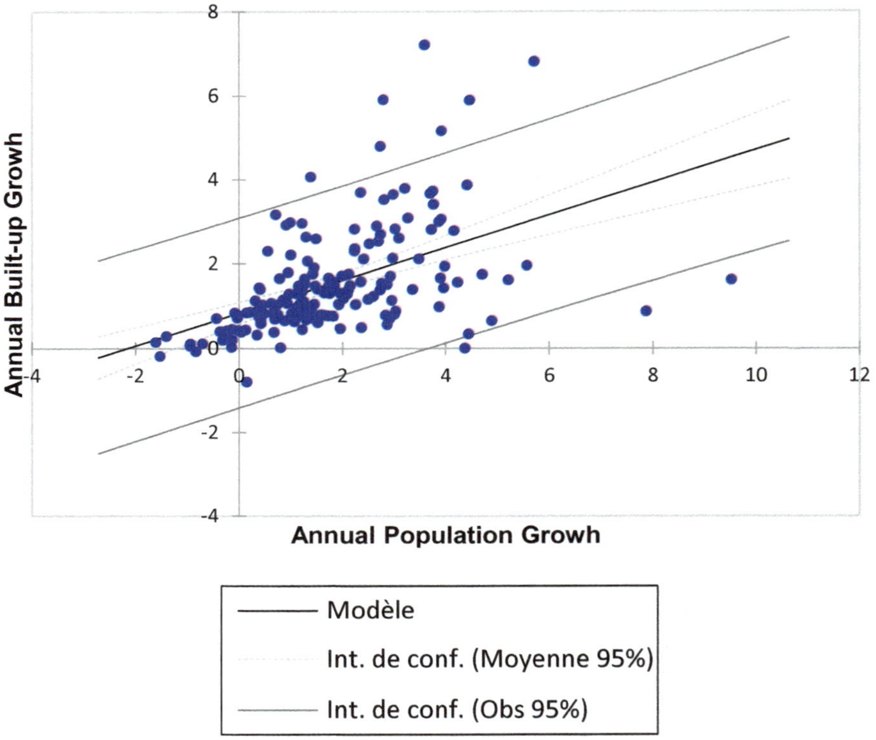

Fig. 13.2 Regression of annual built-up growth 2000 to 2015 by annual population growth (R = 0.507; $R^2 = 0.257$)

and quest to fix capital (5.8 vs. 4.4). There, surplus pour massively into plots and construction at a rhythm much higher than their population growth demand. Western Africa, notably Burkina Faso (3.7%/y for built-up) and Nigeria (3.5%) are other significant cases of rapid urban landscape metamorphosis, less extreme but yet very high. They are characteristic of a public governance of land that induces a selective redistribution of the public domain toward civil servants, army officers and supporters as a path to preserve the stability of the regime.

Houses, apartments and plots constitute as many tangible and irremovable assets, which protect their owners in the context of high inflation and against the poor return and lack of security of financial wealth. In those configurations, we found a majority of lower-income countries with a limited banking system poorly inclusive.

In such situation, we found also most of the Asian countries. Cambodia, Bangladesh and Myanmar represent extreme cases. The sustained pace of 2.2% per year of China physical expansion of cities is of particular significance as it represents almost of quarter of all the physical urban expansion of the world between 2000 and 2015. There, despite a lacklustre demographic growth limited to 0.5%, the built-up

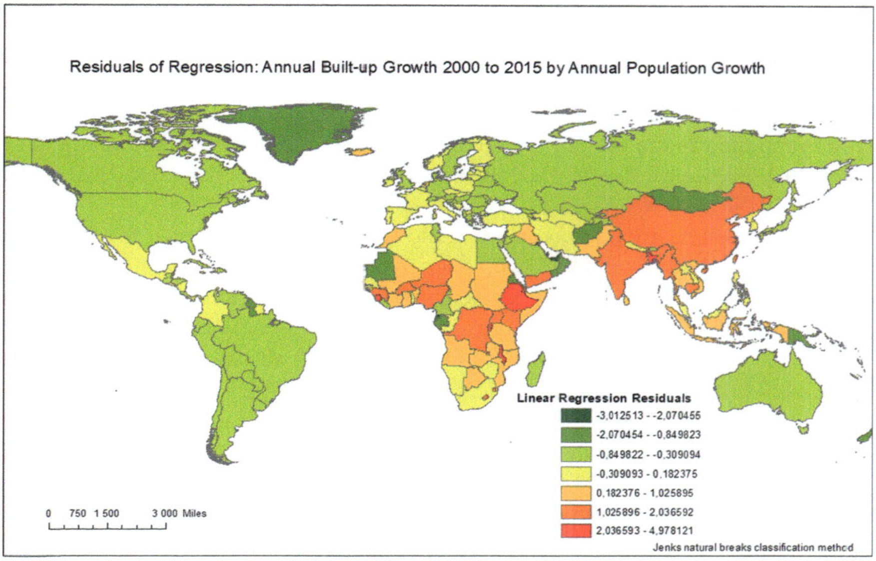

Fig. 13.3 Regression residuals: annual built-up growth by annual population growth (2000–2015)

footprint continues to expand fast at 3.8% per year since 2000. In 15 years, almost 14 m^2 have been added per urban 2015 inhabitant.

This trend exemplifies a group of countries that a Chinese colleague Hsing (2010) called "land-driven development machine". A variant would be the "property state" as coined by Haila (2016) looking at Singapore public management of urban land rent. There, the commodification of land supports the economic growth. It backs money creation and credit development. It is important to understand at the global scale this process has a complex decision game not simply driven by market forces but mediated by state policies vis-à-vis land supply and rate of credit as well by various "cultural ethics" determining the conception of home ownership for household and institutional in their management of assets and embeddedness. Consequently, it leads to an extreme consumption of space. Hence, housing oversupply characterizes today Chinese urban landscape. In 2014, the China Household finance Survey underlined that 40% of the annual housing supply would be enough to meet the additional housing demand. This year (2018), the vacation rate has reached 22%, it was already of 20% in 2011. This represents some 50 million units!

India is not far, with an urban built-up growth at 2.6% per year but associated to a much higher demographic growth (1.6 per year). Beside this consequent land conversion to urban uses, due to market distortions and severe inequalities,[9] the improvement of the urban housing condition remains very limited. Some 65 million

[9]Oxfam reports that in 2017, the richest 1% in India cornered 73% of the wealth generated. Official figures about urban inhabitants living below poverty line is 25.7% (Rangarajan and Mahendra Dev 2015). It represents today more than 100 million persons.

urban Indians or 17% of the urban citizen lives in slums and their number continue to increase. The real estate boom, which constitutes the major component of the new urban frontiers, generates an offer targeting the upper-middle class and the riches who are looking in real investment to fix cash.

It is particularly important to detail the Chinese and Indian cases because between them, they account for nearly a third of the global urban built-up expansion from 2000 to 2015.

Major post-industrial democracies could have also experimented faster built-up footprint spread vis-à-vis their population growth. USA is part of this trend with a built-up expanding at 0.95% per year when the urban population continues to growth at 1.11%. Nevertheless, because urban USA is paradigmatic of the extensive urban with its suburbs and yet represents the second share of the world built-up growth (14%), its trend is of a particular importance. Without the subprime crisis, it could have been even bigger. To some extent, it is comparable to several major Latin American countries and Canada as well: Brazil stands at 1.33 versus 1.03 and Argentina at 0.61 versus 1.07.

Europe is representing an average relation between urban built-up expansion and population growth: only could we distinguish a set Western and Southern Europe vis-à-vis Northern and Eastern Europe. The second exposes a slower built-up rate of change in view of their demographic growth. It is undoubtedly necessary to link this time lag to a process of catching up between the countries of the South and Northern Europe, but also to less control over land use—in other words, to regulations that are more flexible.

The most deficit-oriented countries in term of built-up growth vis-à-vis their population change are, for a part, rich petroleum monarchies (UAE, Qatar, Kuwait and Oman) with an extremely high demographic rate of growth (between 5 and 7% per year) and a built–up growth below one percent per year.[10] In the same group are countries often poor or/and affected by conflict such as Afghanistan where demographic growth remains strong.

13.5 Economic Drivers of Land Conversion and Urban Built-Up Growth

If, globally, the recent urban physical expansion (2000–2015) did not respond linearly to the variability of demographic growth from country to country, the link to the distribution of income by country points to a dissimilarity in trend. There is not much difference between high-income (HIC) and low-income countries[11] (LIC): 0.475 versus 0.501, except that the relation is decreasing in HIC (0.596 previously)

[10]Here we could face a form of underestimation of the built-up change and of densification by verticalisation.

[11]We use the UN income group classification: https://esa.un.org/unpd/wup/cd-rom/WUP2014_DOCUMENTATION/WUP2014_DEFINITION_OF_MAJOR_AREAS_AND_REGIONS.pdf.

and increasing in LIC (0.370). The tendency to a lower correlation of the urban built-up expansion and the population demand/pressure goes with an accomplished demographic transition, the stabilisation of rural to urban migration and an increase of wealth generated.

The population, its growth, did not explain totally and uniformly the intensity of the built-up imprint expansion. Nevertheless, it constitutes the best proxy expressing the demand for housing, productive activities and infrastructures. The artificialisation of soil and new constructions within and around cities answers to multiple goals that the demographic growth is not covering fully: (i) Buildings constitute also places for multiple economic functions which are not strictly proportional to population and (ii) housing is answering to very different segments of demand depending of wide spectrum of inequalities. Furthermore, edifices incorporate a financial value detached of their use value. Fixation of richness in plots and built assets tends to generate a dynamic of construction more and more de-correlated of the population demand. It certainly reflects in the degree of financialisation of economies, wealth formation and inequalities.

Before going further in the macro-economic analysis, we should notice that the family size weight and change by country considered as a proxy of the level of nuclearisation that should affect the housing demand is not correlated to built-up weight nor its change (Table 13.7). Correlations are also non-existent with proxies related to construction, whether it is the share in world concrete production or the weight of fixed capital in GDPs. The weight of the real estate and financial sector in the GDP could perhaps give a better result, but it is not easy to gather for all the countries of the world a proper harmonized and updated indicator.

Significant correlation between economic and urban built-up indicators opposes urbanized land "stock" world distribution by country and its evolution to growth rate of change and densities of urban built-up by country. It means that, when the GDP per capita is negatively correlated with the annual growth rate of the built-up (R = −0.317) as well to the average urban density per country, it is positively correlated with the distribution of the world built-up by country (R = 0.456); in other words, with the largest and richest nations. A reverse dichotomy characterizes the link between annual GDP growth by country from 2000 to 2015 and built-up share (R = −0.343) versus annual growth rate of built-up (R = 0.336) and density (R = 0.420). Here we have a first indication of how the growth of housing by country is localized, in an inversely proportional way, to the various national economic weight; built-up growth responds in priority to a demand linked to strong demographic pressure, hence the link with densities also. It should be noticed here that the relationship is not very simple since the weight of countries in the world GDP in 2015 is very weakly correlated to their share in the global urban built-up growth (R = 0.180), nor to the share of wealth (R = 0.128), but their growths are correlated.

If there is no link with inequality indicators such as the repartition of millionaires or Gini indices, there are significant correlations with the distribution of wealth in general and by quantile. The relationship is more significant than with the global distribution of the GDP. Wealth reflects more the material capital encapsulated in the built-up footprint. The distribution of global urbanized territories by country is

Table 13.7 Correlation (Pearson) between urban change indicators and economic proxies

Variables	Annual growth rate urban population 2000–2015	Annual growth rate urban built-up 2000–2015	Urban density 2015	Urban density change 2000–2015	Urban built-up share 2015	Variation urban built-up share 2000–2015
[e]Family size—2015	−0.033	−0.059	−0.025	0.062	0.082	0.088
[e]Family size change 2000–2015	−0.076	0.013	−0.094	−0.160	0.012	0.006
[b]Share of World GDP—2015	−0.143	−0.095	−0.137	−0.064	0.180	0.098
[b]Annual growth rate GDP 2000–2015	**0.359**	**0.336**	**0.420**	0.101	**−0.343**	−0.231
[b]GDP Per Capita—2015	−0.098	**−0.317**	−0.301	0.146	**0.456**	0.298
[b]Annual growth rate per capita GDP 2000–2015	−0.281	0.029	0.068	−0.273	−0.189	−0.150
[c]2015 share immobilized capital in GDP	0.127	0.100	0.213	−0.047	−0.194	−0.166
[d]Variation cement prod.—2000–2012	−0.085	−0.018	−0.113	0.062	0.020	−0.016
[d]Share World cement prod.—2012	0.030	−0.012	0.037	0.027	−0.034	−0.034
[a]Share World millionaire per country	−0.096	−0.090	−0.153	−0.028	0.149	0.070
[a]Wealth to GDP per adult—2015	−0.265	**−0.325**	−0.235	−0.021	**0.399**	0.246
[a]Share of adult under 10,000 wealth—2015 (USD)	0.111	**0.402**	0.228	−0.151	**−0.383**	−0.213
[a]Share 10,000–100,000—2015	−0.234	**−0.443**	−0.222	0.176	0.254	0.122
[a]Share 100,000–1 million—2015	−0.079	−0.247	−0.256	0.119	**0.537**	**0.373**
[a]Share over 1 million—2015	−0.083	−0.190	−0.271	0.046	**0.367**	0.219
[a]Gini per country—2015	0.006	0.127	−0.017	0.018	−0.038	0.025
[a]By country share of total wealth 2015	−0.107	−0.081	−0.148	−0.044	0.131	0.060
[a]Share change world total wealth 2010–2015	0.035	0.037	−0.049	0.002	−0.195	−0.128
[a]Wealth per adult—2015	−0.102	−0.236	−0.274	0.077	**0.458**	**0.297**
[a]Financial wealth per adult—2015	−0.084	−0.212	−0.284	0.075	**0.478**	**0.354**

(continued)

Table 13.7 (continued)

Variables	Annual growth rate urban population 2000–2015	Annual growth rate urban built-up 2000–2015	Urban density 2015	Urban density change 2000–2015	Urban built-up share 2015	Variation urban built-up share 2000–2015
[a]Non-financial wealth per adult—2015	−0.098	−0.231	−0.259	0.087	**0.320**	0.188
[a]Debt per adult 2015	−0.071	−0.189	−0.243	0.092	**0.403**	**0.315**
[a]Var share of nonfinancial wealth 2010–15	0.089	−0.002	0.099	0.118	−0.067	−0.060
[a]Debt to wealth in 2015	−0.128	−0.254	−0.186	0.137	0.211	0.175

The values in bold are different from 0 to a level of meaning alpha = 0.05; *Sources* [a]Wealth data Credit Suisse Wealth Report 2016; [b]The World Bank, [c]OCDE, [d]U.S. Geological Survey, [e]Global Data Lab

strongly linked to the weight of the high income and capital in national economies: the weight of the upper middle classes (100,000 to 1 million wealth) and the richest (over 1 million) (R = 0.537).

Conversely, the rate of change of built-up imprint is correlated with the weight of the most modest adult populations, below 10,000 USD (R = 0.402). It exposes the materiality of a meta-trend of convergence. The least developed countries are experiencing an urban sprawl that corresponds to an expansion of popular housing and thus to an overall improvement in housing conditions. Multi-storey apartment buildings remain often extremely limited accenting the sprawl during the first cycle of land conversion.

The analysis of the same set of correlations between building growth and economic indicators by countries gathered by income groups confirms the contrast between countries whose building expansion is insufficient vis-à-vis the demographic pressure they experiment and advanced and rich countries where building expansion is more and more uncorrelated to the immediate final user demand (inhabitants). In this last, the demand is mediated by complex market rationalities. Land transformation and real estate fuel, both, the physical production of cities and the enlargement of the financial sector.

Lower-income countries (LIC) expose a significant link between the share of debt vis-à-vis the wealth and the built-up growth rate: R = 0.465 versus R = 0.105 for high-income countries (HIC), but the debt share per adult link to built-up growth is more a characteristic of HIC (R = 0.308). It contrasts the limited capital of LIC with HIC. In HIC, credit could expand individual wealth including built capital. In other words, the built-up growth in HIC is more dependent of the debt level.

In lower-income countries, the density in 2015 is inversely correlated to the degree of inequalities (R = −0.434) and positively to the weight in the world built-up (R = 0.449). Dedensification, notably in larger LIC countries, could have a link with increasing inequalities. The strong correlation between density and the share of adult below 10,000 USD in LIC confirms it (R = 0.526). There, physical expansion has to do with the capture of the urban rent by the upper class. In LIC the growth of GDP per capita is significantly correlated to built-up growth (R = 0.420) while it is the opposite for HIC (R = −0.329).

We know after Piketty (Alvaredo et al. 2017) in particular for LIC and medium low-income countries including BRICS that the increase of GDP per capita goes with strong upsurge in inequalities. The capture of wealth including urban land and their rent by a very limited and rich minority is at stake. There, wealth grows at the rate of return to capital, a rate that normally exceeds the economic growth rate. The expansion of built-up environment encompasses the current tendency of fast growing economies to have ever-increasing ratios of wealth to income. The shaping of new peripheries involves a permanent reconversion of capital from financial (monetary) to nonfinancial (commodity) assets, and back again. Here we confirm the intuition of a growing disconnection between the production of new urban spaces and the final demand of users, inhabitants, productive activities, services and infrastructures. Nevertheless, in general (R = 0.498) it is indeed where the population growth more rapidly that the built-up is expanding faster but it is not meeting straightforward this rising demand.

The multifactorial unsupervised clustering provides a typology of urban growth articulated to the distribution of wealth (Fig. 13.4). Its mapping clearly contrasts advanced and rich economies in majority slowing down in term of population growth as well built-up footprint change (type 3) and developing and poor nations where urban population and built-up rate of growth are the highest (type 5). It consists

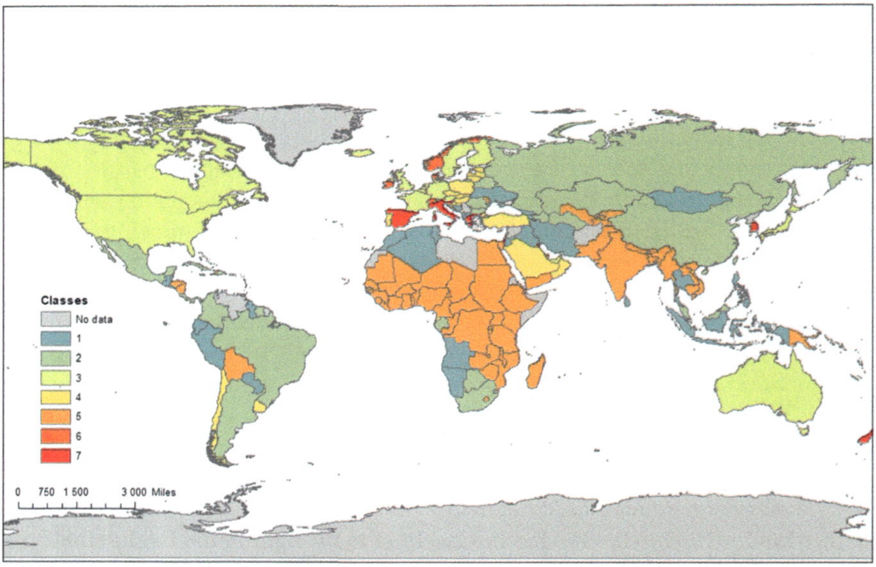

Fig. 13.4 City growth and economic factors: k-means unsupervised clustering

mainly of sub-Sahelian and South-Asian countries that are experimenting the fastest expansion of their urban footprint (Table 13.8).

The other classes represent various intensity of built-up growth articulated to GDP per capita and wealth's quartile distribution. The class 1, for instance, gathers transitional countries relatively poor but with a significant middle class and medium-rich. There, the urban population pressure is less intense than in poorest countries (class 5) but built growth is high as well the GDP growth. The class two is more gathering BRICS type of countries.

The typology suggested by this K-Means clustering confirms a trend towards convergence through the catching-up effect of the poorest countries and a slowdown in built-up growth in the richest countries, due to the densification and structural slowdown in economic growth. In the poorest countries, the high intensity of urban built-up expansion reflects both continuing high population pressure and a structural trend towards housing improvement. Built-up intensity of construction is often all the stronger because buildings constitute a safe haven for fixing their economies or accessing mortgage credit. In these often fast-growing economies, the construction and ownership of buildings is also the driving force behind growing inequality.

13.6 Conclusion: The Urban Built-Up Expansion Tends to Diverge from Population Growth

This first exploration of the dynamics of building expansion by country using a harmonised geodatabase for the entire world reveals very contrasting trends. It is clear that population growth in cities remains the most important factor determining urban built-up spread, but countries differ. In particular, China, where urban built-up expansion continues at a sustained pace despite the significant decline in population growth. The extent of built-up growth in and around Chinese cities accounted for a quarter of the world's urban built-up increase between 2000 and 2015.

Americans are the world's second largest contributors to the global urban built-up expansion. It remains disproportionate to the low population growth within urban areas. However, the pace has slowed. The trend would even be towards a re-densification of urban areas and in any case a decline in construction, which must be linked to the 2008 subprime crisis.

Here we understand that the built-up growth is not simply a reflection of the final demand for housing and therefore of demographic pressure or even driven by the nuclearisation of families. Building is a material good whose value is also interdependent with economic growth. The universal relationship between size and density specific to laws of scale is certainly verified once again in the case of cities. However, economic density or concentration of wealth tends to take precedence over population mass to explain urban built-up expansion. As such, soil conversion is increasingly associated with economic trends and wealth distribution. We show here that there is a significant correlation between urban built-up expansion and

Table 13.8 Detailed legend of map 2 classes with value of variables for each class barycentre and weight of classes

Classes	Density	Population	Built-up	GDP Constant	GDP Per Capita 2015	Under 10,000 USD/Y	10,000–100,000	100,000–1 million	Over 1 million	Weight	Intra-class variance
	2015	Annual growth rate 2000–2015			2015 (wealth)						
1	2538	1.26	1.13	4.16	4334	65.2	17.7	9.7	0.9	31	2,024,606
2	2499	1.09	0.95	4.79	8481	66.4	18.7	5.8	0.4	23	3,842,214
3	2003	1.27	0.84	1.82	47,445	61.1	29.4	9.1	0.4	14	24,864,587
4	2263	0.88	0.72	3.54	16,758	67.7	16.5	8.2	1.4	16	10,160,750
5	3303	2.66	2.56	4.97	1188	68.1	19.9	7.9	0.5	56	1,967,539
6	2037	2.90	1.20	3.91	72,370	75.7	3.9	0.3	0.0	5	133,594,953
7	2560	1.31	0.80	2.14	29,999	59.9	26.4	12.9	0.8	8	31,444,633

Intra-class variance: 3.16%; Inter-classes: 96.84%

the weight of the upper middle classes, particularly in South Asia. In intermediate economies, increasing inequalities boosts extremely intense activity of construction. The capture of the global manufacturing activities and many IT services by these countries stimulates also there a gigantic dynamic of plant construction.

This first study remains at the global level of countries and compares them with each other. Further study is needed to examine the extent to which the intensity of urban expansion is correlated with the size of cities. Are the largest metropolitan areas expanding faster? Does the size and shape of city systems influence how and where land is converted to urban use? What is the precise share of the sprawl in the built-up growth?

This should lead us to question how to infer populations with measured built-up physical extension and thus how to generalize urban areas as DEGURBA does—today the DEGURBA urban population is overestimated. Other groupings are possible and likely to bring out more small towns at the expense of the extension of large urban areas as well as the extent of rural areas. In the immediate future, the approach tends to underestimate the emergence from below and thus the weight of rural localities that become small towns (see Russo et al. 2017). Nevertheless, the overall measurement of urban changes remains valid and robust, especially the harmonized built-up detection. The analyses and interpretations that we are receiving undoubtedly open up new avenues for qualifying and understanding, and even predicting, the spread of cities.

References

Alvaredo, F., Chancel, L., Piketty, T., Saez, E., Zucman, G.: Global inequality dynamics: new findings from WID.world. Am. Econ. Rev. **107**(5), 404–409 (2017)

Angel, S., Blei, D.M., Civco, P., Lamson-Hall, J., Parent, N., Galarza Sanchez, K.: Atlas of Urban Expansion—The 2015 Edition. Edited by New York: The NYU Urbanization Project. Cambridge MA: The Lincoln Institute of Land Policy. U.N. Habitat, Nairobi, Kenya (2015)

Dijkstra, L., Poelman, H.: A harmonised definition of cities and rural areas: the new degree of urbanisation. Regional Working Paper 2014 WP 01/2014. Bruxelles: European Commission. Regional and Urban Policy (2014)

Dijkstra, L., Florczyk, A., Freire, S., Kemper, T., Pesaresi, M.: Applying the degree of urbanization to the globe: a new harmonized definition reveals a different picture of global urbanization. Paper prepared for the 16th Conference of IAOSOECD Headquarters, Paris, France, 19–21 Sept 2018

Esch, T., Heldens, W., Hirner, A., Keil, M., Marconcini, M., Roth, A., Zeidler, J., Dech, S., Strano, E.: Breaking new ground in mapping human settlements from space—the Global Urban Footprint. ISPRS J. Photogram. Remote Sens. **134**, 30–42 (2017)

Florczyk, A.J., Airaghi, D., Ehrlich, D., Corban, C., Freire, S., Kemper, T., Maffenini, L., Melchiorri, M., Politis, P., Pesaresi, M., Sabo, F., Schiavina, M., Tommasi, P., Zanchetta, L.: Community pre-Release of GHS Data Package (GHS CR2018) in support to the GEO Human Planet Initiative. JRC Technical Report. Luxembourg: Publications Office of the European Union, 20 p. (2018)

Henderson, V.: The urbanization process and economic growth: the so-what question. J. Econ. Growth **8**(1), 47–71 (2003)

Hsing, Y.T.: The great urban transformation: politics of land and property in China (2010)

Jun, C., Ban, Y., Li, S.: China: open access to Earth land-cover map. Nature **514**, 434 (2014). https://doi.org/10.1038/514434c

Leyk, S., Uhl, J.H., Balk, D., Jones, B.: Assessing the accuracy of multi-temporal built-up land layers across rural-urban trajectories in the United States. Remote Sens. Environ. **204**, 898–917 (2018)

Li, X., Zhou, Y.: Urban mapping using DMSP/OLS stable night-time light: a review. Int. J. Remote Sens. **38**(21), 6030–6046 (2017)

Lu, L., Guo, H., Pesaresi, M., Ehrlich, D., Ferri, S.: 11 assessment of fine-scale built-up area mapping in China. In: Global Urban Monitoring and Assessment through Earth Observation, p. 217. CRC Press (2014)

Melchiorri, M., Florczyk, A.J., Freire, S., Schiavina, M., Pesaresi, M., Kemper, T.: Unveiling 25 years of planetary urbanization with remote sensing: perspectives from the global human settlement layer. Remote Sens. **10**(5), 768 (2018)

Pesaresi, M., Huadong, G., Blaes, X., Ehrlich, D., Ferri, S., Gueguen, L., Halkia, M., et al.: A global human settlement layer from optical HR/VHR RS data: concept and first results. IEEE J. Sel. Top. Appl. Earth Obs. Remote Sens. **6**(5), 2102–2131 (2013). https://doi.org/10.1109/jstars.2013.2271445

Pesaresi, M., Melchiorri, M., Alice, S., Kemper, T. (eds.): Atlas of the Human Planet 2016: Mapping Human Presence on Earth with the Global Human Settlement Layer. Science for Policy report by the Joint Research Centre (JRC). The European Commission's science and knowledge service (2016)

Rangarajan, C., Mahendra Dev, S.: "Counting the Poor". Econ. Polit. Weekly. **50**(2), (2015)

Russo, A.P., Serrano Giné, D., Pérez Albert, M.Y., Brandajs, F.: Identifying and classifying small and medium sized towns in Europe. Tijd. economische en sociale geografie **108**(4), 380–402 (2017)

Seto, K.C., Fragkias, M., Güneralp, B., Reilly, M.K.: A meta-analysis of global urban land expansion. PLoS ONE **6**(8), e23777 (2011)

Smith, D.A.: Visualising world population density as an interactive multi-scale map using the global human settlement population layer. J. Maps **13**(1), 117–123 (2017)

Tikoudis, I., Oueslati, W., Dimitropoulos, A., Engelfriet, L.: Rethinking Urban Sprawl: Moving Towards Sustainable Cities. OECD Publishing, Paris (2018)

Part V
Complex Systems Modeling for Better Understanding Urbanization Theories

Chapter 14
Unveiling Co-evolutionary Patterns in Systems of Cities: A Systematic Exploration of the SimpopNet Model

Juste Raimbault

Abstract Co-evolutionary processes are according to the evolutionary urban theory at the center of urban systems dynamics. Their empirical observation or within models of simulation remains however relatively rare. This chapter is focused on the co-evolution of transportation networks and cities and applies high performance computing numerical experiments to the SimpopNet co-evolution model in order to understand its behavior. We introduce specific indicators to quantify trajectories of such models for systems of cities, and apply these to exhibit co-evolutionary regimes of the model. This illustrates how the systematic exploration of a simulation model can qualitatively transform the knowledge it provides.

14.1 Introduction

14.1.1 Exploring Models of Simulation

The development of new knowledge production practices, in particular the use of simulation models to understand complex systems, has been more and more fostered by the increase in computational possibilities. Arthur (2015) has proposed in that sense that these trends would consist in a computational shift of science, moving progressively from analytical-based approaches to simulation-based approaches. The study of complexity is naturally not the only field of science benefiting from these technological advances, as witness for example recent progresses in computer vision thanks to deep learning techniques made efficient by intensive computing (LeCun et al. 2015), or the importance of cloud computing for processing the massive amount of data produced by the LHC detectors (Bird 2011). These new methods, tools and practices are however particularly suited to the study of complex systems, because

J. Raimbault (✉)
Center for Advanced Spatial Analysis (University College London), London, UK
e-mail: juste.raimbault@polytechnique.edu

Institut des Systèmes Complexes Paris-Ile-de-France (CNRS 3611), Paris, France

UMR Géographie-Cités (CNRS 8504), Paris, France

D. Pumain (ed.), *Theories and Models of Urbanization*, Lecture Notes in Morphogenesis, https://doi.org/10.1007/978-3-030-36656-8_14

among other reasons their flexibility to take into account numerous interacting heterogeneous agents composing this kind of systems. In the case of socio-technical systems, several examples of such streams of research can be given such as generative social science (Epstein 2006), geosimulation (Benenson and Torrens 2004), sociophysics (Galam 2008) or econophysics (Mantegna and Stanley 1999). In that case, models are a crucial piece among other knowledge domains (Raimbault 2017a) such as theoretical and empirical investigations. All knowledge domains are however complementary and often necessary, and (Raimbault 2016a) recalls the risks of falling into blind fully computational practices.

The study of urban systems, which are a typical illustration of such complex systems (Batty 2007), has witnessed a significant gain of knowledge from the "*liberation of modeling and simulation practices*" as Banos (2013) puts it. Recent developments occurred around the evolutionary urban theory (Pumain 1997), synthesized in particular in Pumain and Reuillon (2017b). Following (Banos 2017), these efforts are the archetype of the complementarity of knowledge domains mentioned above, and have acted as a "knowledge accelerator" with a true beneficial interdisciplinary exchange between computer science and geography.

In particular, the development of new methods for model exploration such as Pattern Space Exploration algorithm (Chérel et al. 2015) or the Calibration Profile algorithm (Reuillon et al. 2015), particularly designed for the use of high performance computing, have allowed a qualitative shift in the knowledge that could be extracted from a simulation model. Several illustrations can be given. A set of parameters that are necessary and sufficient to obtain targeted stylized facts for the emergence of a system of cities are obtained for the SimpopLocal model (Pumain and Reuillon 2017a). Arduin (2018) obtains confidence intervals for the estimation of parameters of a non-tractable epidemiological model through the use of the Calibration Profile algorithm. Brasebin et al. (2017) facilitate urban planning by exploring the feasible space of building envelopes under the constraints of local regulations. Raimbault (2018b) indirectly quantifies interactions between networks and territories through the calibration of a simulation model with a genetic algorithm. These results are obtained thanks to the use of the model exploration software OpenMOLE (Reuillon et al. 2013), which is built around three complementary axis: (i) the possibility to embed almost any model as a black box whatever the language in which it is written (as soon as it runs on a Linux machine); (ii) the implementation of innovative model exploration and calibration methods; and (iii) a transparent access to high performance computing environments. These features are integrated seamlessly with the use of a domain specific language to compose experiment workflows (Passerat-Palmbach et al. 2017).

We consider thus that a considerable gain in knowledge can be observed, from the conceptual or thematic description of a model, to its mathematical formalization, its implementation, its systematic exploration, up to its exploration in depth with the help of specific meta-heuristics. These changes may furthermore be of a qualitative nature, in the sense that the nature of knowledge follows abrupt transitions during the advance of the investigation in this continuum.

The objective of this chapter is to illustrate the impact of these new methods in the case of a model of co-evolution between cities and transportation networks, the SimpopNet model, introduced by Schmitt (2014). Our contribution is significant on the following points: (i) we provide a supplementary proof-of-concept on the role of new simulation practices, tools and methods; (ii) we introduce a set of indicators to study the behavior of simulation models for systems of cities; (iii) we establish the behavior of this particular model, in particular we assess its sensitivity to spatial initial configuration and unveil the different regimes for interactions between cities and networks it can produce.

The rest of this chapter is organized as follows: we first briefly review co-evolutionary models for systems of cities and describe the model studied. We then introduce methodological elements to study such kind of models for systems of cities, describe results of the systematic exploration, and finally discuss the implication of these.

14.1.2 Co-evolution Within Systems of Cities

Co-evolution within system of cities, in the sense of complex intricate dynamics in space and time, is a central feature of the evolutionary urban theory (Pumain 2010). Paulus (2004) has for example applied this concept to the study of economic trajectories of French urban areas. Evolutionary economic geography has also developed an extensive literature using this concept for spatial economic systems (Schamp 2010), for example for the location of firms and networks (Wal and Boschma 2011). We use the definition of this concept proposed by (Raimbault, 2018d), which can be synthesized as the statistical existence of causal relationships within spatio-temporal niches, for which a practical characterization method uses a weak causality based on lagged correlations (Raimbault 2017b).

Considering more precisely the co-evolution of transportation networks and territories, which is of particular interest because of potential "structuring effects" of transportation infrastructures (Pumain 2014), some empirical investigations have been proposed by Bretagnolle (2003) and Bretagnolle (2009) for the French system of cities. The validity of these results was however recently questioned by more thorough data analysis in Mimeur et al. (2017) and Raimbault (2018c). For this reason, models of co-evolution are crucial to gain further insight into this concept.

These kinds of models are however rare for cities and transportation networks, as Raimbault (2017c) suggested that this could be due to the fact that this object of study is at the crossroad of several disciplines with different interests and underlying questions. We can give a few examples of such models (see Raimbault (2018c) for a more thorough review). At the microscopic and mesoscopic scales, Achibet et al. (2014) describe a model of the co-evolution of buildings and roads, whereas Raimbault (2019) develops a morphogenesis model coupling multi-modeling of road network growth with a reaction-diffusion model for population density. At the scale of systems of cities, Baptiste (2010) proposed a reinforcement co-evolution model.

Blumenfeld-Lieberthal and Portugali (2010) have focused on topological breakdown of the network of cities. The SimpopNet model introduced by Schmitt (2014), is to the best of our knowledge the only co-evolution model in the perspective of the evolutionary urban theory.

This last model was however not systematically explored, and the question remains if it actually produces patterns of co-evolution at an aggregated level. This makes it a good candidate for our approach. We will in the next section briefly recall the structure of this model.

14.1.3 Description of the SimpopNet Model

We reformulate here the SimpopNet model (Schmitt 2014), following the notations for the formalization of the interaction model introduced by Raimbault (2018b), since a certain number of parameters and processes are similar. Cities grow following a specification for their populations $\mu_i(t)$ such that
$\mu_i(t+1) - \mu_i(t) = \mu_i(t) \cdot \frac{d_G^{\gamma_G}}{N} \cdot \frac{\sum_j V_{ij}}{V_{ij>}}$ where the potential V_{ij} is of the form:

$$V_{ij} = \mu_j / d_{ij}^{\gamma_G}$$

such that $V_{ii} = 0$, γ_G is a parameter for the distance decay (which gives indeed a level of hierarchy as a function of distance) and d_G a shape parameter for the decay function which gives the typical distance of interaction.

The network grows at each time step through a process that can be seen as a potential breakdown (as described by Raimbault (2019)):

1. two cities are chosen, the first according to populations with a hierarchy γ_N (i.e. with a probability proportional to $\mu_i^{\gamma_N}$) and the second following interaction forces $\mu_i \mu_j / d_{ij}^{\beta}$ with the same hierarchy γ_N;
2. a link is then created if the network is not efficient enough given a threshold parameter θ_N, i.e. if $d_{ij}/d_{ij}^{(N)} > \theta_N$;
3. the links created at a date t have a speed $v(t)$, which will depend on current transportation technologies;
4. a creation of new intersections to yield a planar graph is done, but only for links with a similar speed.

In order to study a stylized version of the model, we consider a configuration such that $v(t > 0) = v_0$ and $v(0) = 1$ (the initial model considers three values for speed that corresponds to the reality of transportation technologies between 1830 and 2000). Indeed, the initial precision in the parametrization of dates and speeds makes it a hybrid model, and should correspond to an application on a real spatial configuration. In a synthetic configuration as used in the model, these parameters have a sense only if we know the behavior of simulated dynamics, and in particular the role of the spatial configuration, i.e. if we are able to differentiate effects linked to the dynamics from effects linked to the initial spatial configuration.

14.2 Methods

14.2.1 Generation of Synthetic Setup Configurations

An important aspect for studying the behavior of such a simulation model is the role of the initial spatial configuration in emerging patterns observed. We therefore apply the methodology developed by Cottineau et al. (2017), which allows extending the sensitivity analysis of a model to spatial meta-parameters (in our case a meta-parameter is a parameter allowing to generate an initial configuration upstream of the model).

A synthetic system of cities is constructed the following way (see Raimbault (2016b) for the notion of synthetic data, calibrated at the first and the second order). A fixed number N of cities is uniformly distributed in space, under the constraint of a minimal distance between each, and their population is attributed following a rank-size law which parameters P_m and α can be adjusted (the distribution of city sizes in the original model corresponds to $\alpha \simeq 0.68$ with $R^2 = 0.98$).

A skeleton of network is created by progressive connection (similar to a percolation algorithm): the algorithm connects cities two by two by closest neighbors in terms of Euclidian distance, and then iteratively selects randomly a cluster and connects it perpendicularly to the closest link outside the cluster. The network is then extended by the creation of local shortcuts, through a repetition n_s times of the random selection of a city with a probability proportional to populations, and its connection to a random neighbor in a radius r_s under conditions of a maximal degree d_s. The final network is then made planar.

This process creates networks that visually correspond (in terms of the order of magnitude of the number of loops, and their spatial range) to the initialization of the model, knowing that a single instance of the network does not allow to determine distributions of topological parameters for which a more precise calibration could be done. We show in Fig. 14.1 examples of synthetic setups compared to the original model setup.

14.2.2 Indicators for Trajectories of Systems of Cities

Context

A crucial aspect of the study of simulation models is the definition of relevant indicators, particularly in the case of synthetic models where it is not possible to produce outputs that are directly linked to data for example. Very general stylized facts, as

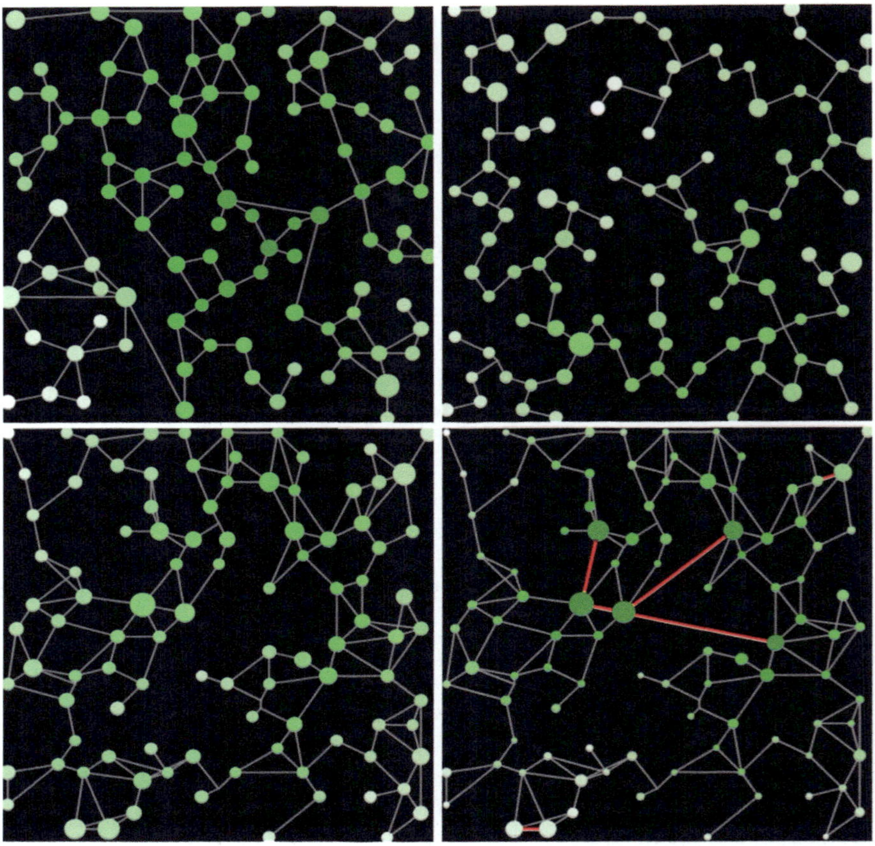

Fig. 14.1 Examples of model setup. (*Top Left*) Unique stylized setup provided in the original implementation of the model (Schmitt 2014); (*Top Right*) Example of a synthetic setup (generator parameters $N = 100$, $\alpha = 0.8$, $P_m = 1.57e5$, $n_s = 10$, $r_s = 7$, $d_s = 5$) with a network close to a tree, obtained with a very low number of shortcuts; (*Bottom Left*) Example of a synthetic setup (generator parameters $N = 100$, $\alpha = 0.8$, $P_m = 1.57e5$, $n_s = 70$, $r_s = 7$, $d_s = 5$) with a higher number of local connections; (*Bottom Right*) Outcome of the model after $t_f = 100$ time steps on the second synthetic configuration. City size gives the population and color the closeness centrality

aiming at producing an urban hierarchy or a network hierarchy, are relatively limited. Moreover, the hierarchy is mechanically produced by most models including aggregation processes. We therefore need more elaborated indicators to understand the dynamics of the system. These indicators must in particular give elements of answer to the following questions:

- types of systems of cities produced by the model;
- change in time of the organization of the system of cities;
- typical profiles of trajectories;
- ability to "produce some co-evolution".

In order to concentrate on the ability of the model to produce trajectories that are both diverse and complex, and for example its ability to produce bifurcations that would manifest as inversions in ranks, and also its ability to capture different aspects of co-evolutive dynamics, we propose a set of indicators, including for example lagged correlation measures in the spirit of causality regimes introduced by Raimbault (2017b), or a correlation measure as a function of distance, to understand the role of spatial interactions in the coupling of trajectories.

Indicators

Given a variable $X_i(t)$ defined for each city and in time (that will be the population or centrality measures for example), we define the following indicators.

Summary statistics

Simple but crucial indicators characterize the distribution of X_i in time:

- hierarchy (slope of the least squares adjustment of X_i as a function of rank) $\alpha(t)$
- entropy of the distribution $\varepsilon(t)$
- descriptive statistics (average $\mathbb{E}[X](t)$ and standard deviation $\sigma(t)$)

Rank correlation

The rank correlation between the initial time and the final time, which translates the quantity of change in the hierarchy during the evolution of the system, and is defined by

$$\rho_r = \hat{\rho}\left[rg(X_i(t=0)), rg\left(X_i\left(t=t_f\right)\right)\right]$$

where $rg(X_i)$ is the rank of X_i among all values (similar to a Spearman rank correlation).

Diversity of trajectories

The diversity of trajectories $\mathcal{D}[X_i]$ captures a diversity of time series profiles for the considered variable. With $X'_i(t) \in [0;1]$ the trajectories that have been individually rescaled, it is defined by $\mathcal{D}[X_i] = \frac{2}{N\cdot(N-1)} \cdot \sum_{i<j}\left(\frac{1}{T}\int\left(X'_i(t) - X'_j(t)\right)^2\right)^{\frac{1}{2}}$

The L2-norm can be generalized by any Minkovski distance, as done in Raimbault et al. (2014). More elaborated indices for this aspect could imply the use of specific time-series clustering techniques (Liao 2005).

Shape of trajectories

We quantify the shape of temporal trajectories through the changes in direction of these $\mathcal{C}[X_i]$, that we take as the number of local extrema, detected by a change of sign of the derivative. In the context of such a type of model, which mainly produces monotonous trajectories, this indicator witnesses in a certain way of a "complexity" of trajectories. In the case of more elaborated shapes, measures such as permutation entropy would be better candidates (Scarpino and Petri 2017).

Distance correlations

We also introduce the correlations as a function of distance, to understand the way the effect of distance is translated at the macroscopic scale. The profile of this function, regarding interaction distance parameters included in the model, will translate the tendency of the model to lead to the emergence of one level of interaction or the other. It is computed as

$$\rho_d = \hat{\rho}\left[(X(\vec{x}_k), Y(\vec{x}_{k'}))\right]$$

where X_i, Y_i are the two variables considered and $\left(k, k'\right)$ the set of couples such that $|\|\vec{x}_k - \vec{x}_{k'}\| - d| \leq \varepsilon$ with ε a tolerance threshold (in practice taken to regroup couples by distance deciles).

Lagged correlations

Lagged correlations between the variations of variables, to identify causality patterns between variables X and Y. The patterns ρ_τ for all variables, and for τ lag or anticipation, must be understood in the sense of potential regimes, explored by Raimbault (2017b).

$$\rho_\tau = \hat{\rho}[\Delta X(t-\tau), \Delta Y(t)]$$

Variables

These indicators are used in our case on the following variables:

- populations $\mu_i(t)$,
- $c_i(t) = \frac{1}{N-1}\sum_{i \neq j} \frac{1}{d_{ij}(t)}$ closeness centralities which capture the position within the urban system,
- $X_i = \frac{1}{\sum_k \mu_k}\sum_{i \neq j} P_j \cdot \exp\left(-d_{ij}(t)/d_G\right)$ accessibilities which capture the insertion within the urban system.

They capture both city trajectories, network trajectories, and the coupling of both with accessibility. The application of above operators to these state variables will thus inform on trajectories of cities, trajectories of the network and trajectories of their coupling, whereas operators based on correlations will inform on interactions between the two aspects.

14.3 Results

14.3.1 Model Implementation

We modified and extended the NetLogo implementation of the model provided by Schmitt (2014), to include in particular (i) methods for the synthetic setup; (ii) indicators described above; (iii) methods for the inclusion within OpenMOLE experiments. The modified code with exploration scripts are available on the open git repository of the project at https://github.com/JusteRaimbault/CityNetwork/tree/master/Models/Reproduction/SimpopNet.

14.3.2 Experience Plan

Given an initial spatial configuration (i.e. a value of meta-parameters for the initial city system and network generator), we establish the behavior of indicators by exploring a grid of the parameter space. The number of parameters being relatively low and the objective being a first grasp of the model behavior, in particular if it is able to produce co-evolution dynamics, we do not use more elaborated exploration methods. The parameters are $(d_G, \gamma_G, \gamma_N, \theta_N, v_0)$ and the meta-parameters $(N_S, \alpha_S, d_S, n_S)$. We take also the meta-parameters into account in order to understand the sensitivity of the model to space.

We explore a grid of 16 configurations of meta-parameters (see Table 14.1 for all values), 324 configurations of parameters (such that $d_G \in [0.001, \ldots, 0.016]$ by 0.005, $\gamma_G \in [0.5, \ldots, 2.5]$ by 1.0, $\gamma_N \in [0.5, \ldots, 2.5]$ by 1.0, $\theta_N \in [1.0, \ldots, 21.0]$ by 10.0 and $v_0 \in [10.0, \ldots, 110.0]$ by 50.0), and 30 random replications, what corresponds to 155,520 simulations. They are executed on a computation grid with the intermediary of OpenMole. Simulation results are available at http://dx.doi.org/10.7910/DVN/RW8S36.

14.3.3 Convergence

Since the model is stochastic, it is important to control the convergence of indicators, that will be more or less easy depending on their variability. To quantify the variability of an indicator X regarding stochasticity, we use a measure similar to the one used by Raimbault (2018a), given by $v[X] = \mathbb{E}[X]/\sigma[X]$ with basic estimators for the expectancy and the standard deviation. On the full set of replications, we obtain for all indicators given previously, a median for the ratio $v[X]$ estimated within the 30 replications, estimated on all parameter values, which takes a minimal value of 3.94, for the average accessibility at final time, what witnesses a low stochastic variability. We can furthermore use this value to estimate the level of convergence:

Table 14.1 Sensitivity to space of the SimpopNet model

N_S	40	40	40	40	40	40	40	40	80	80	80	80	80	80	80	80
α_S	0.5	0.5	0.5	0.5	1.5	1.5	1.5	1.5	0.5	0.5	0.5	0.5	1.5	1.5	1.5	1.5
d_S	5	5	10	10	5	5	10	10	5	5	10	10	5	5	10	10
n_S	10	30	10	30	10	30	10	30	10	30	10	30	10	30	10	30
d'	0	0.05	0.26	0.21	1.79	1.80	1.79	1.72	0.44	0.36	0.42	0.42	2.25	2.23	2.24	2.21

Each column corresponds to an instance of the phase diagram, for which meta-parameters are given, with the relative distance to an arbitrary reference diagram. As inputs we have the meta-parameters N_S, α_S, d_S, n_S and as outputs of simulations the distance d'

it corresponds to a 95% confidence interval around the mean of relative size 0.18 (under the assumption of a normal distribution), i.e. a good convergence. This aspect is crucial for the robustness of results, as this experiment shows that working with this number of repetitions and aggregate averages is consistent.

14.3.4 Sensitivity to Spatial Initial Conditions

We quantify the sensitivity to spatial initial conditions by using the definition of the relative measure of sensitivity, given by Cottineau et al. (2017). This measure is for two phase diagrams f_1, f_2 and d euclidian distance,

$$d' = \frac{2 \cdot d(f_1, f_2)}{(V[f_1] + V[f_2])}$$

Table 14.1 gives values of d' for the 16 configurations of meta-parameters. These are given in comparison to an arbitrary reference configuration (first column). The hierarchy within the initial system of cities appears as the stronger determinant of variability, since all configurations with $\alpha_S = 1.5$ give values larger than 1.7, what witnesses a very strong sensitivity relative to this hierarchy.

Then, the number of cities plays a non negligible secondary role, giving the stronger effects of space. Thus, it is crucial to keep in mind this role of the initial configuration during the analysis of phase diagrams. To stay within the same spirit than the model that was initially proposed, we will however comment a phase diagram for a given spatial configuration. The study of the extended model with integration of meta-parameters to which it is sensitive at their full extent is however beyond the reach of this first analysis.

14.3.5 Model Behavior

Figure 14.2 reports the behavior of the model according to a selection among the diverse indicators given above. We comment a particular spatial configuration which corresponds to a low hierarchical system with a network having only local shortcuts, given by meta-parameters $N_S = 80, \alpha_S = 0.5, d_S = 10, n_S = 30$, which are the values giving configurations that are the most similar to the one of the original model. More exhaustive plots for this parameter configuration are available in Raimbault (2018d) (Appendix A.7).

The values taken by the entropy for centralities (first panel of Fig. 14.2), as a function of time, for $\gamma_N = 2.5$ and $v_0 = 110$, exhibit different regimes depending on d_G and γ_G. A low hierarchy leads to an entropy stabilizing in time, what corresponds to a certain uniformization of distances. On the contrary, a strong hierarchy produces a regime with a minimum, and then an increase of disparities in time. More hierarchical

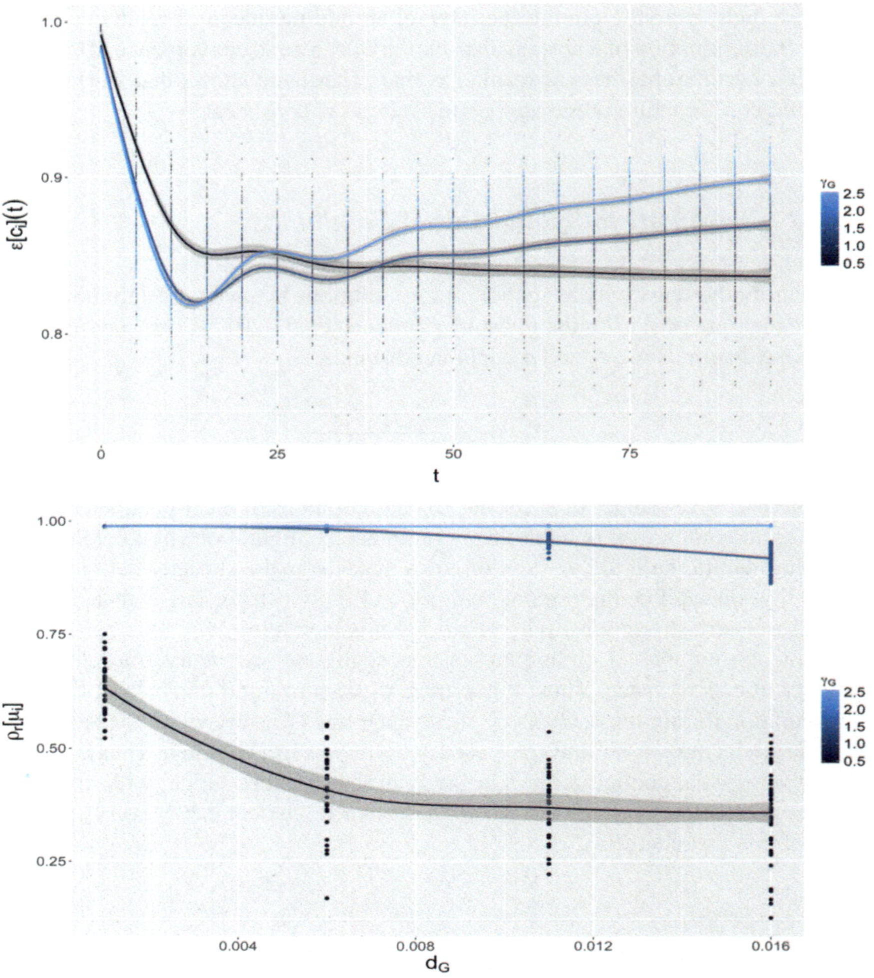

Fig. 14.2 Model behavior for the spatial configuration $N_S = 80, \alpha_S = 0.5, d_S = 10, n_S = 30$. (Top) Temporal trajectories of the entropy for closeness centralities, for $\gamma_N = 2.5$, $v_0 = 110$, $d_G = 0.016$, $\theta_N = 11$, as a function of γ_G (color); (Bottom) Rank correlation for population, as a function of d_G and of γ_G (color), for $\theta_N = 11$, $\gamma_N = 2.5$

interactions produce more hierarchical systems on the long terms, what could have been naturally expected, but with a transient behavior in which the system goes through a point with a maximum of equality between cities in terms of centralities. This confirms that taking into account dynamics in systems of cities is crucial for their understanding.

This variety of behaviors can be found again with the rank correlation ρ_R, that we show here for the population variable, as a function of d_G. It has a low sensitivity to

θ_N and γ_N (see (Raimbault, 2018d), Appendix A.7), but strongly varies as a function of d_G and γ_G as shown in Fig. 14.2 (second panel): interactions at a higher distance induce systematically a larger number of changes in the hierarchy of populations. These can occur when the hierarchy of distance is low. To summarize, the increase of the range of interactions will diminish the inertia of trajectories of the system of cities, whereas the increase of their hierarchy will increase it. This is relatively credible from a thematic point of view: longer and uniform interactions have more chances to make individual trajectories change.

The behavior of correlation indicators is shown in Fig. 14.3. Concerning the effect of distance on correlations between variables, i.e. the evolution of ρ_d, it is interesting to note that an increase of d_G systematically diminishes the levels of correlation, what corresponds to the complexification that we previously showed. As expected, $\rho_d[d]$ decreases as a function of distance, and exhibits non zero values for the correlation between population and centrality for a high hierarchy γ_G, what shows that simultaneous adaptation regimes are rare in this model.

14.3.6 Causality Regimes

Finally, by studying ρ_τ (Fig. 14.3, bottom panel), we observe that causality regimes in the sense of Raimbault (2017b) are relatively restrained (see Raimbault (2018d), Appendix A.7, for the confirmation for a broader range of parameters). The population is systematically caused by the centrality, but there exists no regime in which we observe the contrary. This is a logic of an effect of reinforcement of hierarchy by centrality. This exploration does not provide a configuration with circular causalities, and thus not a co-evolution properly speaking as we defined in the statistical sense.

14.3.7 PSE Algorithm

The last conclusion is crucial regarding the thematic questions that one can ask to the model, and was obtained with a limited experiment (simple grid sampling). However, because precisely of non-linearities, such simple sampling may miss regions of the parameter space in which dramatic changes occur in the phase diagram of the model, as Chérel et al. (2015) showed on toy models and on the Marius model. We apply here therefore the Pattern Space Exploration algorithm, which is precisely designed to unveil such unexpected behavior, what can be abstracted as a sampling of the image space of the model instead of its parameter space.

We use here the following indicators as targets on which the algorithm must find diverse patterns. Given all couples of variables (X, Y) and for $\tau > 0$ and $\tau < 0$, we consider $\max_\tau \rho_\tau[X, Y]$ and $\min_\tau \rho_\tau[X, Y]$, and report the one with the largest absolute value which absolute value is larger than correlation at the origin, and 0 otherwise. This adaptation of the method of Raimbault (2017b) has been proposed

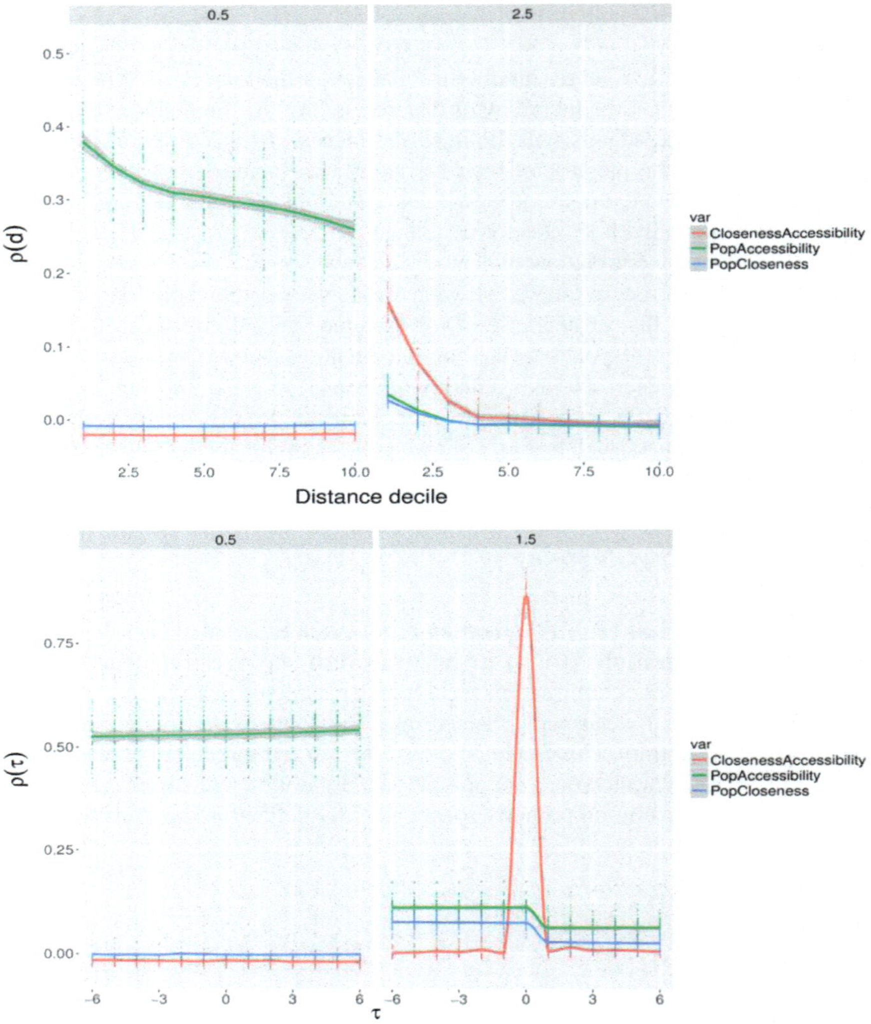

Fig. 14.3 Correlations in the model for the spatial configuration $N_S = 80, \alpha_S = 0.5, d_S = 10, n_S = 30$. (Top) Correlations as a function of distance, for couples of variables (color), for $\gamma_N = 2.5$, $\theta_N = 21$, $v_0 = 10$, and for d_G (columns) and γ_G (rows) variables; (Bottom) Lagged correlations for the same parameters

by Raimbault (2018c) for a similar model, for which simultaneous correlations are high as a consequence of model structure. The algorithm was run on 100 islands, with 500 parallel instances, and stopped at generation 16,000.

Figure 14.4 gives the results as a scatterplot of diversity targets of the algorithm. Several regimes are located on the axis, corresponding to one-directional relations

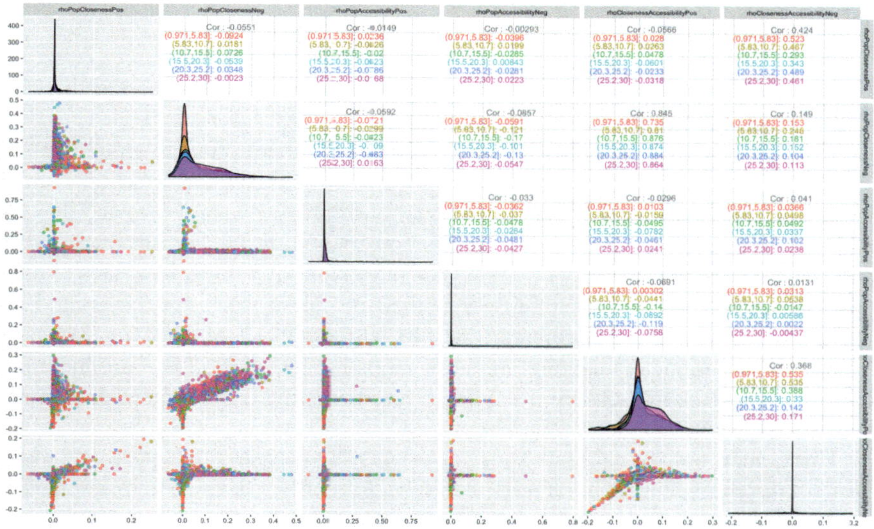

Fig. 14.4 Feasible space of lagged correlation obtained with the PSE algorithm. We give as a scatterplot the six objectives of the algorithm. Color level gives θ_N, which does not seems to be a simple driver of correlation values

between the variables. We obtain however a point cloud within positive correlations for both negative and positive delays for population and centrality, what actually witnesses a co-evolution (circular causality between the two). We have a similar pattern in negative correlations for centrality and accessibility, the point cloud being however more constrained on the diagonal, what may due to a direct repercussion of one variable on the other as they are structurally linked. We also observe points with a positive correlation between population and centrality for positive delays and negative delays, what corresponds to a direct circular causality, and between population and centrality and centrality and accessibility. This analysis reveals thus that the model is actually able to produce co-evolutive regimes in the sense of Raimbault (2018d).

Although the algorithm seems to have converged (less patterns discovered after generation 10,000), a more thorough investigation of the role of stochasticity would be needed since several points have few repetitions, to ensure the robustness of these results. These remain out of the scope of this proof-of-concept exploration, in which we show how a qualitative change can occur in knowledge about a model when using specific exploration methods.

14.4 Discussion

First of all, some thematic observations about the question of interactions between networks and territories can be formulated.

This model could be a useful tool to study the "tunnel effect", which is as we recall is the absence of interaction of an infrastructure traversing a territory with it (Raimbault, 2018d). Indeed, the rules allowing variable values for $v(t)$ and the non-planarity mechanism (when a new link is constructed, it does create intersections only with links of similar speed), allows the introduction of this effect. This remains however exogenous since explicitly specified in model rules, on the contrary to the interaction model of Raimbault (2018b) with feedback of flows, in which the variations of parameters capture an endogenous tunnel effect. The introduction of specific indicators to measure it would be an interesting development direction in the case of this model.

This model could be calibrated on real systems, but the use of the interaction model without the endogenous Gibrat term would be difficultly adaptable to an application of the model on real data because of the values for the calibrated parameter with endogenous growth for example by Raimbault (2018b). The application to a real system would thus require to study the complete model first.

In comparison, the co-evolution model introduced by Raimbault (2018c) seems to be less constrained for network dynamics and to produce more varied interaction regimes. We can not however compare the two in such different contexts. Future work to better understand the role of co-evolution in urban systems shall require multi-modeling approaches (Cottineau et al. 2015) and more systematic model benchmarks.

Regarding the modeling side, this work provides a supplementary proof-of-concept of the importance of the use of new tools and methods to extract knowledge from simulation models, since we indeed showed that the conclusion on the ability of the SimpopNet model to produce co-evolutive regimes would not have been obtained without the use of the PSE algorithm.

14.5 Conclusion

This chapter has illustrated the systematic exploration of a simulation model for the co-evolution of cities and transportation networks. Our contribution covers several dimensions, including a proof-of-concept of the importance of new model exploration tools and methods, a set of indicators to study similar models for systems of cities, and thematic knowledge about the model such as its sensitivity to spatial initial conditions and its ability to produce co-evolutive regimes.

Acknowledgements Results obtained in this paper were computed on the vo.complex-system.eu virtual organization of the European Grid Infrastructure (http://www.egi.eu). We thank the European Grid Infrastructure and its supporting National Grid Initiatives (France-Grilles in particular) for providing the technical support and infrastructure. This work is part of DynamiCity, a FUI project funded by BPI France, Auvergne-Rhône-Alpes region, Ile-de-France region and Lyon metropolis. This work was also funded by the Urban Dynamics Lab grant EPSRC EP/M023583/1.

References

Achibet, M., Balev, S., Dutot, A., Olivier, D.: A model of road network and buildings extension co-evolution. Procedia Comput. Sci. **32**, 828–833 (2014)

Arduin, H.: Modélisation mathématique des interactions entre pathogènes chez l'hôte humain: Application aux virus de la grippe et au pneumocoque. Ph.D. thesis, Université Paris-Saclay (2018)

Arthur, W.B.: Complexity and the shift in modern science. In: Conference on Complex Systems, Tempe, Arizona (2015)

Banos, A.: Pour des pratiques de modélisation et de simulation libérées en géographie et shs. Ph.D. thesis, Université Paris 1 Panthéon Sorbonne (2013)

Banos, A.: Knowledge accelerator' in geography and social sciences: further and faster, but also deeper and wider. In: Urban Dynamics and Simulation Models, pp. 119–123. Springer (2017)

Baptiste, H. (2010). Modeling the evolution of a transport system and its impacts on a French urban system. In: Graphs and Networks: Multilevel Modeling, 2nd edn, pp. 67–89

Batty, M.: Cities and Complexity: Understanding Cities with Cellular Automata, Agent-Based Models, and Fractals. The MIT Press (2007)

Benenson, I., Torrens, P.M.: Geosimulation: object-based modeling of urban phenomena. Comput. Environ. Urban Syst. **28**(1–2), 1–8 (2004)

Bird, I.: Computing for the large hadron collider. Annu. Rev. Nucl. Part. Sci. **61**, 99–118 (2011)

Blumenfeld-Lieberthal, E., Portugali, J.: Network cities: a complexity-network approach to urban dynamics and development. In: Geospatial Analysis and Modelling of Urban Structure and Dynamics, pp. 77–90. Springer (2010)

Brasebin, M., Chapron, P., Chérel, G., Leclaire, M., Lokhat, I., Perret, J., Reuillon, R.: Apports des méthodes d'exploration et de distribution appliquées à la simulation des droits à bâtir. In: Spatial analysis and geomatics 2017 (2017)

Bretagnolle, A.: Vitesse et processus de sélection hiérarchique dans le système des villes françaises. Données Urbaines **4** (2003)

Bretagnolle, A.: Villes et réseaux de transport: des interactions dans la longue durée, France, Europe, États-Unis (HDR). Université Panthéon-Sorbonne, Paris I (2009, June)

Chérel, G., Cottineau, C., Reuillon, R.: Beyond corroboration: strengthening model validation by looking for unexpected patterns. PLoS ONE **10**(9), e0138212 (2015)

Cottineau, C., Reuillon, R., Chapron, P., Rey-Coyrehourcq, S., Pumain, D.: A modular modelling framework for hypotheses testing in the simulation of urbanisation. Systems **3**(4), 348–377 (2015)

Cottineau, C., Raimbault, J., Le Texier, M., Le Néchet, F., Reuillon, R.: Initial spatial conditions in simulation models: the missing leg of sensitivity analyses? In: 2017 International Conference on Geocomputation: Celebrating 21 Years of Geocomputation (2017)

Epstein, J.M.: Generative Social Science: Studies in Agent-Based Computational Modeling. Princeton University Press (2006)

Galam, S.: Sociophysics: a review of galam models. Int. J. Mod. Phys. C **19**(03), 409–440 (2008)

LeCun, Y., Bengio, Y., Hinton, G.: Deep learning. Nature **521**(7553), 436 (2015)

Liao, T.W.: Clustering of time series data—a survey. Pattern Recogn. **38**(11), 1857–1874 (2005)

Mantegna, R.N., Stanley, H.E.: Introduction to Econophysics: Correlations and Complexity in Finance. Cambridge University Press, Cambridge (1999)

Mimeur, C., Queyroi, F., Banos, A., Thévenin, T.: Revisiting the structuring effect of transportation infrastructure: an empirical approach with the French Railway Network from 1860 to 1910. Hist. Methods J. Quant. Interdisc. Hist. **51**, 65–81 (2017)

Passerat-Palmbach, J., Reuillon, R., Leclaire, M., Makropoulos, A., Robinson, E.C., Parisot, S., Rueckert, D.: Reproducible large-scale neuroimaging studies with the openmole workflow management system. Front. Neuroinform. **11**, 21 (2017)

Paulus, F.: Coévolution dans les systèmes de villes: Croissance et spécialisation des aires urbaines françaises de 1950 à 2000. Ph.D. thesis, Université Panthéon-Sorbonne-Paris I (2004)

Pumain, D.: Pour une théorie évolutive des villes. L'Espace Géographique 119–134 (1997)

Pumain, D.: Une théorie géographique des villes. Bull. Soc. Géogr. Liège **55**, 5–15 (2010)

Pumain, D.: Les effets structurants ou les raccourcis de l'explication géographique. Espace Géogr. **43**(1), 65–67 (2014)

Pumain, D., Reuillon, R.: Evaluation of the SimpopLocal model. In: Urban Dynamics and Simulation Models, pp. 37–56. Springer, Cham (2017a)

Pumain, D., Reuillon, R.: Urban Dynamics and Simulation Models. Springer International (2017b)

Raimbault, J.: For a cautious use of big data and computation. In: Royal Geographical Society-Annual Conference 2016-Session: Geocomputation, The Next 20 Years (1) (2016a)

Raimbault, J.: Génération de données synthétiques corrélées. In: Rochebrune 2016, journées d'Etude sur les systèmes complexes naturels et artificiels (2016b)

Raimbault, J.: An applied knowledge framework to study complex systems. In: Complex Systems Design & Management, pp. 31–45 (2017a)

Raimbault, J.: Identification de causalités dans des données spatio-temporelles. In: Spatial analysis and geomatics 2017 (2017b)

Raimbault, J.: Models coupling urban growth and transportation network growth: an algorithmic systematic review approach. Plurimondi (17) (2017c)

Raimbault, J.: An urban morphogenesis model capturing interactions between networks and territories. In: The Mathematics of Urban Morphology, pp. 383–409. Birkhäuser, Cham (2019)

Raimbault, J.: Calibration of a density-based model of urban morphogenesis. PloS. One. **13**(9), e0203516 (2018a)

Raimbault, J.: Indirect evidence of network effects in a system of cities. Environ. Plann. B Urban Anal. City Sci. 2399808318774335 (2018b)

Raimbault, J.: Modeling the co-evolution of cities and networks. ArXiv E-Prints (2018c)

Raimbault, J.: Characterizing and modeling the co-evolution of transportation networks and territories. Theses, Université Paris 7 Denis Diderot (2018d). Retrieved from https://tel.archives-ouvertes.fr/tel-01857741

Raimbault, J., Banos, A., Doursat, R.: A hybrid network/grid model of urban morphogenesis and optimization. In: 4th International Conference on Complex Systems and Applications, pp. 51–60 (2014)

Reuillon, R., Leclaire, M., Rey-Coyrehourcq, S.: OpenMOLE, a workflow engine specifically tailored for the distributed exploration of simulation models. Future Gener. Comput. Syst. **29**(8), 1981–1990 (2013)

Reuillon, R., Schmitt, C., De Aldama, R., Mouret, J.-B.: A new method to evaluate simulation models: the calibration profile (cp) algorithm. J. Artif. Soc. Soc. Simul. **18**(1), 12 (2015)

Scarpino, S.V., Petri, G.: On the predictability of infectious disease outbreaks (2017). arXiv Preprint arXiv:1703.07317

Schamp, E.W.: 20 on the notion of co-evolution in economic geography. In: The Handbook of Evolutionary Economic Geography, p. 432 (2010)

Schmitt, C.: Modélisation de la dynamique des systèmes de peuplement: De simpoplocal à simpopnet. Ph.D. thesis, Université Panthéon-Sorbonne-Paris I (2014)

Wal, A.L.J.T., Boschma, R.: Co-evolution of firms, industries and networks in space. Reg. Stud. **45**(7), 919–933 (2011)

Chapter 15
Modelling Hierarchy and Specialization of a System of Cities from an Evolutionary Perspective on Firms' Interactions

Mehdi Bida and Céline Rozenblat

Abstract Despite their great diversity, most systems of cities show remarkably similar patterns when comparing the size distribution and the economic specialization of their constitutive cities. The universality of these patterns sparked the interest of geographers, economists and physicists. However, until now, no economic model has relied on a micro-based and evolutionary approach to reproduce these regularities. In this chapter, we intend to fill this gap by proposing a model where the micro dynamics of localized firms generate the two macro regularities of size distribution and economic specialization. The model is based on boundedly rational firms' competition and path dependent innovation. We discuss the possible emergence of macro properties from these micro behaviors of firms.

15.1 Introduction

The modelling of cities' systems has got now several experiences through the seminal SIMPOP model (Bura et al. 1996; Pumain 2012; Pumain and Reuillon 2017; Sanders et al. 1997) and other innovative efforts (Batty 2007; Portugali 2011). These models permitted to better understand the driving forces leading to the formation of cities' system hierarchies based on their interactions. However, none of these models is based on the interactions between micro agents. They all start from some assumptions of interactions between cities, thus between already established groups of agents. The question remains open on how cities and cities' systems that are characterized by hierarchical and diversity properties, emerge from the micro-processes of interacting micro economic agents (Lane et al. 2009; Pumain 2006a).

The complex approach of cities through micro economic agents needs to be grounded both in evolutionary theory of urban systems (Pumain 2006b) and in the

M. Bida (✉) · C. Rozenblat
University of Lausanne, Lausanne, Switzerland
e-mail: mehdi.bida@unil.ch

C. Rozenblat
e-mail: celine.rozenblat@unil.ch

D. Pumain (ed.), *Theories and Models of Urbanization*, Lecture Notes in Morphogenesis, https://doi.org/10.1007/978-3-030-36656-8_15

theories of evolutionary economic geography (Boschma and Martin 2010), in order to integrate together the general meso and macro properties of co-evolving cities and the assumptions on self-organized economic agents (firms), such as bounded rationality and path dependence (Nelson and Winter 1982).

The paper proposes a first elaboration of such bottom-up model where cities' systems represent a general meso and macro level constraining environment for micro-agents (firms), while interactions between evolving innovative firms transform this cities' systems environment. The bottom-up approach of cities is firstly questioned (part 1), enabling to introduce the built model (part 2) and to interpret and discuss a set of simulations exploring several variations of competition and distance effects (part 3).

15.2 Bottom-Up Approach to Systems of Cities

Even if there is no clear definition of complex systems (Ladyman et al. 2013), cities and the system they form are generally taken to be a good illustration of what complex systems are. Indeed systems of cities are seen as the result of a self-organization of many interacting individual agents (Pumain et al. 2009). These are for example, firms, workers, consumers, etc. when looking at systems of cities through the economics' lenses. Cities and the system they form are thus structures of individual agents that interact in a particular pattern of which spatial agglomeration and other structural properties are the mere manifestation.

15.2.1 Cities as Multilevel Complex Systems

Usually three levels of organization are identified in the study of systems of cities (Pumain 2006b): the micro level at which individual agents are considered, the meso level at which individual cities are considered, and the macro level at which a whole system of cities is considered. In terms of a pure set approach, one can designate any element of high level (meso or macro) as a subset of intensely interacting elements of the micro level. The obvious incompleteness of this approach shows the importance of the interaction processes and the resulting organization in defining the levels of observation as more than arbitrary levels of observational aggregation. As a starting point, elements of the meso and macro level have their constituents that interact more with each other than with the constituents of other elements. Moreover, the observed organization at a higher level shows that these local interactions generate the patterns of the higher level interactions thanks to which this higher level organization is observed. Thus, higher level elements are not only constituted by more intensely interacting elements, but their interaction has a particular pattern that creates the higher level organization. In the case of systems of cities, the local interactions between micro agents (firms, workers, consumer, etc.) create cities, that themselves self-organize, into the systems of cities.

15.2.2 Understanding Urban Systems: The Interest of a Micro-founded Approach

If cities and the systems they form are the result of the self-organization of individual agents, any model to understand their dynamics that does not deduce them from the behavior and interactions of the individual micro level agents is partial at best. Indeed, if cities can be seen as interacting by competing, or exchanging goods or people as in the SIMPOP models (Pumain 2012) or some economics models (e.g. Mills 1967), such interactions remain metaphors as they feature interactions between two abstract entities. The interest of a micro founded approach is to explain such abstractions by the actions of tangible entities that take real actions such as firms or consumers. The result is a more solid grounding in reality of the higher level phenomena on two aspects. The first is an understanding of how these phenomena arise from tangible actions, i.e. through the self-organization of micro agents. And the second is a better understanding of the relations governing the co-evolution of such higher level phenomena, through the unfolding of the chain of downward-upward causation. That is by showing how the change in a higher level property A can, through downward causation, change the lower level properties which in turn re-organize into new structures and change another higher level property B. This can be seen for example in Tabuchi and Thisse (2011) where the authors show how the change in transportation costs (which is a macro property in the model they propose) induces changes in the structure of the system of cities.

In this work, we propose to further examine the micro-macro link with an economic model in line with theories of evolutionary economic geography, particularly concerning regional knowledge. The model is based on micro level interactions of firms and is built using the weakest assumptions we can about the behavior of firms: firms have incomplete information and bounded rationality. The goal is to estimate in which extend the micro-level interactions can make emerge some meso level cities and macro level system of cities observing some specific properties.

15.2.3 The Two Empirical Evidences of Hierarchy and Specialization and Associated Theories

The model focuses on reproducing the two main universal properties of systems of cities: the hierarchical size differentiation and the socio economic specialization (Pumain et al. 2009).

15.2.3.1 Urban Size Hierarchy

Perhaps the most salient empirical regularity that is observed at the level of the system of cities is the hierarchical differentiation of cities' sizes. Within a same system (usually the cities of a same country that are supposed to have strongly interacted for a long time), cities sizes (usually the city's number of inhabitants) have been observed to follow approximatively a Pareto distribution with a great consistency across time and countries (Brakman et al. 2009; Rosen and Resnick 1980). This fact was first observed at least as early as 1913, when Auerbach (1913) observed that the product between the rank of German cities and their population is approximatively constant. Later, Singer (1936) observed for several countries (e.g. France, U.S.A., Germany) and dates, that the distribution of their cities' sizes was close to the Zipf distribution:

$$\text{Probability}(Size = k) = Ak^{-\beta}, \quad \beta > 0 \tag{1}$$

The observation was eventually made famous by Zipf (1949), when he noted the same relation between the size and rank of cities in several countries as part of his more general research on what later became his eponymous distribution. The level of the size hierarchical differentiation is usually examined by looking at the relation between the size of cities and their rank when the sizes are ordered in a decreasing order. The relation should be close to log-linear when the size is approximatively Zipf distributed[1] (Fig. 15.1).

Although Zipf himself insisted on the fact that the city size is proportional to its rank ($\beta = 1$), the empirically observed values of β range actually from 0.48 to 1.22 depending on the country and the used definition of city (Brakman et al. 2009). Higher β coefficients indicate a stronger differentiation between the cities' sizes, that is a more important difference between cities of two different ranks.

Modeling the processes that lead to a Zipf distribution of city sizes have received great attention from geographers and economists. We quickly review here the different categories of models based on the main idea proposed to explain the city sizes distribution, for a deeper discussion of the question, the reader can refer to Pumain (2006), Brakman et al. (2009), and Dimou and Schaffar (2011).

A first category of models is based on purely random processes suggested by Gibrat (1931). Gabaix (1999) was among the first economic models that was shown to generate a Zipf distribution. Gabaix considers cities randomly growing at a common mean rate, with their growths rates being independent across cities and time. Despite the discrepancy between the model assumptions and empirical observations (Pumain et al. 2009), an interesting conclusion can be drawn from Gabaix's work regarding the process that governs the growth of cities. If the growth rate is random and independent

[1] Although more recent models succeed at generating distributions that fit better the empirical observations (Duranton 2007; Giesen and Suedekum 2012), it remains interesting to use Zipf's distribution as a benchmark when modeling processes that seek at reproducing urban hierarchies without confronting it to data about existing urban systems.

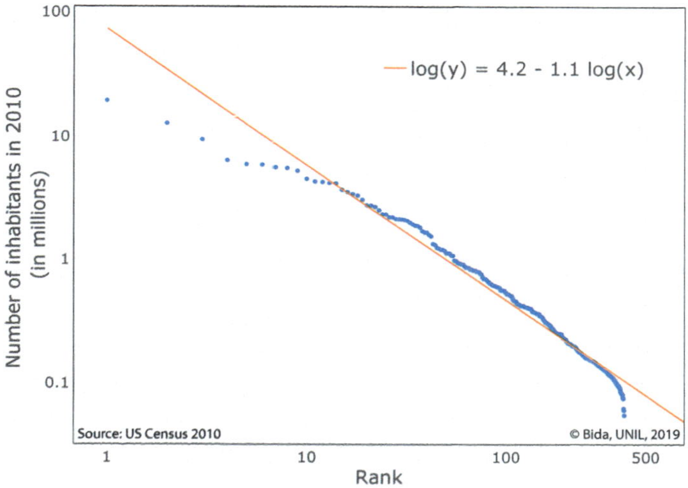

Fig. 15.1 Size distribution of US metropolitan statistical areas in 2010 (*Source* U.S. Census Bureau)

and if there is no mechanism that prevents cities from indefinitely shrinking, then the resulting distribution will necessarily be degenerate. The consequence of this is that either the growth rates are not independent or their mean is not proportional to their size.

A second category of models, from mainstream economics, proposes that Zipf like distributions of cities are the result of an equilibrium between agglomeration and dispersion forces. Brakman et al. (1999) took into account the congestion costs in order to avoid to agglomerate all the population in a single city as in Krugman's model (1991). Eaton and Eckstein (1997) consider dynamic growth of cities of which steady state results in a Zipf like distribution, the model features knowledge spillovers between cities.

A third category of model is based on spillovers or diffusion, where cities' growth is dependent on their own size, but also on the size of the other cities in the system through a spillover mechanism. The spillover avoids the situation where a small subset of cities grow too fast compared to the other cities of the system. Although belonging to two different modeling traditions, the SIMPOPlocal model (Pumain and Reuillon 2017) and Duranton (2007) share the reliance of this common mechanism.

Thus, in order to generate a Zipf like distribution, it seems that the growth advantage due to city size whether it is in the form of agglomeration economies or increasing returns to scale, should be compensated by the presence of dispersion forces or intercity spillovers.

15.2.3.2 Cities' Economic Specialization

A second property of system of cities appears in their division of labor at this macro-level leading to the relative specialization of each city (Aydalot 1985). Larger cities have been observed to be generally more diversified than smaller ones (USA: Henderson 1997; Canada: Marshall 1981; China: Min-rong and Yan-hua 2013; Japan: Mori et al. 2008) (Fig. 15.2). This pattern, also, is consistent in time and by sectoral composition. Not only cities keep the same degree of specialization over time, but also the activities in which they specialize (Duranton and Puga 2000).

If the negative relation between size and specialization has not received as much attention as the distribution of city sizes, several theories give some possible explanations to this observed relation. Static economic theories explaining this facts either emphasize the role of the advantage to economic diversity (Davis and Dingel 2014), or the interaction between transportation costs, internal increasing returns and competitive pressure (Tabuchi and Thisse 2011) as an extension of the central place theory (Christaller 1933; Lösch 1954).

Dynamic theories seem however more appealing, as they exhibit the processes through which cities become more diversified as they grow. In this line, Pumain et al. (2009) propose a model of urban diversification based on the unequal diffusion of innovations throughout the system of cities. New industries are created through successive innovation waves, which are better captured by bigger cities, and then selectively diffuse to smaller cities depending on the availability of the local resources (physical or knowledge). This model appears as a plausible consequence at the cities'

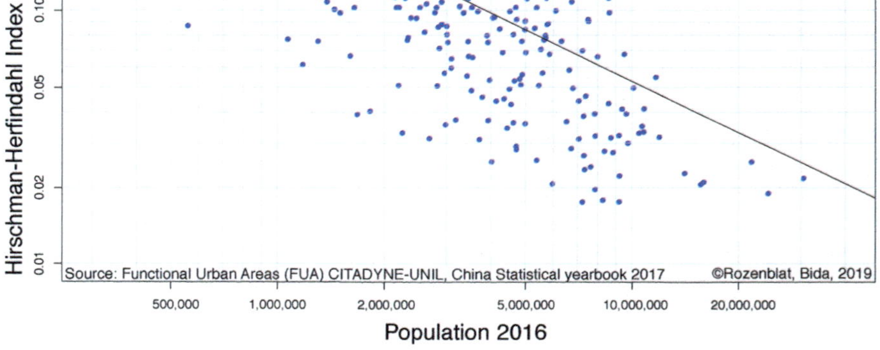

Fig. 15.2 Size and economic specialization for Chinese cities in 2016

system level of the findings of evolutionary economic geography. One of the main findings of latter is that local knowledge is essential in determining regional economic evolution. Indeed, the development of the notion of technological relatedness has shown how the growth of regions happens primarily through related diversification (Boschma 2017). Related diversification stands for the fact that the emergence of new activities in a region is strongly dependent on the nature of the present activities. In particular, empirical studies such as Neffke et al. (2011) have shown that the new industries that develop in a given region are predominantly technologically related to the region's current industries.

15.3 The Model

The micro founded model aims at reproducing the two main observed properties of systems of cities: the size hierarchy and the pattern of economic specialization, through micro interactions between firms. To this end, the model features interactions between micro level agents: firms in the context of the meso level of cities and of the system of cities acting on the micro level of firms' interactions. In return, cities and the system of cities are transformed by the results of the micro interactions between firms.

15.3.1 General Description

The bottom-up model is compatible with the *evolutionary framework* in the sense that at the micro-level, firms are endowed with bounded rationality (*myopic optimization*) and have their capabilities evolve through time with the possibilities of *innovation* (Fig. 15.3).

MESO (CITY)	Inherited Cities' industrial mix	
MICRO (FIRM)	INDIVIDUAL	INTERACTIONS
	Innovation Myopic optimization	Trade Competition Selection

Fig. 15.3 Main concepts of the model

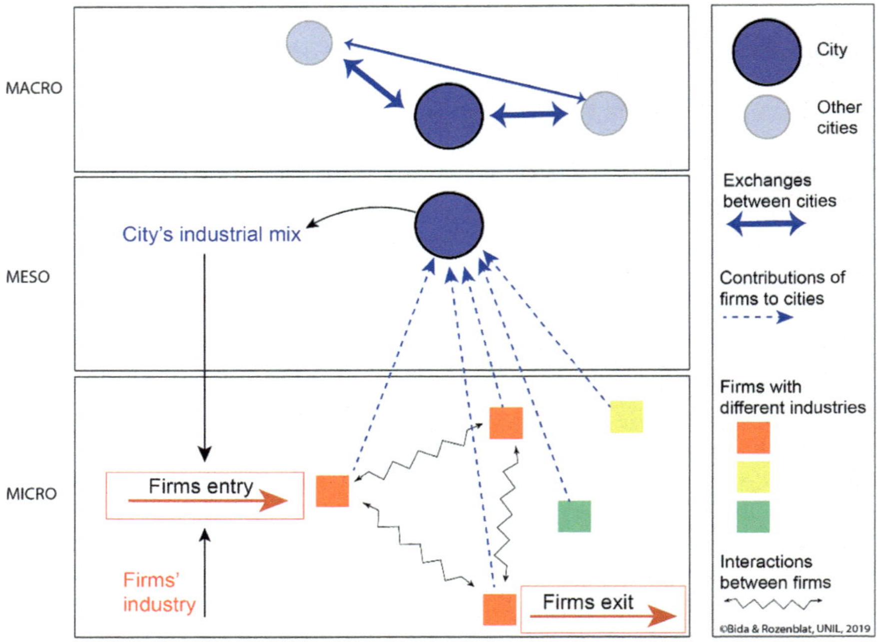

Fig. 15.4 Functioning of the model

Firms exchange each other by *trade*, but also *compete* within the same industry, a *selection* operating among the less competitive ones. At micro and meso-level, the model features *path dependence* by the fact that the presence of firms with certain industries in a city creates an *inherited industrial mix* that changes the possibilities of its future evolution by: (1) influencing its wealth and size and by (2) influencing its future path in the technological space (future entries of firms). The functioning of the model integrates micro, meso and macro levels (Fig. 15.4).

In line with the idea of related diversification, the probability that a firm of a given industry enters (micro level) depends on the current industrial mix of this city. This is because we assume that a firm belonging to a certain industry has to tap into local competences in order to produce, which is impossible if the local competences are inadequate to the firm's need.

We assume that initially, every city is endowed with a particular industry, and that the relatedness between industries is uniformly distributed so that no city is initially advantaged. Every iteration corresponds to a trade phase, a phase where firms improve their production process and a phase where new firms enter the market. Interactions between firms consist here only in market competition between firms of the same industry. Over the iterations, cities enrich their industrial mix (meso-level in Fig. 15.4) through the entry of new local firms that belong to related industries conferring them competitive advantage if the industry is still profitable (otherwise money-loosing firms exit according to the selection process). The hierarchical diffusion of innovations is embodied in the two following processes: largest cities have a higher likelihood to host new undiscovered industries (because they have a more diverse industrial mix). Once they discover a new industry, cities hosting existent related industries can easily host a new firm belonging to the newly discovered industry.

This model reveals sufficient to generate the desired properties of systems of cities (macro-level). The exchanges between cities consist in the sum of the trade flows between firms of cities and other cities' consumers. Such flows also exist inside each city, but does not change the city income (which is consistent with the economic base theory, where local exchanges are only considered as induced activities without any influence on the wealth of cities). In order to implement this model, few variables and parameters are necessary (Fig. 15.5).

At the micro-level:

- Each firm faces a *fixed cost* (its functioning) and *production cost* which depends on the innovative level reached at one moment by the firm (*Production efficiency*).
- The *production efficiency* increases thanks to the *R&D investment* procured by the *profit* made at the previous step. It increases also thanks to the location economies realized through the *industrial proximity* permitted by the *city industrial mix* (meso-level) reached at the previous step.
- The *profit* of each firm depends of the *fixed cost* and *production cost* (negatively) and on the combination of the *price* and the amount of the *demand*.

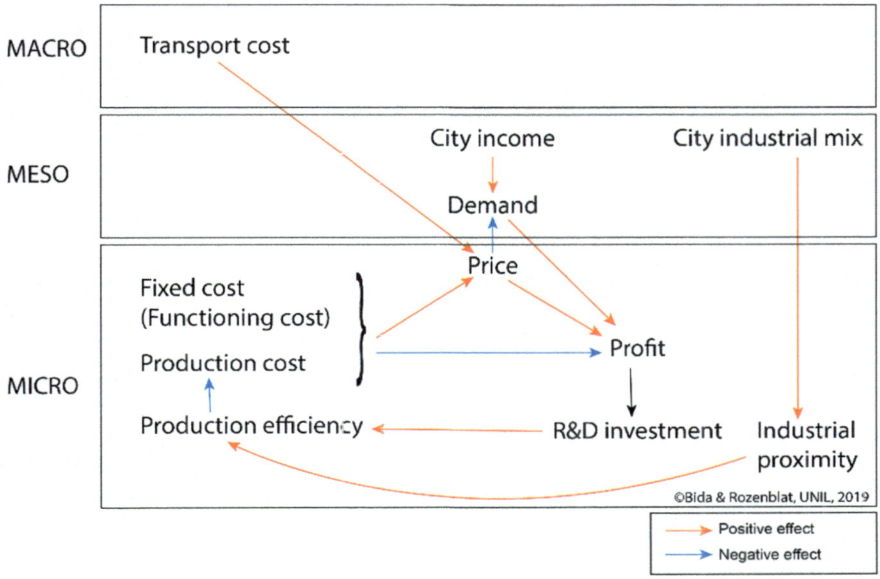

Fig. 15.5 Variables of the model

– The *price* determined by each firm depends on the *costs of the firm* (micro-level) but also on the *transportation costs* which is define at the macro-level.

At the meso-level:

– The *demand* itself depends on one hand on the *price*, but also on the *city income* (meso-level);

At the macro-level:

– Transportation costs are fixed and similar for all firms and across the city system.

As a result at the macro-level:

– Hierarchy of cities size will emerge from the micro/meso interactions;
– Specialization of cities also contributes to qualify the division of labor within the urban system, that will be qualified by heterogenous levels of cities' specialization.

15.3.2 Formal Description

The modelling framework is roughly based on the standard dynamic micro-economic framework of which some assumptions are relaxed. The model features firms and consumers that are located in cities. Firms compete within a same industry to sell their good to consumers located in all cities and function using only local labor. New firms randomly enter each time step. The entry success of new firms depends on the proximity of their industry to the city industrial mix. Nor positive neither negative externalities are accounted for in this model. In particular cities are neither subject to congestion cost nor to external scale economies. However the model rests on the meso-level effect that is the dependence on the current industrial mix, of the probability of entry and the level of efficiency of new firms.

15.3.3 Cities and Consumers

We assume that initially, there is a finite number V of cities. The cities are spatially located in a finite one-dimensional isotropic space. We choose to place the cities on a circle of length L to avoid any boundary effects (Fig. 15.6). We furthermore assume that they are regularly spaced, so that no particular subset of cities can benefit of be disadvantaged by the irregularity of their spatial distribution. We however note that the regular spacing of the cities is not a definite feature of the urban system, as the model allows cities to decay and completely disappear.

Cities are characterized by their *income* that is the sum of their firms' previous revenues, since firms only employ local labor. All this income is used to buy goods from active firms at the next iteration. For the sake of simplicity, we assume that

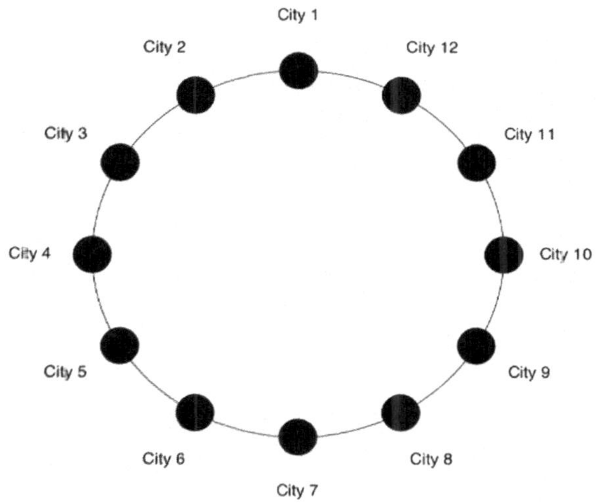

Fig. 15.6 Example of initial configuration with 12 cities

the income of each city is evenly spread over the active industries. On the aggregate level, for a same industry, consumers do not favor the product of a firm over another, and thus the attractiveness of a product within an industry only depends on its price. Given these requirements, we choose the aggregate city-demand to be:

$$q_{ij} = \frac{C_j}{N} \frac{p_{ij}^{-\alpha}}{\sum_{i \in I} p_{ij}^{1-\alpha}}, \alpha > 0 \tag{2}$$

where C_j is the income of city j, N the number of active industries (with operating firms), p_{ij} is the price of firm i to consumers located in city j, and α a parameter that determines the sensitivity of the consumers to the relative price level. The choice of the constant elasticity of substitution function is motivated by the ability of this function to represent a variety of aggregate consumer behavior: from a choice only motivated by the price level (high values of α) to the case where each firm has its own consumption niche (α close to 0).

15.3.3.1 Firms and Industries

There exists a finite number I of industries to which firms can belong. Not all industries need to be active. An industry is considered as active when at least one firm belonging to it is active. Industries determine the type of good that is produced by firms, one can think of it as an abstraction of economic sectors. All the firms that belong to a same industry produce a homogenous good, and are directly in competition to each other, meaning that the entry of a firm of the same industry will directly impact their market share, whereas this is not the case when firms in other active industries enter the market. However, industries also compete in a certain way: every time a firm belonging to an inactive industry enters the market, it takes a proportion of the city total income shared by existing industries.

In order to implement the idea of *related diversification* (Boschma and Frenken 2011), we introduce a level of *technological relatedness* between the industries. It is a proximity measure that indicates the level of technological relatedness between two industries. We build on the empirical findings about industrial evolution at the regional (Neffke et al. 2011) and the international scale (Hidalgo et al. 2007). Both studies use the concept of *industry/product space* to represent the technological proximity between the different industries in order to analyze the dynamics of the evolution of the industrial mix of countries and regions. In spite of the difference of scale, both studies present similar findings about the structure of the industry space and the dynamics of industrial evolution. The industry space has been found to be modular, i.e. with groups of industries of high intra relatedness and low inter relatedness. Moreover, regions' as well as countries' industrial mixes have been shown to evolve consistently with the industries' space structure, by remaining closely related to its previous state.

In light of these results we assume that, at the meso level, the current industrial mix of a city will determine the dynamics of firms' entry. More precisely, the level of technological relatedness between the present industries and the industries of potential entrants will determine their probability of entry at each time step. The firms' entries make evolve the city' industrial mix.

The level of relatedness θ_{kl} ($=\theta_{lk}$) between two industries k and l lies between 0 and 1, where a level of 1 stands for the highest level of relatedness and 0 for the lowest level of relatedness. A modular industry structure in this case would mean that industries would be divided in groups with very high θ for two industries within the same group and very low θ for two industries each belonging to a distinct group (Fig. 15.7a). We will depart from this structure and assume that the industries cannot completely be separated into clearly distinct groups, but that each industry is related to a fixed number of other industries constituting a low density quasi-regular industry space (Fig. 15.7b).

This connection between industry groups will allow cities to *discover* inexistent industries while preserving the idea that new industries can only be discovered through a definite set of existing industries. Finally, we further simplify the structure by assuming that the relatedness between two industries is either 0 or 1.

Each firm is located in only one city, where it only uses local labor to operate. Firms produce and transport goods to consumers in all cities. A constant level of labor is necessary to the functioning of the firm, and goods are produced and transported at constant return.

Selling goods to consumers located in other cities requires (local) labor to transport them that depends on the quantity and the distance over which they carried. Each firm is moreover endowed with a level of production efficiency that determines the amount of labor needed to produce one good. The amount of labor needed by the firm at each time step is given by:

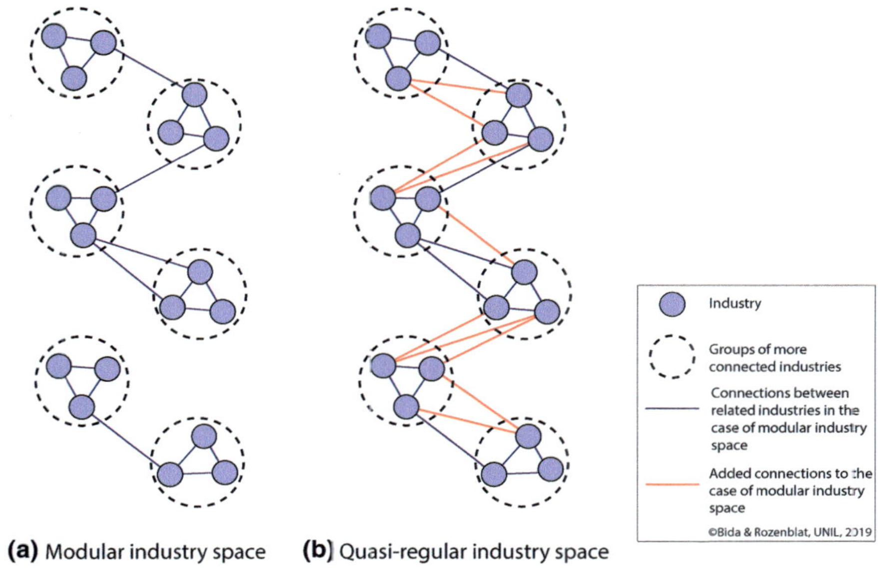

Fig. 15.7 Example of industry space with 20 industries. **a** is a modular pattern, as usually empirically observed. **b** is a more connected version of (**a**), as taken in the model. Each industry is related to 4 others

$$l_i = b_i + \left[\sum_{j=1}^{V} q_{ij}\left(a_i^{-1} + D_{ij}\right)\right], \quad D_{ij} = \left(1 + d_{ij}\right)^{\gamma} - 1, \quad \gamma \geq 0 \tag{3}$$

where b_i is a fixed amount of labor to keep the firm functioning, a_i is the production efficiency and D_{ij} is the necessary amount of labor to transport one unit of good between producer's i city and consumers located in a city j that depends on the distance d_{ij} between the city of producer i and the city of consumers j.

In this model, we do not account for the interaction between workers and firms. For simplicity we assume that firms can find all the necessary amount of labor needed in the city where they are located at a constant wage that we set equal to 1 for all industries and cities. In these conditions, the profit of producer i is given by:

$$\Pi_i = q_i p_i - l_i \tag{4}$$

15.3.3.2 Initial Conditions and Dynamics

In order not to advantage any city, we assume that initially, all cities are endowed with the same income, and host only one firm. All the initial firms belong to the same industry.

At every time step t, for each city j and industry l, a maximum of one firm enters the market with the following probabilities:

– if the industry already has been or is still active in the economy

$$\left[\max_{l \text{ in city } j} \theta_{lk}\right] \tag{5}$$

– if the industry was never active before

$$\left[\max_{l \text{ in city } j} \theta_{lk}\right] p_{discovery} \tag{6}$$

Note here that the more other related industries are present in a city, the higher are the chances of a given industry to enter the city. Over the iterations, we expect this to create a process of related diversification of cities. As cities become more likely to foster an industry as their industrial mix is more related to it. We set two different probabilities for discovered and undiscovered industries to reflect the fact that the emergence of a new industry is much more difficult than the entry of competitors in an established industry. However, the latter still depends on the industrial mix of the city since it requires knowledge resources and supporting institutional structures that have to be available locally (Boschma and Frenken 2011).

Entering firms belonging to a unprecedently discovered industry enter the market with a level of efficiency equal to 1. Otherwise, for an already discovered industry k, a new competitor h belonging to k a that enters the market is initially endowed with a level of production efficiency a_h randomly chosen as:

$$a_h = 1 + A_{kj}, \quad A_{kj} \sim \text{Binom}\left(\left[\max_{\substack{i \text{ in city } j \\ i \text{ in industry } M(k,j)}} a_i\right] - 1, \quad \theta_{M(k,j)k}\right) \tag{7}$$

where $M(k,j)$ is the industry already present in city j that has is the most related to industry k. This means that the more related and efficient the current firms in city j are, the more likely entering firms will also be efficient. Note that a_h is never higher than the most efficient incumbent of industry $M(k,j)$, which is the industry from which we can consider that h "branched". This is because the efficiency of production is partly acquired through learning-by-doing (Thompson 2010) and thus incumbent firms, which are more experimented, are assumed to be more efficient. However, this also implies that cities having very efficient firms in closely related industries can see their entering competitors being more efficient than “unexperimented” incumbents.

Newly entering firms join the market with active firms and trade with the consumers located in all cities according to the equations above. Firms with a negative profit exit the market, and earning profit firms invest a certain constant proportion q of their profit in research and development (R&D), also employing local labor, in order to improve their production process and raise their level of efficiency. Decisions concerning R&D investments in general are conditioned by several factors such as the appropriability of the innovation and market structure (Cohen 2010). Firm size (usually firm revenue) has been empirically shown to explain more than half of the intra-industry variance with which R&D investments grow below but close to proportionally (Cohen and Klepper 1996). In order to avoid unnecessary complexity, we simply assume that the proportion of R&D investment depends proportionally on the firm size, and take it to be equal to a fraction r of the current profit (firms that remain active only have positive profits).

The result of R&D is an increase in the production efficiency of the firm. The efficiency increase is assumed to be equal to the invested amount. For an efficiency level a_t, the efficiency at the next time step will be:

$$a_{t+1} = a_t + \Pi_t \tag{8}$$

Firms then adjust their prices to consumer demand in order to increase their profits. Given the assumption of bounded rationality, firms neither know the demand function of the consumers nor the prices of their competitors. They adjust their price through a myopic optimization process:

$$p_{t+1} = p_t + (p_t - l_t)(1 + m)^{H\left\{\log\left(\frac{p_t}{p_{t-1}}\right)(q_t p_t - l_{t-1} - \Pi_{t-1})\right\}}, \quad 0 < m < 1 \tag{9}$$

where H is a function that determines the sensitivity to profit change. It needs to be increasing an following the sign of its argument. To avoid unstable price adjustment processes, we also require it to be between -1 and 1. This adjustment process means that firms follow the variation of the profit they make to adjust their prices. If a positive or negative price adjustment leads to a profit increase, firms will again adjust their price in the same direction. Note that firms optimize without taking the last efficiency improvement into account. This is in order to separate the effect of efficiency improvement from the effect of price variation of consumer demand on the change of profit.

We finally reallocate consumer’s incomes according to the revenues of local firms and expenditures of local consumers. Given that firms only employ local labor and their owners also belong to the city, all the revenues they generate remains in their cities:

$$C_{j,t+1} = \sum_{i \text{ in city } j} \left(\sum_{\text{cities g}} p_{ig,t} \quad q_{ig,t} \right) \tag{10}$$

15.4 Results and Discussion

Given the difficulty to solve the model analytically, we resort to numerical simulations in order to explore the different outcomes of the model. We only explore the effect of three parameters, the others are left unchanged throughout the exploration, their values are given in Table 15.1.

The focus of the exploration is on the price competition (α), distance friction (γ) and the probability of discovering a new industry (p_{disc}). For each combination of parameters, 100 runs have been performed in order to evaluate the robustness of the result. Each run was performed for the number of steps that was necessary to discover all the industries, plus 50 extra steps to allow the stabilization of the dynamics.

We explore the results for the number of remaining cities at the end of the simulations (Fig. 15.8a), the value of mean slope coefficient of the log-size distribution (Fig. 15.8b), and the value of mean slope coefficient of the relation between city size and specialization (Fig. 15.8c). We preformed preliminary runs in order to determine the intervals over which the exploration was done. We find that for a level of competition that is high enough ($\alpha \geq 1.5$), the evolution of the system ends with a unique city concentrating all active firms and the sum of all the cities' initial incomes. Similarly, for high values of the distance friction ($\gamma \geq 2$), the final number of cities is lower than half of initial number.

The parameters for which the effect of variation is explored are the level of price competition α, the level of distance friction γ, and the probability to discover a new industry p_{disc}. The parameters' values for exploration are chosen so that the cases

Table 15.1 Constant parameter values used in all the simulations

Parameter	Value
Initial number of cities V	30
Total number of industries I	100
Number of related industries	4
Cities' initial income C_0	500
Perimeter of the circle L	2
Necessary quantity of fixed labor b_i	5
Firms' sensitivity to profit m	0.1
R&D (profit) intensity r	0.05
Price guiding function H	$H(x)=2\pi^{-1}$ arctan$(0.5\pi x)$

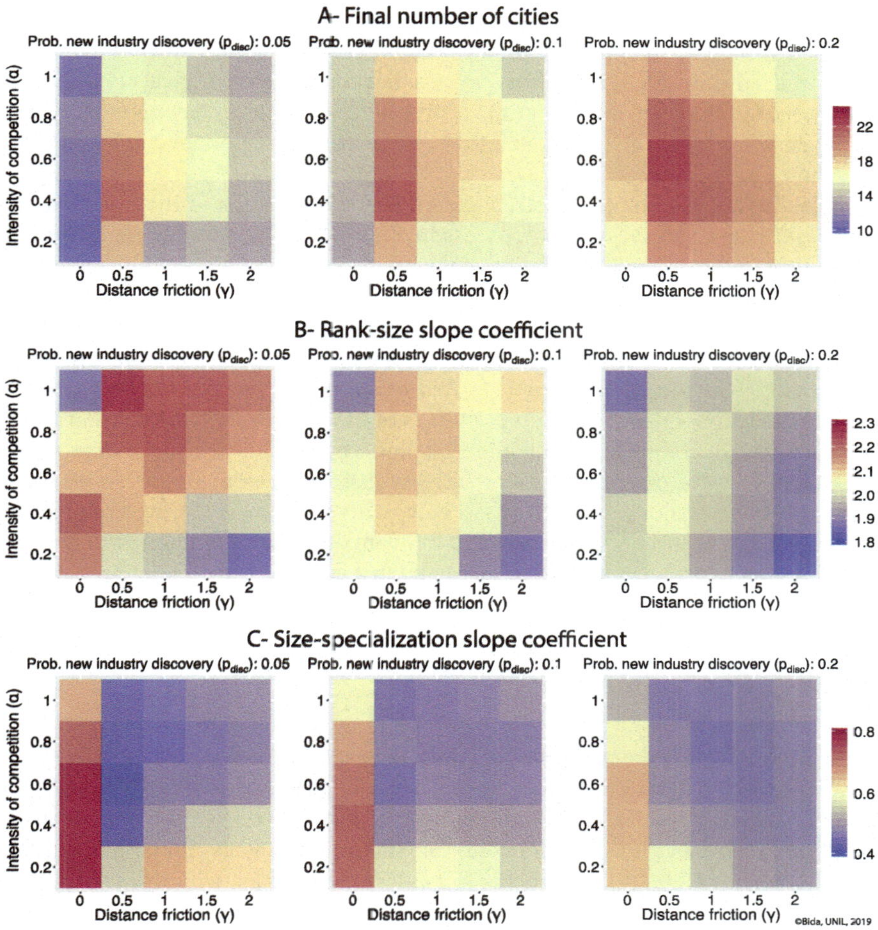

Fig. 15.8 Average results of the cities' system simulations (100 simulations for each parameters combination)

with too few remaining cities are avoided. The explored values can be found in Table 15.2.

Table 15.2 Chosen values for the model exploration

Parameters	Chosen values
Level of price competition α	0.2, 0.4, 0.6, 0.8, 1
Distance friction γ	0, 0.5, 1, 1.5, 2
Probability of discovering a new industry p_{disc}	0.05, 0.1, 0.2

The level of price competition α determines the importance of the relative price for the consumers when they choose among different products offered by firms belonging to a same industry. A High level of α means that consumers will give a considerable importance to the relative price of the different products when buying them. On the contrary, a low value of, means that the price is of little importance in determining the choice of consumers.

The distance friction parameter γ determines the variation of the transportation cost with respect to the distance over which the products are carried. Given that transportation costs are at the charge of the producing firms, higher values of γ raise further located firms' trading costs, decreasing their competitiveness compared to closer located firms. In the exploration, the particular case of the absence of transportation cost ($\gamma = 0$) is considered, along with three other forms of transportation costs: sublinear ($\gamma < 1$), linear ($\gamma = 1$), and super-linear ($\gamma > 1$).

The model is explored for different values of the probability p_{disc} of discovering a new industry. In combination with the relatedness of the industries present in a given city, the p_{disc} value determines the probability that a firm belonging to an inexistent industry enters in a city, conferring it a competitive advantage during the time step it enters. High values of p_{disc} mean more frequent discoveries of industries, however this does not allow cities to host totally unrelated new industries.

15.4.1 Number of Remaining Cities

For all the tested parameters combinations, the average number of remaining cities was computed (Fig. 15.8a). Recalling that the initial number of cities is 30, this number can only decrease through the time steps. For none of the combinations, the final number of cities was found to remain the same as the initial number. The result becomes globally more robust to random fluctuations with the average final number of cities of the 100 runs (for each combination), where we observe a standard deviation from 24% for lowest mean value to 11% for the highest mean value. The final number of cities seems to increase with p_{disc}. However, the pattern seems to be non-monotonic for the combination of the parameters α and γ. The number seems to reach a maximum for low-intermediate values of the level of competition and distance friction. A moderate level of competition and distance friction seems to favor the maintenance of cities, possibly by allowing some profits and a certain level of production efficiency for larger cities, and a level of protection from outside competitors for less efficient firms in smaller cities. The lowest average final number of 10 cities was found for the combination ($\alpha = 0.2$, $\gamma = 0$, p_{disc}=0.05). If a low transportation cost is not surprising, since it exposes less efficient producers to competition, the low value of price competition is less expected. We think that when the level of competition within a same industry is low, it is the interindustry competition that determines the growth of cities. In this case, larger cities, because they are more diversified gain an important advantage and "absorb" smaller, less diversified cities.

15.4.2 City Size Distribution

One of the main goals of the model is to generate the empirically observed Zipf distribution of cities. An example of resulting city size distribution is given with the parameters $\alpha = 0.5$, $\gamma = 1$, and $p_{disc} = 0.2$ (Fig. 15.9).

For all the parameters' combinations, we estimate slope $-\beta_1$ between the logarithm of the city income C and size rank ρ using least squares:

$$\underset{\beta_0,\beta_1}{\operatorname{argmin}} \sum_{C_j>0} \left[\log(C_i) - \beta_0 - \beta_1 \log(\rho_i)\right]^2 \tag{11}$$

For the range of the explored parameters, the model seems to perform qualitatively reasonably well, with an explained variance ranging between 0.9 and 0.97. Moreover, over the 100 runs, for any parameters' combination, the standard deviation of the explained variance is below 6.2%.

The resulting values of the slopes (Fig. 15.8b) range between 1.8 and 2.2, which is high compared to the empirically observed distributions [typical real city size distribution slopes range from 0.48 to 1.22 (Brakman et al. 2009)]. Accounting for congestion costs due to city size could allow to reach more realistic values of slope

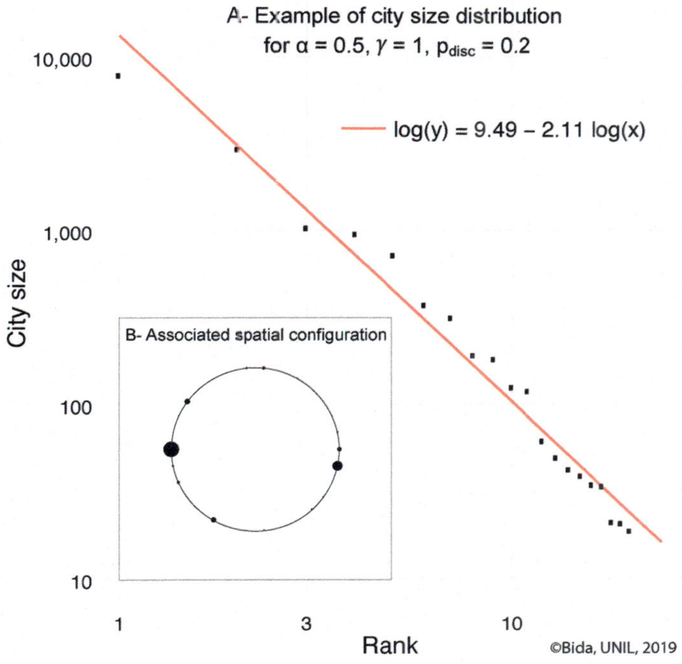

Fig. 15.9 Example of city size distribution for one simulation

of the rank size distribution as in (Brakman et al. 1999). Other options can also be considered, such as the inclusion of a non-traded industry in each city, that would prevent excessive shrinking of small cities as in Sanders et al. (1997), Duranton (2007), or in a more general way in Gabaix (1999).

As for the influence of the parameters on the slope, again, we observe that higher values seem to monotonically change the outcome by decreasing the size hierarchy of cities. The level of competition α seems to have a higher importance than the distance friction γ when the latter has positive values. Higher level of competition allows larger firms and more efficient firms to better compete, enhancing the positive feedback loop between size and efficiency. Interestingly, when transportation costs are absent ($\gamma = 0$), a city size hierarchy is still observed, but the influence of the level of competition seems to have an opposite effect to the case of existing transportation costs. The same cause as the observed effect on size can be the reason: when the level of intra-industry competition is low, inter-industry competition favors the growth of larger, more diversified cities. Thus intra-industry competition seems so play a mitigating effect for the advantage of larger cities, possibly because smaller specialized cities host more efficient firms.

15.4.3 Size-Specialization Relation

The second main objective of the model is to reproduce the observed relation between size and specialization. In order to study the relation between city size and specialization, we use the Hirschmann-Herfindahl index (HHI) to calculate the specialization level of each city *j*:

$$HHI(j) = \sum_{\text{industries } l} \left(\frac{\sum_{\text{firms } i \text{ of } l \text{ in } j} \Pi_i}{\sum_{\text{firms } i \text{ in } j} \Pi_i} \right)^2 \tag{12}$$

The HHI index is widely used as concentration measure in general and as a measure of cities economic specialization in particular (e.g. Henderson 1997). Lower values of the HHI index stand for a more diversified city. Cities with lower HHI index values are hosting a higher number of industries, of which size are more similar. An example of resulting city size distribution is given with the parameters $\alpha = 0.5$, $\gamma = 1$, and $p_{disc} = 0.2$ (Fig. 15.10).

As previously, we find using least squares, the average linear relation between the logarithm of the HHI and the logarithm of the city sizes, given the 100 simulations for each parameter combination (here taken as the sum of profit making local firms). Here again, for the range of the explored parameters, the model seems to perform qualitatively reasonably well. For all parameters' combinations the relation between size and specialization is negative. The negative relation seems to be robust to random fluctuations for all parameters' combinations, with a maximum standard deviation of 24%. The log linear model seems to yield a good fit with a minimum average

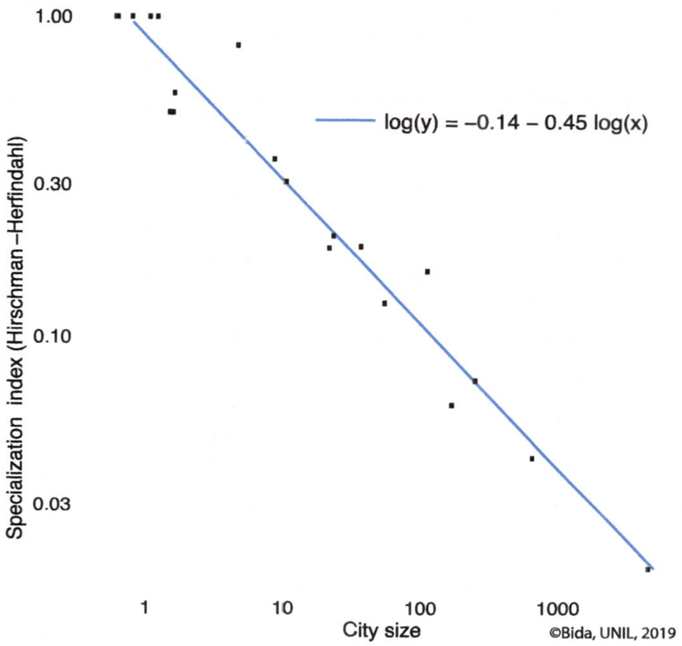

Fig. 15.10 Example of size-specialization relation for $\alpha = 0.5$, $\gamma = 1$, and $p_{disc} = 0.2$

explained variance of 0.85. The model is thus an idealized case, as, to our knowledge, no empirical observation has revealed such a great regularity.

We observe that low levels of intra-industry competition favor a higher differentiation in economic specialization. This is consistent with the previous given explanation concerning the mitigating effect of intra-industry competition. Lower values of intra-industry competition seem to give more importance to inter-industry competition, intensifying the positive feedback between the size and the diversity of a city.

15.4.4 Synthesis

For all the explored parameters' combinations, the generated system of cities qualitatively matches the empirically observed features of real systems of cities concerning the hierarchization in size and economic specialization. The results have shown that the interaction between the three variables of the model can lead to non-monotonic effects when varying them. We note here that the interplay between intra-industry and inter-industry competition seems crucial in determining the final features of the system of cities. Interestingly, we also note that the expected hierarchization according to size and economic specialization is generated for both the cases of existent and absent transportation costs.

An important result is the low negative effect in general of the probability of new industry discovery (p_{disc}) on the Rank-Size coefficient. In the SIMPOP model, the formation of the cities' hierarchy was only with the apparition of new innovations (Bura et al. 1996; Sanders. et al. 1997). In the model we propose, the apparition of new industry is also necessary to create the cities' hierarchy. Our model shows however that raising the likelihood of apparition of new industries does not necessarily result in an increase of the cities' hierarchy. On the contrary, in the model presented here, increasing the likelihood of apparition of new industries benefits to all the existing cities, and thus mitigates their hierarchization. Thus, despite the fact that the model includes in a way, the hierarchical diffusion of innovations from largest cities to the smaller ones (Pumain 2006b) because when a new industry appears in a city, the other cities can catch it during the following steps, the model seems however to overestimate the speed of this diffusion. These spatial waves of diffusion of innovations are the result of economic cycles of which mechanisms and manifestations (Dosi and Nelson 2010; Klepper 1997) need to be better reproduced by the model.

15.5 Conclusion

We proposed in this chapter a micro-founded economic model, able to reproduce two main properties of system of cities considering the size distribution and the economic specialization. Relaxing the less realistic assumptions of main stream economic models, the model rests upon bounded rationality (Myopic optimization) and evolving firms' capabilities. In line with the recent findings of evolutionary economic geography, the model features path dependence through a process compatible with the idea of related diversification. Cities' industrial diversification is the result of past firms' interactions, and this diversification in turn constrains future firms' dynamics.

Starting from undifferentiated cities in terms of size and economic specialization, the model managed to generate a system of unequal cities by their size and their specialization. The obtained rank-size slopes depend largely on the combination of the level of intra-industry competition and the distance friction (but there is a low effect of the average probability of innovation). Simulation results also underlined the role of diversification of cities (several industries), especially when transportation costs are virtually inexistent.

However, the model lacks parts of complexity, given that several parameters that must evolve like in "real" urban system remain constant. For example, the sum of all cities' income is assumed to be constant. This restrains the expansion of the whole system, and thus the growth of cities and the diversification of their industries. Another simplification made in the model that deserves more attention is the structure of the industry space. Our results are valid for a particular simple structure of the industry space. This calls for further improvements to make the results robust to less regular and more conformable to realistic structures of the industry space.

Eventually, despite the qualitative similarity of the cities size distribution with real empirical observations, the range of the slope generated by the model is unrealistic. This might come from the lack of inclusion of several processes like non-traded local services, or congestion. This would allow to reproduce existing cities' systems in order to better understand their properties. In particular, it would be interesting to develop this model to study more applied questions as the effect of the intensity and forms of collaborations between firms on the features of the system of cities. Especially we are concerned by searching how to improve spatial organization of cities compatible with the contemporary challenges of sustainability.

References

Aydalot, P.: Économie régionale et urbaine. Economica, Paris (1985)

Batty, M.: Cities and Complexity: Understanding Cities with Cellular Automata, Agent-Based Models, and Fractals. The MIT Press (2007)

Boschma, R.: Relatedness as driver of regional diversification: a research agenda. Reg. Stud. **51**(3), 351–364 (2017)

Boschma, R., Frenken, K.: Technological relatedness and regional branching. In: Bathelt, H., Feldman, M.P., Kogler, D.F. (eds.) Beyond Territory: Dynamic Geographies of Knowledge Creation, Diffusion and Innovation. Routledge, London (2011)

Boschma, R., Martin, R.: The Handbook of Evolutionary Economic Geography. Edward Elgar Publishing (2010). https://doi.org/10.4337/9781849806497

Brakman, S., Garretsen, H., van Marrewijk, C.: The New Introduction to Geographical Economics, 2nd edn. Cambridge University Press, Cambridge (2009). https://doi.org/10.1017/CBO9780511818745

Brakman, S., Garretsen, H., Van Marrewijk, C., Van Den Berg, M.: The Return of Zipf: Towards a Further Understanding of the Rank-Size Distribution. J. Reg. Sci. **39**(1), 183–213 (1999)

Bura, S., Guérin-Pace, F., Mathian, H., Pumain, D., Sanders, L.: Multiagent Systems and the Dynamics of a Settlement System. Geogr. Anal. **28**(2), 161–178 (1996)

Christaller, W.: Die zentralen Orte in Süddeutschland: Eine ökonomisch-geographische Untersuchung über die Gesetzmässigkeit der Verbreitung und Entwicklung der Siedlungen mit städtischen Funktionen. Gustav Fischer, Jena (1933)

Cohen, W.M.: Fifty years of empirical studies of innovative activity and performance. In: Hall, B.H., Rosenberg, N. (eds.) Handbook of the Economics of Innovation, vol. 1. North-Holland (2010)

Cohen, W.M., Klepper, S.: A reprise of size and R & D. Econ. J. **106**(437), 925–951 (1996)

Davis, D.R., Dingel, J.I.: The Comparative Advantage of Cities, Working Paper No. w20602. National Bureau of Economic Research (2014). https://doi.org/10.3386/w20602

Dimou, M., Schaffar, A.: Les théories de la croissance urbaine, Urban growth theories. Rev. Écon. Polit. **121**(2), 179–207 (2011)

Dosi, G., Nelson, R.R.: Technical change and industrial dynamics as evolutionary processes. In: Hall, B.H., Rosenberg, N. (eds.) Handbook of the Economics of Innovation, vol. 1. North-Holland (2010)

Duranton, G.: Urban evolutions: the fast, the slow, and the still. Am. Econ. Rev. **97**(1), 197–221 (2007)

Duranton, G., Puga, D.: Diversity and specialisation in cities: why, where and when does it matter? Urban Stud. **37**(3), 533–555 (2000)

Eaton, J., Eckstein, Z.: Cities and growth: theory and evidence from France and Japan. Reg. Sci. Urban Econ. **27**(4), 443–474 (1997)
Gabaix, X.: Zipf's law for cities: an explanation. Q. J. Econ. **114**(3), 739–767 (1999)
Gibrat, R.: Les inégalités économiques. Recueil Sirey, Paris (1931)
Giesen, K., Suedekum, J.: The Size Distribution across all 'Cities': A Unifying Approach, Working Paper No. 3730. CESifo Group Munich (2012)
Henderson, V.: Medium size cities. Reg. Sci. Urban Econ. **27**(6), 583–612 (1997)
Hidalgo, C.A., Klinger, B., Barabási, A.-L., Hausmann, R.: The product space conditions the development of nations. Science **317**(5837), 482–487 (2007)
Klepper, S.: Industry life cycles. Ind. Corp. Change **6**(1), 145–182 (1997)
Krugman, P.: Increasing returns and economic geography. J. Polit. Econ. **99**(3), 483–499 (1991)
Ladyman, J., Lambert, J., Wiesner, K.: What is a complex system? Eur. J. Philos. Sci. **3**(1), 33–67 (2013)
Lane, D., Pumain, D., van der Leeuw, S.E., West, G. (eds.): Complexity Perspectives in Innovation and Social Change. Springer, Netherlands (2009)
Lösch, A.: Economics of Location. Yale University Press, New Haven (1954)
Marshall, J.U.: Industrial diversification in the Canadian urban system. Can. Geogr. Le Géogr. Can. **25**(4), 316–332 (1981)
Mills, E.S.: An aggregative model of resource allocation in a metropolitan area. Am. Econ. Rev. **57**(2), 197–210 (1967)
Min-rong, C., Yan-hua, M.: Specialization and diversification in cities: the evidence from China. In: 2013 6th International Conference on Information Management, Innovation Management and Industrial Engineering, vol. 1 (2013)
Mori, T., Nishikimi, K., Smith, T.E.: The number-average size rule: a new empirical relationship between industrial location and city size. J. Reg. Sci. **48**(1), 165–211 (2008)
Neffke, F., Henning, M., Boschma, R.: How do regions diversify over time? Industry relatedness and the development of new growth paths in regions. Econ. Geogr. **87**(3), 237–265 (2011)
Nelson, R.R., Winter, S.G.: An Evolutionary Theory of Economic Change. The Belknap Press of Harvard University Press, Cambridge, MA (1982)
Portugali, J.: Complexity, Cognition and the City. Springer, Berlin (2011)
Pumain, D. (ed.): Hierarchy in Natural and Social Sciences. Springer, Dordrecht (2006)
Pumain, D.: Alternative explanations of hierarchical differentiation in urban systems. In: Hierarchy in Natural and Social Sciences. Springer, Dordrecht (2006b). https://doi.org/10.1007/1-4020-4127-6_8
Pumain, D.: Multi-agent system modelling for urban systems: The series of SIMPOP models. In: Heppenstall, A., Crooks, A., Batty, M. (eds.) Agent-Based Models of Geographical Systems. Springer, Dordrecht (2012). https://doi.org/10.1007/978-90-481-8927-4_36
Pumain, D., Paulus, F., Vacchiani-Marcuzzo, C.: Innovation cycles and urban dynamics. In: Lane, D., Pumain, D., van der Leeuw, S.E., West, G. (eds.) Complexity Perspectives in Innovation and Social Change. Springer, Berlin (2009). https://doi.org/10.1007/978-1-4020-9663-1_9
Pumain, D., Reuillon, R.: Urban dynamics and simulation models. Springer International Publishing (2017). https://doi.org/10.1007/978-3-319-46497-8
Rosen, K.T., Resnick, M.: The size distribution of cities: an examination of the Pareto law and primacy. J. Urban Econ. **8**(2), 165–186 (1980)
Sanders, L., Pumain, D., Mathian, H., Guérin-Pace, F., Bura, S.: SIMPOP: a multi-agent system for the study of urbanism. Environ. Plan. **24**, 287–305 (1997)
Singer, H.W.: The 'Courbe des Populations.' A parallel to Pareto's law. Econ. J. **46**(182), 254–263 (1936)
Tabuchi, T., Thisse, J.-F.: A new economic geography model of central places. J. Urban Econ. **69**(2), 240–252 (2011)
Thompson, P.: Learning by doing. In: Hall, B.H., Rosenberg, N. (eds.) Handbook of the Economics of Innovation, vol. 1. North-Holland (2010)
Zipf, G.K.: Human behavior and the principle of least effort. Addison-Wesley Press (1949)

Chapter 16
Conclusion: Perspectives on Urban Theories

Denise Pumain and Juste Raimbault

Abstract In this concluding chapter, we propose two ways to synthesize the scientific contributions of this book. The first discusses several major principles that can be retained for constructing relevant theories within urban science in relation with the contents of the chapters of the book. The second part identifies communities of scientific discussion by analyzing a citation network capturing the scientific neighborhood of this book, constructed starting from the works quoted in the bibliographies of the chapters and enlarging to the papers citing them. This mining of digital bibliographical data confirms the wide inter-disciplinarity of urban questions justifying the plurality of urban theories and opening a variety of solutions for complementing them and coupling their generic models. We finally suggest directions for the construction of integrated theories, in particular through the coupling of simulation models corresponding to the different approaches to be bridged.

16.1 Introduction

At the end of the five years of work in our GeoDiverCity program,[1] we brought together a diversity of authors from different disciplines. Each person was invited to present an important question about the theories and models of urbanization. They are representative of a variety of currents in urban research. Rather than repeat here the contents of all chapters, we propose two ways to synthesize the scientific contributions of this book. In a first part we replace them in relation to a few principles

[1]Funded by ERC Advanced Grant 269826 (PI Denise Pumain).

D. Pumain (✉)
Université Paris I, UMR Géographie-Cités, Paris, France
e-mail: pumain@parisgeo.cnrs.fr

J. Raimbault
Institut des Systèmes Complexes Paris-Ile-de-France (CNRS 3611), Paris, France
e-mail: juste.raimbault@gmail.com

UMR Géographie-Cités (CNRS 8504), Paris, France

D. Pumain (ed.), *Theories and Models of Urbanization*, Lecture Notes in Morphogenesis, https://doi.org/10.1007/978-3-030-36656-8_16

that were experimented in our program, and in a second part we situate them with respect to a broader view of international literature on these topics.

The first part of this concluding chapter is a selection of salient points from our evolutionary theory of urban systems that are discussed in several of the chapters. As many of our results were already published (Pumain et al. 2015; Cura et al. 2017; Pumain and Reuillon 2017)[2] it was possible to confirm them, or to bring more different evidence or contradictory views. For each of these lively research questions, we report the convergent opinions that emerged from the topics discussed, as well as the open and even controversial perspectives for future work. The second part reports a quantitative analysis which arrives at another form of synthesis from the bibliographies of the chapters of the book and their networks of citations. This requires the use of methods for constructing and exploring large digital bibliographical data. Each part gives an overview of the current state of urban science, first of all according to its reported results and then according to the articulations between the conceptions of those who make it. We can easily imagine that in the near future, the second method will become an essential prerequisite for realizing the first, provided that further semantic analysis of the content of papers could be made.

In the United States a recent controversy has been fueled, both by people claiming either that theories about the city were not available (Brenner and Schmid 2014), or that these theories had been constructed mostly from empirical cases and inspiration from the Western world (Robinson 2016) and demonstrated most of times on the cases of large cities rather than "ordinary" cities and towns. We want to show that the arguments put forward by the instigators of this controversy are in contradiction with the knowledge acquired throughout the course of the history of urban science. We illustrate here a spiral conception of the cumulativeness of knowledge, according to which it is important not to neglect existing theories, which in principle already contain results drawn from a large number of empirical observations, even if their conceptions of society may seem partly out of date. It is therefore just as useful and necessary to propose revisions of old theories as to pretend to bring entirely new ones. Actually, the theories and models presented in this book have revealed themselves widely compatible and complementary.

16.2 Robust Grounds for Theories and Models in an Interdisciplinary Urban Science

We agree with Scott and Storper (2015) when they claim in a recent paper about urban theories that they can identify "dimensions common to all cities without on the one hand, exaggerating the scope of urban theory, or on the other hand, asserting that every individual city is an irreducible special case' (p. 1). Indeed, several points

[2] A series of papers in journals should be added to that short list, as well as Ph.D. Dissertations by Baffi (2016), Cottineau (2014), Finance (2016), Ignazzi (2015), Rey-Coyrehourcq (2015), Schmitt (2014) and Swerts (2013).

of convergence appear in the chapters, regardless of the authors' disciplinary origin. These points mainly characterize the structuration and transformation of systems of cities and summarize what can be called a common urban dynamics. Although this dynamics is complex, and shares properties with other complex systems, it concerns all systems of cities when constituted in quite extensive territories. It admits slightly different but intelligible modalities according to the particular conditions of the historical development of these systems including rather strong path dependence effects. These historical conditions together with the common dynamics constitute the evolutionary theory of urban systems (Pumain 2018). Because of the common urban dynamics and its rather strong path dependence, this theory authorizes a certain predictability for investigating the future of urban systems.

16.2.1 Urban Growth and the Hierarchical Structuration of Systems of Cities

The authors of this book have again found that the distribution of the size of cities is always very dissymmetrical, and it can be described by several types of statistical models, such as Zipf's law or the lognormal distribution, both statistically explained by quasi-stochastic models of urban growth (in which urban growth is proportional to city size). These models allow the comparison between systems of cities located in different regions of the world and at different periods of time. The general relevance of these models reveals a great coherence in the adequacy of the size of the cities to that of the territories in which they are located. The science of cities can therefore rely on this knowledge to make forecasts about the future size and number of cities in the short or medium term, and to look for processes common to the evolution of these distributions, in terms of dynamic growth of cities.

Whatever the country and period of observation, there is a complete imbrication of urban dynamics, economic development, innovation waves and the diversity of urban growth trajectories. As a first approximation, cities belonging to the same territory, thus in principle sharing the same rules of socio-economic and cultural functioning, are growing on the long-term at about the same rate, even if their growth rates fluctuate sharply on shorter time intervals. A stochastic model like Gibrat's (1931) is a good description of this process. When one observes not only the quantitative evolution of the population of the cities, but also their qualitative transformations (modification of the production and services, occupations, levels of education and skills, cultural and social practices…), there is also a general propensity for these transformations to occur fairly quickly in all parts of the system (Pumain and Saint-Julien 1979; Paulus 2004). According to an evolutionary theory or urban systems, *there is a generic co-evolution process that structures the hierarchical and functional differentiation of interdependent cities within systems of cities.* In terms of the theory of complex systems, this strong coupling in city trajectories is similar but not equivalent to the co-evolution processes observed in biology (Raimbault 2018;

Schamp 2010). However it is well explained by the numerous relational networks that are maintained and continuously renewed between cities that interact through the various types of exchanges of their multiple agents and stakeholders. In such well-connected systems, innovations propagate very often according to a hierarchical process that is reactivated each time a new large wave of novelty appears in the modes of economic production or kinds of social organization. Large cities are adapting first to the changes, which on average reach later the medium and smaller towns. This process corresponds to the variations of the exponents of the scaling laws between size of cities and urban attributes over time (see Sect. 16.2.4 below). Chapter 6 brings a novel confirmation of such a process, showing that even in the networks generated by last cutting edges information technologies, larger cities as effectively more unequal or concentrated in terms of their social relations, as they become more diverse in terms of the structures of each individual's social networks.

It is now recognised that the urbanization process is self-reinforcing, because it is generated from social interactions of all kind taking place in dense urban milieus. Chapter 13 provides evidence from an unprecedented global vision of the physical growth of cities in relation to their demographic expansion showing that this material expansion and its financial value have grown much faster than population. If economic growth is essential to the progress of urbanization, the latter is making an essential contribution to local and global economic development. This expansion is not only a measure of "agglomeration economies" but also the expression of the multiplying power of networking activities. Actually, each discipline tends to formalize this complex evolutionary process in its own terms. Whether named "agglomeration economies" or "increasing returns to scale" or "accelerating pace of life with city size" (West 2017) or "multiplicative power of networks" as in Chap. 15, the somehow auto-catalytic incentive and trend to a more or less continuous increase of urban population has generated at the same time increasing inequalities in city sizes. Rather than focusing its theories on explaining the growth of "the" city, geography insists we pay attention to their mutual relationships as well as with their embedding territory. From the observation of the quantitative and qualitative co-evolution of urban trajectories, it appears that urbanization is mainly driven by the exploitation of unequal quality and quantity of resources and costs that may vary widely according to city size. For a long time the process that substitutes and updates new products and services more rapidly in large cities than in small towns occurred within small regions or national territories, whereas since the second half of 20th century it has become a global process exploiting the differentials in resource prices and wage levels according to an "international division of labor" (Aydalot 1976).

16.2.2 General Predictability of Urban Growth and Decline

As a result of the lengthy urbanization process there is a huge diversity of urban trajectories, which however obey two main logics. The largest cities are the ones which repeatedly succeed in adopting successive innovations. In this development, they also extend the spatial scope of their activities. In expanding the networks of their activities, they both help and hamper the development of smaller towns and nearby cities by bypassing their customer networks. This last trend is especially reinforced when the acceleration of transport communication accentuate the historical trend to "space-time contraction" leading to a systematic "spatial reorganization" (Janelle 1969). Thus in the dynamics of systems of cities, there is a tendency to accentuate hierarchical inequalities which has two origins: from the top, large cities develop on average a little more rapidly at the beginning of each innovation wave, and each time the small towns lose relative importance. Thus, at the same time, there is a strengthening of the urban hierarchy from the top and a "simplification from below". In the long run, this dynamics therefore produces population growth in small towns, which tends to be slower than that of the larger ones, with less diversification of their activities. For a long time, the effects of this trend remained not too visible due to the general population growth, which allowed a wide spread of urban growth throughout the system of cities. However, in recent decades in some developed countries, the demographic decline brings out the phenomenon of "shrinking cities", which alarms local officials, observing the devitalization of these cities and fearing for their future evolution. The theory of urban dynamics, however, ensures that such an evolution is quite predictable. In fact, the decline in population growth does not prevent the concentration dynamics from continuing, increasingly opposing growing metropolises and declining small towns.

Some urban trajectories may deviate from these regularities, depending on certain specializations by economic activities in which they have a comparative advantage based on specific deposits (minerals, energy sources are classics, but nowadays cheap labor pools, research capacities or touristic sites are also selective factors of population concentration). As long as the activity remains prosperous, these cities can grow faster than the entire system, but when the activity enters into recession, it may cause these specialized cities to decline faster than others.

The authors of this book share knowledge that may reassure the actors who receive too often some alarmist messages about the urbanization process. Although as in all complex systems the prediction is in theory impossible, we affirm that there is a certain statistical predictability in city growth and city size over short time periods due to the relatively slow time scale of the dynamics of cities. Urban growth is rarely totally explosive and can be rather well anticipated. This relative predictability does not mean that the intervention of urban actors is vane; proactive adaptive strategies (safe imitation of successful ones in similar contexts, or anticipation and risk to find new development niches) are always necessary, in a pervasive context of emulation (or co-opetition). The common critics about the largest metropolises that are too often described as "monstruopolises" are not accurate. In fact, there is a correspondence between the population size of the cities and the size of population in the territories where they are located. Our comparative analyses have demonstrated the robustness and sustainability of most urban systems, even if the rather large variability in their spatial, political and economic organization demonstrates that none of them can be erected as a norm nor optimum. Against a unified alignment on urban planning for competitiveness, Chap. 8 advocates for the promotion of diversity as a more fundamental incentive for sustainable growth. Indeed, through both processes of hierarchical differentiation and functional specialization the resulting geodiversity of cities is perhaps the most historically secure engine of social change. It is in that sense that cities and systems of cities are resilient and can be considered as a product of collective territorial intelligence of humankind. Big cities have more inequality but more resources for the poorest. The geodiversity of cities also means that it accommodates for the various stages of individual trajectories through migration and relocation within cities. Homogeneity of housing stock, demographics or economic activities on the other hand tend to lock city in short cycles of prosperity.

16.2.3 Historical Transitions and Ecological Constraints

The urbanization that has everywhere transformed housing patterns on the planet by grouping populations in increasingly large agglomerations is a process that seems so far irrepressible and irreversible. This process has undergone historically two major acceleration phases, which can be represented as "transitions" that are the social equivalent of the physical bifurcation processes (or phase transition) (Sanders 2017). The first of these "urban transition" characterizes the emergence of cities, which occurred in all the regions of the world that experienced the Neolithic Revolution, and were sufficiently dense in population, vast in extent and open in terms of circulation (Bairoch 1985). Archaeologists are sometimes reluctant to accept the terms of "revolution" or "emergence" about the apparition of cities, because of the slowness and progressivity of the associated social changes that were rooted in rural communities. This evolution lasted a few millennia, with many new arrivals and disappearances of cities, though not exceeding a proportion of about 10% of the total world population in the pre-industrial ages (Bairoch 1985). A second transition took place with the great industrial revolution of the 19th century, which has enormously increased the size of cities (tens of millions of inhabitants instead of just one million for the largest ones before then), and raised the proportion of urban population to more than 50%, 80% are expected at the end of the 21st century according to the UN. In the first post-Neolithic phase, cities were strongly constrained by the resource limitations of their immediate environment, even though they had already invented a way of multiplying their capacity to take advantage of these resources by creating new artifacts and forms of social organization and by exchanging these innovations between different places. Their high vulnerability to natural disasters and the vagaries of conflict, as well as the weakness of their technical means, are sufficient to explain the slowness of this first urban development. In the second phase, cities grew into networks, earlier in the richest and most technologically advanced territories than in the countries colonized by them. In these dominated countries, urbanization occurred partly spontaneously, partly under the influence of colonizing countries, then more rapidly under the effect of strong demographic growth and subsequent economic growth. It is now in some of these territories as large as China and India that the largest cities in terms of population are the most numerous. However the economic weight of cities in developed countries remains predominant. In less than two centuries, dominant habitat patterns across the planet have shifted from small, relatively uniform and spatially dispersed rural cores to considerably larger concentrations of much higher densities and extremely differentiated sizes. During this second phase, urban development has meshed the entire planet with a variety of communication networks of very different natures. All in all, these new forms of habitat seem to cover only a small part of the globe, the built-up areas and the networks that connect them occupy no more than 15% of the earth surface. But their footprint in terms of mineral and energy resources and amount of soil mobilized is much larger, to the point that in terms of ecological footprint it has been estimated that the equivalent of three to eight planets would be needed to raise the level life of all urban dwellers to that of the richest lot at present.

The urbanization process is therefore entering a third stage in its development, during which the environmental constraint is gaining importance. The novelty is that its expression is no longer local, restricting the development capabilities of this or that city, but global. Fortunately the constraint can now be controlled thanks to the existence in all territories of these solidly constituted urban networks, very coherent in their hierarchical organization and their functional complementarities. Such an organization allows at the same time to circulate top down the new international or national regulations for promoting the environmental transition, and bottom up for collecting and disseminating the multiple initiatives and inventions that emerge locally for its concrete realization. However, the knowledge that we currently have in terms of what is called the "urban metabolism" is very largely insufficient to give clear indications as to the urban planning policies that would be able to drive this transition most effectively. Do we need more compact cities, smaller or bigger? The many measures that are being developed to test scaling laws applied to cities can no doubt help to strengthen recommendations.

16.2.4 Scaling Laws Have Societal Grounds

Among the recent impulses given to urban research, three have been very much invested by the community of physicists, who proposed to apply their formalisms to cities. Fractal geometry has proved itself to account, much better than classical density measurements, for the morphology of built-up surfaces, spatial distributions of urban activities and city networks, as was mentioned in Chap. 2. The second impulse was caused by the avalanche of massive data collected by mobile or immobile sensors. It has not really given rise yet to the emergence of new theoretical propositions, still mostly describing, with other models, well-known regularities concerning urban mobility, for example. However, Chap. 7 demonstrates that some new model of commuting based on these data can help predicting CO_2 emissions with a rather strong accuracy and therefore can bring useful information to urban planners. This work, as well as the one presented in Chap. 6, clearly show that cellphone connections may be a useful proxy of real social interaction, for which until now effective sources of information were too scarce. The fact remains that current approaches are still "socially blind", meaning that people are not differentiated according to the social group to which they belong.

But the third impulse, which consists in seeking the expression of scaling laws in the urban world, has led to new theoretical propositions (Bettencourt and West 2010). Since the authors tend to present this theory as universal, the discussion on this subject remains open. The authors of this book have rejected the idea of the universality of the exponent values of urban scaling laws, by showing their dependence on the data used, and by even identifying intelligible regularities from their variation over time and space. Thus regarding the linear scaling of income level per capita in British cities, Chap. 4 demonstrates that the global statistical picture actually hides the well-known geographical North-South divide between Northern regions having large cities and lower income due to their heavy past of 19th century manufacturing, while Southern cities have not so many large cities but higher income levels. This would be another confirmation of the strong linkage between urban scaling laws and the economic evolution of cities in urban systems, as developed in the evolutionary theory of urban systems. This also suggests that investigations in scaling should be enlarged to those of geographical self-similarities for a better examination of local processes of interactions into the model. In Chap. 5 the authors arrive to the same conclusion with different motives and methods. They also demonstrate the fruitfulness of using the residuals of a model of scaling laws for detecting anomalies or local specificities in location strategies of investment.

Regarding the universality of the interactional urban theory that sustains the model of scaling laws at the level of its fundamental principles (West 2017), it is more difficult to conclude. At this stage universality cannot signify that urban processes are of the same nature as processes in the physical or living worlds. Institutional regulations may always change (and hopefully they will do) for solving the problems linked with the most harmful consequences of exacerbated urban growth and the difficult challenges of all kind generated by this process. At least, research and debates about physics-driven scaling laws have allowed for extended communication about the long-term effects of city size that were studied by economists, historians and geographers (Bairoch 1977) and to provide formalized expressions that are useful for comparisons. This trend in multiplying quantified research could ultimately lead to clearer recommendations for urban public policies, for instance by providing more precise quantified measures, as for smarter cities. Moreover, it may become part of a computational shift in social science that will enable a qualitative shift in the knowledge that could be extracted from simulation models as suggested in Chaps. 12, 14 and 15.

16.2.5 *Do We Need Urban Different Theories for Each Culture or Region of the World?*

In this book we are referring to the long term evolution of cities as well as to its many global variations, first to throw light on some current debates about urban theories and second because these two requirements are necessary when theorizing about social facts and processes. Actually, all things being equal in terms of the size of a country's surface and population, the models that have been presented show a variability in the organization of urban hierarchies, which are generally more contrasted in recent settlement countries than in those where urbanization has a longer history. In addition, the authors of this book have shown that the results of these comparisons can be unstable and sometimes even contradictory. Indeed, they depend heavily on the number of cities that have been selected to represent a system, and in the way chosen to define them. The solution to this problem appears in Chaps. 4, 5, 12 and 13: more stable results are obtained and more robust conclusions can be made to test their robustness when choices appropriate to their goals are made for defining cities and systems of cities (see Sect. 16.2.6 below).

We have demonstrated the usefulness of complexity theories and models for comparing a large variety of urban systems all over the world in another recent book (Rozenblat et al. 2018). It is indeed legitimate to start analyzing any urban evolution by filtering from data the evidence of the common urban dynamics. This does not mean that dynamic models are constructed from the Western point of view only, and historians of the urbanization process have since long established that the temporal delay between urban growth in industrialized countries and emerging ones did not mean that the latter would encounter exactly the same sequence as the former (Bairoch 1985). But the same kind of adaptive processes leading to co-evolution of cities are at work in all systems of cities, as exemplified in Chap. 12 with urban cases as different as the Former Soviet Union and South Africa, the first with a background of planned interdependencies, the second marked by a dual system that extend networks because of so many invisible boundaries of the white colonial cities. It may be that in a highly planned country like China, the national and local policies that develop a system of cities that was organized on the same territorial basis for centuries continue its development by obtaining similar patterns as those observed elsewhere. But Chaps. 10 and 11 demonstrate each on their own grounds to what extent this reflexive development relies on a specificity of interactions that challenge the current urban and financial theories in their more detailed expression.

16.2.6 Relevant Definition and Delineation of Cities and Systems of Cities Are an Essential Component of Urban Science

It is no coincidence that Michael Batty devotes the first part of his Chap. 2 to the question of limits. Many of the controversies and misunderstandings in the literature about the size of cities would be resolved if the authors had been more attentive to the measurement of their size, so to the seemingly ancillary questions of the definition and the delimitation of the objects under study, as in any scientific investigation (Cottineau 2017). Overall, the authors of this book recognize themselves in a two-level ontology that identifies "cities" in "systems of cities" from two scales of relationships in the space-time of societies, that of daily interactions and that of evolutionary interdependencies in the longer term (Pumain 2017). It is true that with the increasing range and diversity of interactions among cities over the last decades, clearly identifying these two levels has become more difficult and their nesting is no longer so strict as in long-established urban theories (Reynaud 1841; Christaller 1933; Berry 1964; Pred 1977).

Cities are complex objects whose multiplicity of definitions refers to the diversity of the interactions that constitute them. These interactions always have a social component, even if they take into account material constraints in terms of the building materials, energy resources or technical and service networks. Some authors are anxious to appear to further objectify their delimitation, based on satellite images of the built space, by naming the forms thus constructed as "natural cities" (Jia and Jiang 2010) or "city proper" (Rosen and Resnick 1980), which could be considered as a conceptual oxymoron. In fact, the spatial expansion of urban buildings is always linked to social organizations, which impose more or less severe constraints and regulations on the occupation of space by buildings. Cities are objects whose definition is always linked to the political organization of the social groups that build and inhabit them, sometimes identified in administrative definitions and delimitations. Populations and sometimes activities are often enumerated within such limits, as in census data bases. In some cases, some geographical overlaps occur in these urban political definitions, as in the Chinese censuses, which may overestimate the urban populations concerned, while on the contrary the selection of eligible population groups (owning urban residence permits, or urban hukou) leads to their underestimation. This example of a floating population significant in terms of its proportions in urban population typically challenges the "classical" definitions of cities, and is a subject of research in itself (Swerts 2013, 2017).

Researchers often prefer building their own delineation rather than using official boundaries whose speed of evolution may be slower than that of the spatial expansion of the cities. The concept of *urban agglomeration* has been invented and sometimes documented by statistical institutions to better observe this reality. But the motorization of transport has often led, in all parts of the world, to disrupt the continuity of the urban limits drawn on the ground by connecting very strongly by daily relations with the historic heart of the cities sometimes more distant places, leading to define *functional urban areas*. Sometimes they form groups of intensely connected cities, called *megalopolises*, such as the one identified in 1957 by Jean Gottmann between Boston and Washington (Gottmann 1957), or as those developing in China, between Nanjing and Shanghai and in the Pearl River region around Guangzhou and Shenzhen. These highly integrated *mega-city-regions* are suggested by some researchers such as Le Néchet (2017) as the latest transition of human settlements. Even if the interactions between cities are spreading more and more at multiple scales, even sometimes connecting a small, very specialized town to the whole world, it is not possible to deduce that the delimitation of cities becomes totally obsolete. The fact that technologies, lifestyles and representations tend to become widely influenced by urban cultures everywhere, as Brenner and Schmid (2014) translate into the idea of a "planetary urbanization" cannot support urban theories that would no longer take into account this meso level of geographical organization that we call the city. In this respect, the systematic explorations of different urban boundaries that are made possible today by algorithms applied to detailed georeferenced databases help to better understand the diversity of spatial configurations of populations and urban activities and to bring nuances to the theories which would be too much generalizing (Cottineau 2017; Cottineau et al. 2017, 2018).

There is no question of advocating for a single definition. The important thing is to know how to harmonize those used and to choose those that are best adapted to the problem studied. This recommendation is also valid to identify what can be considered as a "system" of cities. That definition means a strong evolutionary interdependence between member cities, still observed with relevance in the context of the current world states, but increasingly uncertain as globalization weaves long-distance networks between places. It has to be reminded that this uncertainty has existed for a long time in the case of the largest cities, since the scope of urban interactions is generally quite strongly correlated with their size. Chapter 13 in this book in a good example of how a harmonized data base with its carefully segmented analysis is able to provide safe results on which urban theories can develop and be tested. And the way is now open for applying in research a real multi-level concept of cities (Rogov and Rozenblat 2018).

16.2.7 Why It Is Necessary to Maintain a Plurality of Theories and Models

Science is a continuous process of creating and revising theories. Theories are proposed for summarizing large sets of empirical observation within a simpler description that is guided by a coherent interpretation, validated in a more or less wide consensus and experience by many scholars and considered at a given moment as a good candidate for a possible explanation. Regarding urban sciences, the existence of a plurality of theories is necessary even if not always well understood.

The plurality of theories may receive a first justification from the epistemology of social sciences. It appears as reflecting the life of science, where a diversity of possible explanations is provided. When thinking in a "realistic" way, the co-existence of different interpretations of the same facts may be attributed to uncertainties in knowledge and ambiguity in facts that are justifying a diversity of explanations, while thinking as the adepts of philosophical constructivism may link this situation to the diversity of opinions, experience or social positions of researchers in science. But one of the major reasons for admitting a plurality of theories in social sciences is that their objects are usually multi-level and highly complex. Regarding cities, the multi-level character is obvious and usually lead to identify three major levels of inquiry within the apparently continuous scales of urban settlements and networks in space and time.

However, the explanatory factors that have to be considered may be different at each level due to their categorization through "emerging properties". There are not yet integrative theories that can provide satisfying explanation for all possible levels of organization. Indeed, multi-level organization is a characteristic feature of complex systems. It is often admitted that if distinct "levels" can be observed in their organization it is because emerging properties arise when the scale of observation is changing, leading to investigate different kinds of factors and processes.

Another source of plurality of theories is that each discipline in social science did elaborate consistent sets of interpretation of the urban realm according to a specific perspective and the particular processes it usually investigates. Building a meaningful explanation of a particular urban case often consists in borrowing explanatory concepts from sociology, economics, geography, history, urban planning and political economy –each of them embedded in their own pattern of understanding of complex systems- and combining them according to an explanatory hierarchy where their weights may vary. We have thus suggested that the urban complexity may also receive a definition including the number of disciplinary concepts that are required to understand a case study.

Urban theories are not enough developed to provide yet irrefutable recipes to urban planners and developers. But urban knowledge does exist, and should not be neglected in advancing hazardous theories. This reflection is a prerequisite in the current race to develop "smarter cities" that are more respectful of the rights of humanity to live in peace and harmony with their environment.

16.3 A Citation Network Analysis to Synthesize Urban Theories

16.3.1 Method and Data

We now turn to a quantitative analysis of the relative positioning of disciplines and approaches discussed above. We propose to use citation network analysis as a proxy to understand the structure of that scientific environment, what captures a single dimension of practices but contains relevant information on endogenous disciplinary structures. We use the method and tools of Raimbault (2019a) to construct a citation network from the references cited by chapters of this book. The rationale is to reconstruct from the bottom-up the scientific legacy in which each approach situates itself (a citation is a subjective and positioned asset to provide a basis for further knowledge), what is indeed not fully overlapping with the actual content (e.g. captured by semantics, as Raimbault et al. (2019) show how the two quantifications are complementary).

The bibliography of each chapter was manually indexed to ensure correct citing references retrieval during the data collection process. Furthermore, for performance purposes, but also to ensure a focus of the network content on urban issues, references clearly out of the scope and which would yield a significant part of the initial network totally unrelated to urban theories (the paper on morphogenesis by Turing (1990) is a typical example, being anecdotally cited by papers relating to urban issues, but also massively cited by several branches of biology).[3]

The initial corpus contains $N = 402$ references, and from it the backward citation network at depth two is reconstructed. This means that all papers citing the initial corpus, and a significant proportion of papers citing these citing papers, are collected. This yields a network with $V = 596{,}318$ nodes and $E = 1{,}000{,}604$ links. While for performance of data collection reasons, the network is not full (44% of nodes with positive in-degree have all their entering links), the balance between chapters is good (between 39 and 42% when considering chapter subnetworks separately) so this sampling does not bias the analysis. Regarding the language of papers in the networks, running a language detection algorithm on titles (using the python package polyglot) confirms that most of the corpus is in English (80.9%), the second language being Mandarin (4.2%) followed by Spanish (2.4%), German (2.3%), French (2.0%) and Portuguese (2.0%).

16.3.2 Network Analysis

We then keep the largest connected component (covering 99.98% of the network) and work on the higher order core of the network, obtained by removing nodes with degree one until no such node is present anymore in the network. The resulting network is smaller (159,648 nodes and 563,956 links) but expected to contain important information in terms of topological structure. A community detection algorithm (Louvain method at fixed resolution of 1) on the symmetrized network is used to reconstruct endogenous disciplines from the viewpoint of citation practices. We

[3]Code and results are on the open git repository of the project at https://github.com/JusteRaimbault/Perspectivism/tree/master/Models/QuantEpistemo. The raw dataset of the corpus is available on the dataverse at https://doi.org/10.7910/DVN/QCSAKT.

Table 16.1 List of largest citation communities (covering more than 90% of the network)

Community	Relative size (%)	Representative papers
Regional science	18.00	Cooke and Morgan (1999) [935], Porter (2000) [990]
Planning/governance	12.48	Bulkeley (2013) [541], Healey (2006) [737], McCann (2011) [538]
Urban economics	12.33	Gabaix (1999) [957], Henderson (1974) [831]
Social geography (health, public space, built environment, mode choice)	11.54	Handy et al. (2002) [666], Gehl (2011) [722]
Complexity/urban simulation/geosimulation	8.94	Batty (2013) [642], Waddell (2002) [893], Benenson and Torrens (2004) [535]
Pattern design	7.7	Alexander (1977) [993]
Microdemographics	6.1	Bongaarts (2002) [370]
Mobility	4.6	Cresswell (2006) [701]
Transport Networks	4.2	Rodrigue et al. (2016) [438]
Spatial analysis	4.0	Anselin (2013) [493]

The name of each was given after inspection of papers of highest degree within the community. We give for each some representative papers among these (degree in brackets)

obtain 27 communities which have a directed modularity of 0.71. Their size distribution is particular: 16 of them have a cumulated size of less than 1% and can be ignored in the analysis, while the remaining have a rather low hierarchy (rank-size exponent of -0.68 ± 0.08 with an adjusted r-squared of 0.88). This means that communities are rather balanced, confirming that this book covers a broad range of topics with no topic particularly dominating. The main communities are described in Table 16.1, with their name given after inspection of highest degree papers, their relative size, and some representative papers (chosen among the ones with the highest degree).

The content of communities obtained corresponds to some extent to broad disciplinary trends, but also to some thematic structure with some being apparently rather "interdisciplinary". The largest community (called "Regional science") contains works on innovation, firms, clusters, and regions that we attribute to regional science, which is not far but separated from urban economics working on these particular objects and scales. A second community includes work in planning, but also on governance structure and impacts of these (on climate change for example (Bulkeley 2013). The next cluster is Urban Economics as an expected strongly disciplinary cluster. Then comes works related to social issues, on very different topics (from health to the use of public space, built environment, or transportation mode choice) but all related to the study of the human and social component of the city. An important cluster is then related to complexity and simulation approaches, which can be interpreted more as a "methodological" community. Finally smaller communities can be thematic (Microdemographics, Mobility) or methodological (Spatial analysis). The smallest communities have more chance to being contingent to particular choices or subjects chosen by authors of the book, but the largest components can be seen as a broad overview of urban theory in general. Note that these results remain a partial mapping of urban theories and that many entries remain out of our analysis (urban climate or hydrology for example, population microsimulation models, or purely architectural or urban design approaches, to give a few).

We visualize the network in Fig. 16.1 in order to have an overview of how the different communities relate to each other. Without surprise, regional science, urban economics, and planning interact strongly and form a very compact triangle. Social geography (in which we can include mobility), and complexity connect also strongly to this core, social geography being mostly connected with planning and complexity, while complexity makes the bridge between urban economics, planning and social geography. Finally, some communities are more isolated at the periphery, such as design or demographics. This visualization confirms that the theories considered in

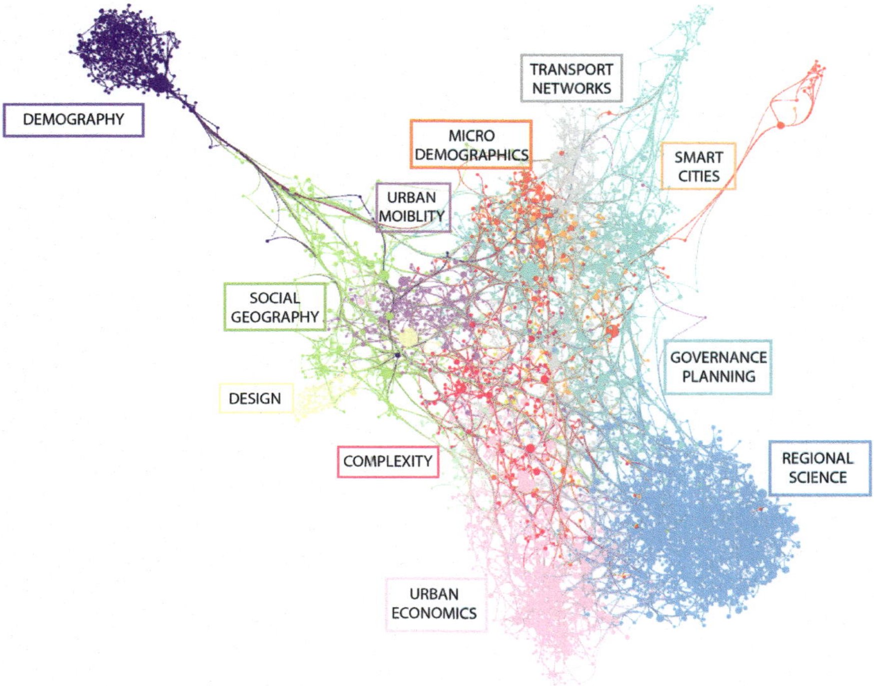

Fig. 16.1 Visualization of the core citation network. The network is visualized using the TULIP software, with Fast Multipole Embedder of Martin Gronemann, Curve edges and Edge Bundling algorithms for the layout. In the Fast Multipole Embedder algorithm, attraction forces are mediated in a multilevel way by iterative processing. The color of nodes and their edges reflects communities of nodes. Thanks to Céline Rozenblat who helped in simplifying and improving the readability of the figure

this book are well balanced and relatively well integrated, at least at such a scale of the full citation network. The landscape we get appears broader in its interdisciplinary scope than the graph obtained by Peris et al. (2018) who analyzed a graph of citations including a smaller number of publications (less than 1500) that were more focused on the topic of systems of cities. Thus, the clusters they identified (regional system, world city network, simulation and complexity, economic geography and city size distribution) are only partially similar to the communities we have found.

The content of largest communities can be studied more precisely, what can also give a better grasp on their level of interdisciplinarity. Therefore, a second community detection can be run within each. The level of modularity then informs whether each community is itself well integrated (low index value) or if it can be decomposed into subfields. As expected, subnetworks are still relatively modular, but with different strengths. Regional science is the least modular community with a modularity of 0.49, urban economics is also relatively low (0.59), while planning (0.63), social geography (0.66) and complexity (0.63) are the most modular communities. This can be interpreted as, for example, urban economics and regional science being more homogeneous in their citation choices. To illustrate how subfields organize, we show in Fig. 16.2 a visualization of the sub network obtained by keeping the "complexity" community only. We observe a continuum between practical approaches (urban sprawl (Nechyba and Walsh 2004) and urban growth (Seto et al. 2011) at the bottom), dominating applied simulation approaches (largest communities in the middle, corresponding to models such as Land-use transport interaction models (LUTI) on the left (Waddell 2002), and cellular automata models on the right (Clarke and Gaydos 1998), and more methodological and theoretical approaches at the top, including geosimulation (Benenson and Torrens 2004), agent-based modeling (Schelling 1971), urban complexity (Batty 2007), and urban systems (Pumain 1997). It is noteworthy to observe the diversity of these "sub-disciplines", but also their complementarity since applied models rely on theoretical and methodological investigations on one side, but also on data-driven investigations on the other side. Furthermore, to connect this diverse community with the rest of the full networks, each sub-community will play its own role in introducing bridges (e.g. applied models will connect to planning, while complexity approaches can connect with economics).

Until now, we performed a mostly visual and descriptive analysis, but it is also possible to quantify the relation between the endogenous disciplines identified, to understand the effective bridges existing or potential integrations. We use for this a basic indicator of inter-citation proportions. Given a total number of citation links made by a given community, we evaluate the proportion of these links made to a paper in another given community. The corresponding matrix for the 5 largest communities is shown in Table 16.2. The values confirm highly clustered communities, with all having an internal citation rate higher than 77%, the largest being regional science with a rate of 89%. This suggest potential for more bridges (although we quantify here only "direct bridges"; which may miss some intermediate role that would be revealed by centralities e.g.—such an advanced analysis is however out of the scope of this descriptive analysis) between urban theories. One can also distinguish "self-centered" disciplines, in particular regional science, for which the balance of given

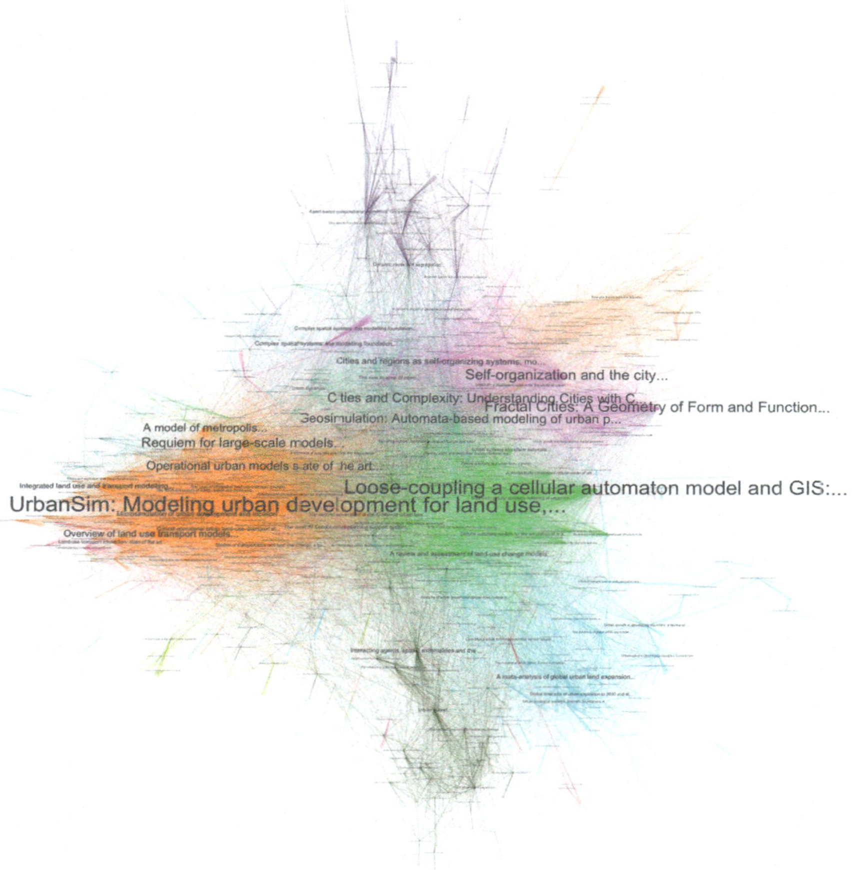

Fig. 16.2 Visualization of the subnetwork for complexity/simulation models. The network is visualized using the Gephi software, with a Force-Atlas 2 algorithm for the spatialization. In this algorithm, attraction forces are mediated through links, thus the spatial proximity between two communities relates to a proximity in number of links, and the relative positioning but also the compactness of communities can be interpreted. The color of nodes reflects their communities. We observe here communities ranging from data analysis to theoretical and methodological complexity approaches, with applied models in the middle

Table 16.2 Citation links between main communities

	Reg. sci.	Planning	Eco. geo.	Social geo.	Complexity	Others
Reg. sci.	89.25	2.59	4.91	0.42	0.27	2.55
Planning	5.15	80.32	1.96	4.35	1.71	6.51
Urb. eco.	5.77	1.90	84.12	1.08	3.18	3.96
Social geo.	1.26	5.66	2.05	78.86	4.27	10.49
Complexity	0.88	2.69	6.87	4.79	77.38	13.19

For the 5 largest communities, proportion (in %) of outcoming citation links in each other community

citation against received citations is always strongly negative, from more open disciplines such as complexity for which it is exactly the contrary. We also confirm the relative positioning discussed with the spatialisation of the network (for example social geography being mostly related to planning and complexity).

Another important insight into the content of this book is then how each chapter is positioned within the network, i.e. how each contributes to the emergence of each different endogenous community. First of all, one can consider subnetworks associated to each chapter. Starting from the references cited by a given chapter, one can reconstruct its subnetwork by getting iteratively citing papers. This produces a subset of the total network as only a subset of the initial corpus was considered. We find that subnetwork sizes range between 113,269 and 139,393 nodes, which corresponds respectively to 71% and 87% of the network, confirming the very high connectivity of branches sprout from different initial seeds. This also confirms a global robustness of the urban theories considered, i.e. that the corresponding scientific practices do refer to a broad common ground. Subnetworks have a high overlap between chapters, as the number of common nodes ranges from 113,133 to 134,467. Focusing on relative overlaps gives some information on the proximity between chapters. The relative overlap is taken as a Jaccard similarity index between sets, that is if N and N' are two sets of nodes, their similarity is given by $J = 2\,|N \cap N'|/(|N| + |N'|)$. We show in Fig. 16.3a (above panel) the relative similarity matrix between all chapters. We observe non-intuitive results, as for example (Samaniego 2020) working on transportation network scaling which relatively does not share much citations with the other chapters on scaling laws and in urban economics. The epistemological chapter dealing with complexities (Raimbault 2020a) is the farthest from most others, reflecting the difficulty to link meta considerations with applied urban theories. The two chapters on scaling (Arcaute and Hatna 2020; Finance and Swerts 2020) intersect mostly between themselves and with the definition of urban complexity (Batty 2019, 2020) and urban economics, but surprisingly not that much with the econophysics chapter (Barthelemy 2020) which does not refer to a large part of work done on scaling in the field of physics methods applied to urban systems. All in all, we find an absolute high integration, and some unexpected patterns in relative integrations, recalling the contingency of the citation practices that are intrinsic to each scientist with a culture and preferences beside its disciplinary affinities.

Finally, we can study the composition of chapter subnetworks in terms of endogenous communities. Considering a given subnetwork i, we compute the probabilities p_{ij} of its nodes to belong to the community j. This probability matrix, normalized by taking $p'_{ij} = p_{ij} - \langle p_{ij} \rangle / std(p_{ij})$ where average and standard deviation are computed over columns, gives patterns of under or overrepresentation of the different themes within chapters. This normalized matrix is visualized in Fig. 16.3b (bottom panel). We can understand the origin of some communities: for example, demographics mostly come from the chapter on emerging urban systems (Baffi and Cottineau 2020), while a community on settlement data comes from the chapter on urban sprawl (Denis 2020). This also highlights missing entries in some chapters, such as Raimbault (2020a) which has a very low proportion of urban economics, which is natural given that complexity theories are rather antagonist with the mainstream economics. This also allows finding subtle differences in content, such as the two chapters on scaling, Finance and Swerts (2020) invoking more spatial analysis in a geography tradition, while Arcaute and Hatna (2020) have a relatively higher link to complexity and urban economics. Finally, studying Herfindhal concentration index on composition probability as a measure of "interdisciplinarity" of each chapter does not give significant results (values ranging from 0.84 to 0.86) to differentiate them, and further analysis would be necessary to study this particular aspect [for example using more elaborated indices such as the Rao-Stirling index (Leydesdorff and Rafols 2011)] but remains out of the scope of this chapter.

This analysis allows to better situate each chapter in a global picture of the literature and thus better understand their complementarity. Possible bridges, or new points of view, can also emerge from considering interactions between communities and chapters.

16.4 Modeling and Simulation as a Medium to Couple Approaches

In the previous two sections, we gave an overview of how different urban theories can be complementary in theory and in practice. We now discuss why the coupling of heterogeneous approaches is relevant for future urban research and how modeling and simulation could be a powerful medium to do so.

16.4.1 Coupling Theories Through Models

This main proposal is based on general principles for modeling and simulation in the social sciences introduced by Banos (2013), which develops general guidelines to extract knowledge from simulation models. These relate to and draw on widely established practices in diverse disciplines using modeling and simulation, such as

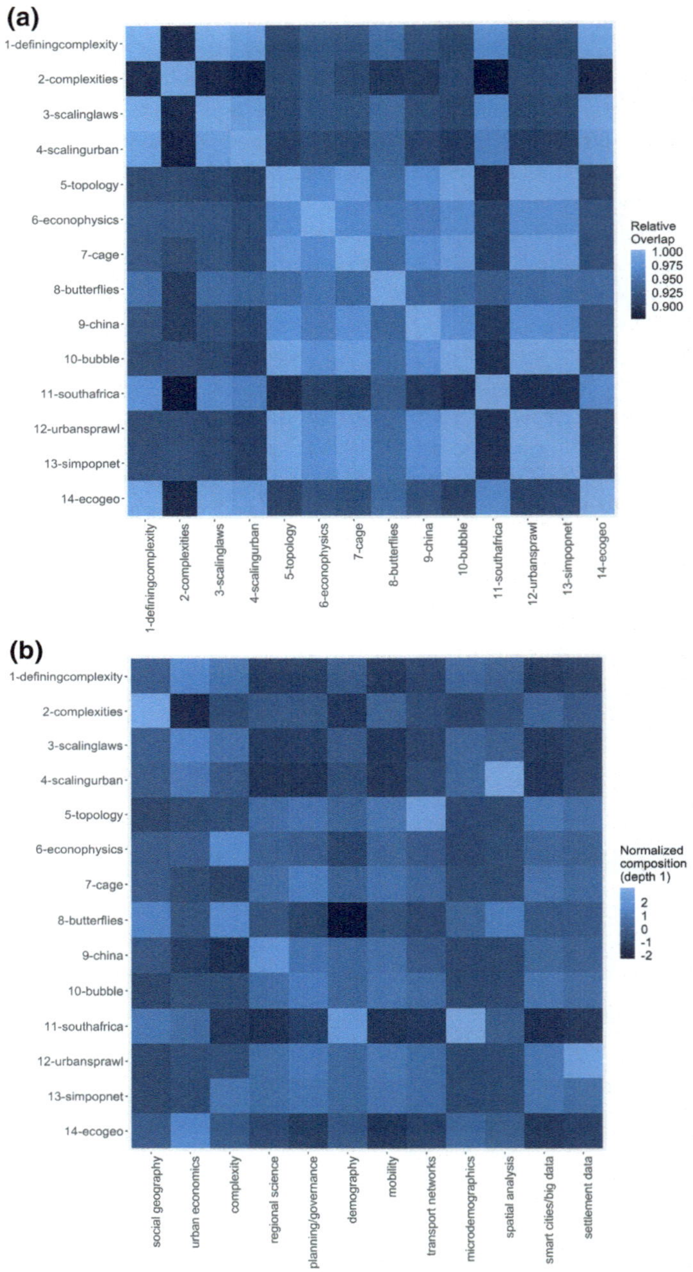
(a)
1-definingcomplexity
2-complexities
3-scalinglaws
4-scalingurban
5-topology
6-econophysics
7-cage
8-butterflies
9-china
10-bubble
11-southafrica
12-urbansprawl
13-simpopnet
14-ecogeo
Relative Overlap
1.000
0.975
0.950
0.925
0.900
(b)
social geography
urban economics
complexity
regional science
planning/governance
demography
mobility
transport networks
microdemographics
spatial analysis
smart cities/big data
settlement data
Normalized composition (depth 1)
2
1
0
-1
-2

◂**Fig. 16.3** Network coverage and composition of chapters, given by a similarity matrix between chapters (**a**), and a composition matrix of chapters in terms of communities (**b**). Proximities between chapters given by a Jaccard similarity index between subnetworks corresponding to each. Chapters are coded the following way: "definingcomplexity" (Batty 2020); "complexities" (Raimbault 2020); "butterflies" (Sanders et al. 2020); "cage" (Bouba-Olga 2020); "topology" (Samaniego et al. 2020); "econophysics" (Barthelemy 2020); "scalingurban" (Finance and Swerts 2020); "scalinglaws" (Arcaute and Hetna 2020); "china" (Wu 2020); "southafrica" (Baffi and Cottineau 2020); "bubble" (Aveline 2020); "urbansprawl" (Denis 2020); "ecogeo" (Bida and Rozenblat 2020); "simpopnet" (Raimbault 2020). (*Right*) Composition of chapters in terms of relative share of subnetworks (considering citing papers at the first level only, i.e. papers directly citing the initial corpus) in each community, normalized as center and reduced variables. Negative values correspond to an underrepresentation of the theme while positive values correspond to an overrepresentation

ecology (Grimm and Railsback 2012), computational social science (Epstein 2006), and general methodological contributions on agent-based modeling for example (Sun et al. 2016). These include in particular that (i) models have different objectives and functions; (ii) they thus must be shared in an open way for their benchmarking and comparison; (iii) models must be reused and coupled; (iv) behavior of models must be known in a precise way with extensive sensitivity analyses. Other principles include for example the need for a strong interaction between models and empirical data, or the fact that problems are most of the time multi-objective and models cannot provide unique optimal solutions, but these have less direct impact on our question. These different aspects are interlinked and form altogether a consistent framework in the spirit of complementary simulation models in an open science and reproducible context. The use of simulation models in itself, beyond all the advantages of being a medium to produce indirect knowledge on processes of a system, is furthermore justified as models are more and more part of the system studied, as Batty (2019, 2020) puts it when considering the concept of a "digital twin".

The case of geographical systems, and more particularly urban systems, furthermore justifies the application of these principles, because of their multi-dimensionality, spatio-temporal non-stationarity, multiple aspects of complexity, multi-scalarity. Some aspects of this complexity of urban systems can be specified and linked to Banos' principles. The "ontological complexity" proposed by Pumain (2003) as a new alternative to define the complexity of a system, which would be based on the number of viewpoints required to grasp most of system processes, is always high for urban systems, which is equivalent to their high multi-dimensionality. Therefore, the principle of various model objectives and functions is intrinsic to urban systems. The high spatio-temporal non-stationarity (Raimbault 2019b) and the non-ergodicity (Pumain 2012) of urban systems directly justify the importance of knowing the model behavior and performing a sensitivity analysis: if the model trajectories are path-dependent or dependent on the application context, an extensive knowledge of model dependency to initial conditions and parameters is essential to extract robust knowledge from it.

Model complementarity and coupling is at the core of Banos (2013) system of principles. We furthermore argue here that model coupling, in the sense of the construction of integrated models, can be a robust way to couple theories. This can be understood as a sort of "transfer postulate" between theories and associated models. Following Livet et al. (2010), ontology in the sense of an explicit specification of object and processes studied, is a powerful mediator to build agent-based models of social systems. In this context, different theories would then be mapped to different ontologies, i.e. models, in the modeling domain, and possibly to different methods, tools, data, and empirical analysis. While the latest can be coupled but do not necessarily induce a new knowledge component (coupling two methods is not necessarily a new method, as coupling two empirical analysis does not imply a new one, or it requires generally new models), the coupling of models is particular as elaborating a coupling of models corresponds to constructing a new model: it indeed requires an ontology for coupling processes, even in the case of sequential coupling which is the case where outputs of a first model are used as inputs of a second model (Voinov and Shugart 2013). The newly created model should correspond to a new theoretical entity that would then be the coupling of theories. In the epistemological framework of perspectivism proposed by Giere (2010) as an alternative to the opposition between realism and constructivism, this relevance of coupling is furthermore justified: as cognitive agents (scientists) have each their own perspectives, including a purpose, the coupling of perspectives is nothing more than collaborative scientific work. If their disciplinary background strongly differs, such a coupling is a tentative to construct interdisciplinary knowledge. Therefore, our proposal can be linked to "applied perspectivism" described in the third chapter of this book (Raimbault 2020).

The construction of such bridges should yield more integrated knowledge, in terms of horizontal integration through the couplings, but also possibly vertical integration if two approaches are at different scales, and furthermore a higher integration of knowledge domains since an increased interaction between them is necessary in the coupling process. Note that a maximum integration is not desirable and would make not much sense, since the practice of deepening knowledge is intrinsically modular and consists in a complex interplay between "disciplinarity and interdisciplinarity" (the virtuous spiral advocated by Banos (2017). The construction of integrative approaches is thus assumed to participate to a wider context of knowledge production, reinforcing both specific and integrated knowledge. Coupling models only for the sake of it can indeed be counter-productive as pointed by Voinov and Shugart (2013), which differentiate between integration of previously existing models and integration of knowledge from existing models into a new model. This echoes the need to construct a new perspective with its own purpose when coupling two perspectives.

16.4.2 Challenges

We have developed why the coupling of urban theories would be fostered by the coupling of simulation models stemming from these. This would yield integrated approaches, in the sense of a horizontal integration (transversal questions) and a vertical integration (towards multi-scale models) as emphasized by the complex systems roadmap (Chavalarias et al. 2009). We postulate that such approaches are crucial to reach higher standards in evidence-based social sciences, in the sense that they are a path among others to more systematic and evidence-based approaches.

New technical tools and methods will play a crucial role in these integrations. Indeed, as suggested in the previous section, if models are used as intermediaries to couple theories, they however must be well known in terms of behavior, using for example sensitivity analysis methods. In that context, a specific tool and associated methods were developed within the OpenMOLE platform (Reuillon et al. 2013) which provides a workflow engine allowing streamlined model embedding, exploration and distribution of computation on high performance computing environment. These new paradigms and methods are particularly suited to urban issues, as they furthermore arose in the context of the development of urban theories.

We suggest that emerging disciplines in urban science may have a key role to play as integrating approaches. For example, the field of Urban Analytics and City Science coined by Batty (2017) when renaming the journal Environment and Planning B Planning and Design, which captures quantitative approaches to urban and territorial systems (with a preferential focus on data analysis methods), is one of these. The new generation of Theoretical and Quantitative Geography inheriting from a long European tradition (Cuyala 2013) is another branch of these approaches. Geosimulation (Benenson and Torrens 2004) is also a hybrid and interdisciplinary field which already provided many integrating approaches. The positioning of studies of urban systems by physicists, described as a part of a "physics of society" by Caldarelli et al. (2018) is not clear yet, as they only claim the application of methods from statistical physics to social data and problems, but neither provide directions for such a transfer to be relevant and efficient, nor clarify the elements that would lay the basis for this "new discipline" (for example should they be methodological, with all associated issue of method transfer, or should they be thematic in the sense of object studied, in which case the relation with e.g. urban analytics is not thought).

Many open questions remain, such as the transfer possibilities towards decision making and planning, which can be very different depending on the fields. To what extent confronting approaches can foster the applicability of some is an issue that still has to be investigated. Besides that, it remains impossible to know if some approaches are missed while they could enlighten the particular issues tackled by a candidate integrative approach. The use of quantitative epistemology methods, such as the one used here with citation networks, or multi-dimensional methods (Raimbault et al. 2019), can however help lowering such risks.

16.5 Conclusion

We have in this concluding chapter provided a synthesis of urban theories overviewed through the whole book, by first recalling the most important issues and questions common to most theories of urban systems, which suggested a necessary plurality of such theory. The general explanation of the urbanization of the world is to be found just as much in the capacities of social organization as in economic growth. These two processes are strongly correlated over the long term, even if the process of emergence of innovations who directs the impulses is still difficult to predict, in terms of the conditions of its appearance and its qualitative content. The spatial organization of urban forms is beginning to be better understood, provided we recognize that the explanations and models that account for them are to be conceived as an open dynamics rather than as static equilibrium. The persistence of urban hierarchies, as well as the scaling laws that appear between various attributes of cities, are the product of a dynamic diffusion of innovations that exploits quantitative inequalities and qualitative differences between cities to build complex networks of complementarity and interdependencies, at all levels of city organization and systems of cities. It is in this sense that all people, businesses and local authorities are concerned by the needs of the next important transition for the future of the cities and the urbanization process, which consists of adaptation to climate change and the reduction of fossil resources consumption. More research is needed to construct bridges between theories, in particular the use of simulation models as a powerful medium for interdisciplinary dialogue, as is suggested through the citation network analysis of the scientific landscape around the chapters of the book.

If we may try to convey a specific message from this to the urban citizens, urban planners and stakeholders seeking for general ideas about cities and urbanization, it is rather clear: the current state of urban knowledge results from the collaboration of many disciplines. Urban challenges cannot be solved by a single disciplinary approach or by technologies only. Quantified models can help for solving local problems as well as providing an easier visualization of possible broader scenarios. The new massive observations captured by all kinds of sensors will help for a better local urban management rather than opening entirely new theoretical issues. As cities are fundamentally adaptive complex systems, anchored in a variety of geographical territories and historical contexts, there is not a single norm or model to recommend. On contrary, as in biology, the wide "geodiversity" which is driving urban evolution should be preserved as much as possible in order to maintain an open evolution. Qualitative changes in urbanization cannot be predicted but the future of cities is to be handled with care, precisely because they soon represent our quasi only way of inhabiting our single planet, in a world made more and more interdependent by multiple networks. While the long term effects of accelerated exchanges as in finance and information are not well known, the developing urban interactions could help in sharing solutions rather than exacerbating tensions.

References

Alexander, C.: A Pattern Language: Towns, Buildings, Construction. Oxford University Press, Oxford (1977)

Anselin, L.: *Spatial Econometrics: Methods and Models*, vol. 4. Springer Science & Business Media (2013)

Arcaute, E., Hatna, E.: Scaling laws: insights and limitations. In: *Theories and Models of Urbanization*. Springer Lecture Notes in Morphogenesis (2020)

Aveline, N.: China's housing booms, a challenge to the bubble theory. In: *Theories and Models of Urbanization*. Springer Lecture Notes in Morphogenesis (2020)

Aydalot, P.: Dynamique spatiale et développement inégal. Economica, Paris (1976)

Baffi, S.: *Le chemin de fer et la ville dans les processus de territorialisation en Afrique du Sud: de la séparation à l'intégration territoriale?* University Paris 1 Panthéon Sorbonne, Doctoral dissertation (2016)

Baffi, S., Cottineau, C.: What is emerging? Understanding urbanisation dynamics in BRICS countries through a geographical approach, the case of Russia and South Africa. In: *Theories and Models of Urbanization*. Springer Lecture Notes in Morphogenesis (2020)

Bairoch, P.: De Jéricho à Mexico. Éditions Gallimard, Paris (1985)

Bairoch P.: Taille des villes, conditions de vie et développement économique. Editions de l'EHESS, Paris (1977)

Banos, A.: Pour des pratiques de modélisation et de simulation libérées en géographie et SHS. HDR, Université Paris 1 (2013)

Banos, A.: Knowledge accelerator' in geography and social sciences: Further and faster, but also deeper and wider. In: Urban Dynamics and Simulation Models. Springer, Berlin (2017)

Barthelemy, M.: Revisiting urban economics for understanding urban data. In: Theories and Models of Urbanization. Springer Lecture Notes in Morphogenesis (2020)

Batty, M.: Cities and Complexity: Understanding Cities with Cellular Automata, Agent-Based Models, and Fractals. The MIT Press (2007)

Batty, M.: The New Science of Cities. MIT Press (2013)

Batty, M.: The future journal. Environ. Plan. B (2017)

Batty, M.: A map is not the territory, or is it? Environ. Plan. B (2019)

Batty, M.: Which complexity for urban systems? In: Theories and Models of Urbanization. Springer Lecture Notes in Morphogenesis (2020)

Benenson, I., Torrens, P.M.: Geosimulation: Automata-Based Modeling of Urban Phenomena. Wiley (2004)

Berry, B.J.: Cities as systems within systems of cities. Papers in regional science **13**(1), 46–163 (1964)

Bettencourt, L., West, G.: A unified theory of urban living. Nature **467**(7318), 912 (2010)

Bida, M., Rozenblat, C.: Modelling hierarchy and specialization of a system of cities as a result of an evolutionary perspective on firms' interactions. In: Theories and Models of Urbanization. Springer Lecture Notes in Morphogenesis (2020)

Bongaarts, J.: The end of the fertility transition in the developed world. Popul. Dev. Rev., 419–443 (2002)

Bouba-Olga, O., Grosseti, M.: The French version of the CAGE mythology (Competitiveness–Attractiveness–Globalization–Excellence) and some ideas on how to escape from it. In: Theories and Models of Urbanization. Springer Lecture Notes in Morphogenesis (2020)

Brenner, N., Schmid, C.: The 'Urban Age' in question. Int. J. Urban Reg. Res. **38**(33), 731–755 (2014)

Bulkeley, H.: Cities and Climate Change. Routledge (2013)

Caldarelli, G., Wolf, S., Moreno, Y.: Physics of humans, physics for society. Nat. Phys. **14**(9), 870 (2018)

Chavalarias, D., Bourgine, P., Perrier, E., Amblard, F., Arlabosse, F., Auger, P., … Amor, S.B.: French roadmap for complex systems 2008–2009 (2009)

Christaller, W.: Die Zentralen Orte in Süddeutschland : eine Ökonomisch-Geographische Untersuchung Über die Gesetz Massigkeit der Verbreitung und Entwicklung der Siedlungen mit Städitschen Funktionen. Fischer Verlag, Jena (1933)
Clarke, K.C., Gaydos, L.J.: Loose-coupling a cellular automaton model and GIS: long-term urban growth prediction for San Francisco and Washington/Baltimore. Int. J. Geogr. Inf. Sci. **12**(7), 699–714 (1998)
Cooke, P., Morgan, K.: The associational economy: firms, regions, and innovation. OUP Cat. (1999)
Cottineau, C.: L'évolution des villes dans l'espace post-soviétique. Observation et modélisations. University Paris 1 Panthéon Sorbonne, Doctoral dissertation (2014)
Cottineau, C.: MetaZipf. A dynamic meta-analysis of city size distributions. PLoS ONE **12**(8), e0183919 (2017)
Cottineau, C., Hatna, E., Arcaute, E., Batty, M.: Diverse cities or the systematic paradox of urban scaling laws. Comput. Environ. Urban Syst. **63**, 80–94 (2017)
Cottineau, C., Finance, O., Hatna, E., Arcaute, E., Batty, M.: Defining urban clusters to detect agglomeration economies. Environ. Plan. B Urban Anal. City Sci., 2399808318755146 (2018)
Cresswell, T.: On the Move: Mobility in the Modern Western World. Taylor & Francis (2006)
Cura, R., Cottineau, C., Swerts, E., Ignazzi, C.A., Bretagnolle, A., Vacchiani-Marcuzzo, C., Pumain, D.: The old and the new: qualifying city systems in the world with old models and new data. Geogr. Anal. **49**(4), 363–386 (2017)
Cuyala, S.: La diffusion de la géographie théorique et quantitative européenne francophone d'après les réseaux de communications aux colloques européens (1978–2011). Cybergeo Eur. J. Geogr. (2013)
Denis, E.: Population, land, wealth and the global urban sprawl. Drivers of urban built-up expansion across the world from 1990 to 2015. In: Theories and Models of Urbanization. Springer Lecture Notes in Morphogenesis (2020)
Epstein, J.M.: Generative Social Science: Studies in Agent-Based Computational Modeling. Princeton University Press (2006)
Finance, O.: *Les villes françaises investies par les firmes transnationales étrangères: des réseaux d'entreprises aux établissements localisés.* University Paris 1 Panthéon-Sorbonne, Doctoral dissertation (2016)
Finance, O., Swerts, E.: Scaling laws in urban geography. Linkages with urban theories, challenges and limitations. In: Theories and Models of Urbanization. Springer Lecture Notes in Morphogenesis (2020)
Gabaix, X.: Zipf's law for cities: an explanation. Q. J. Econ. **114**(3), 739–767 (1999)
Gehl, J. (2011). Life Between Buildings: Using Public Space. Island Press
Gibrat, R.: Les inégalités économiques. Sirey, Paris (1931)
Giere, R.N.: Scientific perspectivism. University of Chicago Press, Chicago (2010)
Gottmann, J.: *Megalopolis* or the urbanization of the northeastern seaboard. Econ. Geogr. **33**(3), 189–200 (1957)
Grimm, V., Railsback, S.F.: Pattern-oriented modelling: a 'multi-scope' for predictive systems ecology. Philos. Trans. R. Soc. B Biol. Sci. **367**(1586), 298–310 (2012)
Handy, S.L., Boarnet, M.G., Ewing, R., Killingsworth, R.E.: How the built environment affects physical activity: views from urban planning. Am. J. Prev. Med. **23**(2), 64–73 (2002)
Healey, P.: Urban Complexity and Spatial Strategies: Towards a Relational Planning for Our Times. Routledge (2006)
Henderson, J.V.: The sizes and types of cities. Am. Econ. Rev. **64**(4), 640–656 (1974)
Ignazzi, A.C.: Coevolution in the Brazilian system of cities. University Paris 1 Panthéon Sorbonne, Doctoral dissertation (2015)
Janelle, D.G.: Spatial reorganization: a model and concept. Ann. Assoc. Am. Geogr. **59**(2), 348–364 (1969)
Jia, T., Jian, B.: Measuring urban sprawl based on massive street nodes and the novel concept of natural cities (2010). arXiv preprint arXiv:1010.0541

Le Néchet, F.: De l'étalement urbain aux régions métropolitaines polycentriques: formes de fonctionnement et formes de gouvernance. Peupler la terre-De la préhistoire à l'ère des métropoles. Presses Universitaires Francois Rabelais, Tours (2017)
Leydesdorff, L., Rafols, I.: Indicators of the interdisciplinarity of journals: diversity, centrality, and citations. J. Inf. **5**(1), 87–100 (2011)
Livet, P., Müller, J.P., Phan, D., Sanders. L., Auatabu, T.: Ontology, a mediator for agent-based modeling in social science. J. Artif. Soc. Soc. Simul. 13(1) (2010)
McCann, E.: Urban policy mobilities and global circuits of knowledge: toward a research agenda. Ann. Assoc. Am. Geogr. **101**(1), 107–130 (2011)
Nechyba, T.J., Walsh, R.P.: Urban sprawl. J. Econ. Perspect. **18**(4), 177–200 (2004)
Paulus, F.: Coévolution dans les systèmes de villes : croissance et spécialisation des aires urbaines françaises de 1950 à 2000. Université Paris 1, thèse de doctorat (2004)
Peris, A., Meijers, E., van Ham, M.: The evolution of the systems of cities literature since 1995: schools of thought and their interaction. Netw. Spat. Econ. **18**(3), 533–554 (2018). https://doi.org/10.1007/s11067-018-9410-5
Porter, M.E.: Location, competition, and economic development: local clusters in a global economy. Econ. Dev. Q. **14**(1), 15–34 (2000)
Pred, A.: City systems in advanced societies. Hutchison, London (1977)
Pumain, D.: Vers une théorie évolutive des villes. L'Espace Géographique **2**, 119–134 (1997)
Pumain, D.: Une approche de la complexité en géographie. Géocarrefour **78**(1), 25–31 (2003)
Pumain, D.: Urban systems dynamics, urban growth and scaling laws: The question of ergodicity. In: Complexity Theories of Cities Have Come of Age, pp. 91–103. Springer, Berlin, Heidelberg (2012)
Pumain, D.: An evolutionary theory of urban systems. In: Rozenblat, C., Pumain, D., Velasquez, E. (eds.) International and Transnational Perspectives on Urban Systems. Springer Nature, Advances in Geographical and Environmental Sciences, 3–18 (2018)
Pumain, D., Saint-Julien, T.: Les transformations récentes du système urbain français. L'espace géographique 203–211 (1979)
Pumain, D., Swerts, E., Cottineau, C., Vacchiani-Marcuzzo, C., Ignazzi, A., Bretagnolle, A., Delisle, F., Cura, R., Lizzi, L, Baffi, S.: Multi-level comparison of large urban systems. Cybergeo 706 (2015)
Pumain, D., Reuillon, R.: Urban Dynamics and Simulation Models. Springer, International. Lecture Notes in Morphogenesis, 123 p (2017). ISBN: 978-3-319-46495-4. https://doi.org/10.1007/978-3-319-46497-8_3
Raimbault, J.: Caractérisation et modélisation de la co-évolution des réseaux de transport et des territoires (Doctoral dissertation, Université Paris 7 Denis Diderot) (2018)
Raimbault, J.: Exploration of an interdisciplinary scientific landscape. Scientometrics **119**(2), 617–641 (2019a)
Raimbault, J.: An urban morphogenesis model capturing interactions between networks and territories. In: The Mathematics of Urban Morphology, pp. 383–409. Birkhäuser, Cham (2019b)
Raimbault, J.: Relating complexities for the reflexive study of complex systems. In: Theories and Models of Urbanization. Springer Lecture Notes in Morphogenesis (2020a)
Raimbault, J.: Unveiling co-evolutionary patterns in systems of cities: a systematic exploration of the SimpopNet model. In: Theories and Models of Urbanization. Springer Lecture Notes in Morphogenesis (2020b)
Raimbault, J., Chasset, P.-O., Cottineau, C., Commenges, H., Pumain, D., Kosmopoulos, C., Banos, A.: Empowering open science with reflexive and spatialised indicators. Forthcom. Environ. Plan. B (2019)
Reuillon, R., Leclaire, M., Rey-Coyrehourcq, S.: OpenMOLE, a workflow engine specifically tailored for the distributed exploration of simulation models. Future Gener. Comput. Syst. **29**(8), 1981–1990 (2013)
Rey-Coyrehourcq, S.: Une plateforme intégrée pour la construction et l'évaluation de modèles de simulation en géographie. Doctoral dissertation, Paris 1-Panthéon-Sorbonne (2015)

Reynaud, J.: Villes, in Encyclopédie nouvelle. In: C. Gosselin (ed), Paris, t. VIII, (ed.) pp. 670–687(1841)
Robinson, J.: Comparative urbanism: new geographies and cultures of theorizing the urban. Int. J. Urban Reg. Res. **40**(1), 187–199 (2016)
Rodrigue, J.P., Comtois, C., Slack, B.: The Geography of Transport Systems. Routledge (2016)
Rosen, K.T., Resnick, M.: The size distribution of cities: an examination of the Pareto law and primacy. J. Urban Econ. **8**(2), 165–186 (1980)
Rogov, M., Rozenblat, C.: Urban resilience discourse analysis: towards a multi-level approach to cities. Sustainability **10**(12), 4431 (2018)
Rozenblat, C., Pumain, D., Velasquez, E. (eds.): International and Transnational Perspectives on Urban Systems. Springer Nature, Advances in Geographical and Environmental Sciences (2018)
Samaniego, H.: The Topology of communicating across cities in increasing sizes, or the complex task of "Reaching Out" in larger cities. In Pumain (ed) Theories and Models of Urbanization. Springer Lecture Notes in Morphogenesis (2020)
Samaniego, H., Franco-Cisterna, M., Sotomayor-Gómez, B.: The topology of communicating across cities of increasing sizes, or the complex task of "reaching out" in larger cities. In: Theories and Models of Urbanization. Springer Lecture Notes in Morphogenesis (2020)
Sanders, L.: Peupler la Terre. De la Préhistoire à l'ère des métropoles. Tours, Presses Universitaires François Rabelais (2017)
Sanders, L., Thomas, I., Vacchiani-Marcuzzo, C.: On the survival of butterflies in the jungle of urban systems. In: Theories and Models of Urbanization. Springer Lecture Notes in Morphogenesis (2020)
Schamp, E.W.: On the Notion of Co-evolution in Economic Geography. Edward Elgar Publishing (2010)
Schelling, T.C.: Dynamic models of segregation. J. Math. Sociol. **1**(2), 143–186 (1971)
Schmitt, C.: Modélisation de la dynamique des systèmes de peuplement: de SimpopLocal à SimpopNet. Doctoral dissertation, Université Paris I Panthéon-Sorbonne (2014)
Scott, A.J., Storper, M.: The nature of cities: the scope and limits of urban theory. Int. J. Urban Reg. Res. 39, 1–15 (2015). https://doi.org/10.1111/1468-2427.12134
Seto, K.C., Fragkias, M., Güneralp, B., Reilly, M.K.: A meta-analysis of global urban land expansion. PLoS ONE **6**(8), e23777 (2011)
Sun, Z., Lorscheid, I., Millington, J. D., Lauf, S., Magliocca, N. R., Groeneveld, J., … Buchmann, C.M.: Simple or complicated agent-based models? A complicated issue. Environ. Model. Softw. **86**, 56–67 (2016)
Swerts E.: Systèmes de villes en Inde et en Chine. Doctoral Dissertation, University Paris 1, Panthéon-Sorbonne (2013)
Swerts, E.: A data base on Chinese urbanization: ChinaCities. Cybergeo Eur. J. Geogr. Data papers 830 (2017)
Turing, A.M.: The chemical basis of morphogenesis. Bull. Math. Biol. **52**(1–2), 153–197 (1990)
Voinov, A., Shugart, H.H.: 'Integronsters', integral and integrated modeling. Environ. Model Softw. **39**, 149–158 (2013)
Waddell, P.: UrbanSim: modeling urban development for land use, transportation, and environmental planning. J. Am. Plan. Assoc. **68**(3), 297–314 (2002)
West, G.: Scale. The universal laws of life and deaths in organisms, cities and companies. Weidenfeld and Nicholson, London (2017)
Wu, F.: Emerging cities and urban theories: a Chinese perspective. In: Theories and Models of Urbanization. Springer Lecture Notes in Morphogenesis (2020)

Printed by Printforce, the Netherlands